OXFORD MATHEMATICAL MONOGRAPHS

Series Editors

E. M. FRIEDLANDER   I. G. MACDONALD
L. NIRENBERG   R. PENROSE   J. T. STUART

OXFORD MATHEMATICAL MONOGRAPHS

Oscillation Theory of Delay Differential Equations

With Applications

I. GYÖRI
Computing Centre of A. Szent-Györgyi
Medical University, Szeged, Hungary

and

G. LADAS
Department of Mathematics,
University of Rhode Island, USA

CLARENDON PRESS · OXFORD
1991

Oxford University Press, Walton Street, Oxford OX2 6DP
Oxford New York Toronto
Delhi Bombay Calcutta Madras Karachi
Petaling Jaya Singapore Hong Kong Tokyo
Nairobi Dar es Salaam Cape Town
Melbourne Auckland
and associated companies in
Berlin Ibadan

Oxford is a trade mark of Oxford University Press

Published in the United States
by Oxford University Press, New York

A catalogue record of this book is available from the British Library

Library of Congress Cataloging in Publication Data
Györi, I.
Oscillation theory of delay differential equations: with
applications / by I. Györi and G. Ladas.
p. cm. — (Oxford mathematical monographs)
Includes bibliographical references and index.
1. Differential equations—Delay equations. 2. Oscillations.
I. Ladas, G. E. II. Title. III. Series.
QA371.G96 1991 515'.35—dc20 90-20897

ISBN 0-19-853582-1

Typeset by Integral Typesetting, Gorleston, Norfolk
Printed in Great Britain by
Courier International Ltd., Tiptree, Essex

PREFACE

In recent years there has been much research activity concerning the oscillation of solutions of delay differential equations. To a large extent, this is due to the realization that delay differential equations are important in applications. New applications which involve delay differential equations continue to arise with increasing frequency in the modelling of diverse phenomena in physics, biology, ecology, and physiology.

Our aim in this monograph is to present in a systematic way the most important recent contributions in oscillation theory of delay differential equations. We will also apply the oscillation theory to several equations in mathematical biology to obtain the oscillatory character of their solutions.

There is no doubt that some of the recent developments in oscillation theory have contributed a beautiful body of knowledge in the field of differential equations that has enhanced our understanding of the qualitative behaviour of their solutions and has some nice applications in mathematical biology and other fields.

This monograph contains some recent important developments in the oscillation theory of delay differential equations with some applications to mathematical biology. Our intention is to expose the reader to the frontiers of the subject and to formulate some important open problems that remain to be solved in this area.

Chapter 1 contains some basic definitions and results which are used throughout the book. In this sense, this is a self-contained monograph. Chapter 2 deals with the basic oscillation theory of linear scalar delay differential equations. In Chapter 3 we introduce a generalized characteristic equation and then use it to establish comparison theorems, to prove the existence of a positive solution and also to obtain general oscillation results for delay equations with variable coefficients and variable delays. In Chapter 4 we present a linearized oscillation result and then apply it to some models in mathematical biology. Chapter 5 deals with the oscillation of linear and nonlinear systems of delay differential equations. In Chapter 6 we develop the oscillation of solutions of neutral differential equations. Chapter 7 deals with the oscillation of delay difference equations. The oscillation of equations with piecewise constant arguments is treated in Chapter 8. In Chapter 9 we present some oscillation results for integrodifferential equations. Chapter 10 contains some oscillation results for equations of higher order. In Chapter 11 we study the asymptotic behaviour of both the oscillatory and non-oscillatory solutions of some equations, mostly from mathematical biology, and obtain results about the global attractivity of their respective steady states. Finally, Chapter 12 contains some miscellaneous topics, including some results on slowly oscillating periodic solutions, rapidly oscillating solutions, and oscillations of solutions of differential equations with periodic coefficients.

At the end of every chapter we have included some notes and references about the material presented and we briefly discuss other related developments. We also present some open problems which are worth investigating and which will stimulate further interest in this subject.

Szeged   I. G.
Kingston   G. L.
1991

ACKNOWLEDGEMENTS

This monograph is the outgrowth of seminars and lecture notes which were given at the University of Rhode Island during the last ten years and in particular during the period September 1987 to May 1990. We are thankful to Professors R. D. Driver, D. A. Georgiou, K. Gopalsamy, E. A. Grove, A. Ivanov, V. Lj. Kocic, M. R. S. Kulenovic, L. Pakula, Z. Wang, and J. Yan, and to our graduate students W. Briden, R. C. DeVault, A. M. Farahani, K. Farrel, J. H. Jaroma Jun., S. A. Kuruklis, S. C. Mosher, E. C. Partheniadis, M. R. Petrarca Jun., C. Qian, I. W. Rodrigues, S. W. Schultz, and P. N. Vlahos for their enthusiastic participation and constructive criticism which helped to improve the exposition. Special thanks are due to E. A. Grove, D. A. Georgiou, V. Lj. Kocic, L. Pakula, S. A. Kuruklis, C. Qian, S. W. Schultz, P. N. Vlahos, and Z. Wang for proofreading in detail parts of this monograph. Finally, we express our appreciation to Edit Györi for her excellent typing of the original manuscript.

CONTENTS

1

PRELIMINARIES

Oscillation theory begins in Chapter 2. The aim in this chapter is to present some preliminary results which will be used throughout the book. In this respect, this is almost a self-contained monograph. The reader may glance at the material covered in this chapter and then proceed to Chapter 2.

Section 1.1 contains a detailed description of all possible global existence and uniqueness results that are needed in our treatment of the oscillation theory of delay and neutral delay differential equations. Of particular importance is to understand what we mean by a solution of an equation. See Definitions 1.1.1(c) and 1.1.2(c). In Section 1.2 we prove that the solutions of linear autonomous differential equations are exponentially bounded. We need this property of solutions because in Sections 2.1, 5.1, and 6.3 we take their Laplace transforms. In Section 1.3 we present some basic properties of Laplace transforms. In Sections 2.1, 5.1, and 6.3 we use Laplace transforms to present some basic necessary and sufficient conditions for the oscillation of all solutions of linear autonomous differential equations. In Section 1.4 we introduce the z-transform, which is the discrete analogue of the Laplace transform. This transform is used in Section 7.1 to give a powerful necessary and sufficient condition for the oscillation of all solutions of linear autonomous systems of difference equations. Section 1.5 contains several basic lemmas from analysis that we use on several occasions throughout this monograph. In Section 1.6 we include some basic results on differential inequalities whose use will simplify considerably the proofs of several theorems. Section 1.7 states some useful theorems from analysis that are needed in this monograph.

1.1 Some basic existence and uniqueness theorems

This section is written for the reader who is not familiar with the existence and uniqueness theory of delay and of neutral delay differential equations which is discussed in specialized books such as Bellman and Cooke (1963), Driver (1977), Hale (1977), and Myskis (1972). Here we emphasize the global existence and uniqueness theorems, because in oscillation theory the solutions are assumed to exist on an infinite interval $[t_0, \infty)$. See Definitions 1.1.1(c), 1.1.2(c), and Remark 1.1.1.

Consider the delay differential system

$$\dot{x}(t) + f(t, x(t), x(t - \tau_1(t)), \ldots, x(t - \tau_n(t))) = 0, \tag{1.1.1}$$

where for some $\tilde{t}_0 \in \mathbb{R}$ and some positive integer m,

$$f \in C[[\tilde{t}_0, \infty) \times \mathbb{R}^m \times \cdots \times \mathbb{R}^m, \mathbb{R}^m] \quad \text{and} \quad \tau_i \in C[[\tilde{t}_0, \infty), \mathbb{R}^+]$$

$$\text{for } i = 1, 2, \ldots, n, \tag{1.1.2}$$

with

$$\lim_{t \to \infty} [t - \tau_i(t)] = \infty, \quad \text{for } i = 1, 2, \ldots, n. \tag{1.1.3}$$

For every 'initial point' $t_0 \geq \tilde{t}_0$ we define $t_{-1} = t_{-1}(t_0)$ to be

$$t_{-1} = \min_{1 \leq i \leq n} \left\{ \inf_{t \geq t_0} \{t - \tau_i(t)\} \right\}. \tag{1.1.4}$$

As we see, t_{-1} depends on the delays $\tau_i(t)$ of the differential equation as well as the point t_0. The interval $[t_{-1}, t_0]$ is called the 'initial interval' associated with the initial point t_0 and the delay differential equation (1.1.1).

With eqn (1.1.1) and with a given initial point $t_0 \geq \tilde{t}_0$ one associates an 'initial condition'

$$x(t) = \phi(t) \quad \text{for } t_{-1} \leq t \leq t_0 \tag{1.1.5}$$

where $\phi \colon [t_{-1}, t_0] \to \mathbb{R}^m$ is a given 'initial function'.

Definition 1.1.1.

(a) *A function x is said to be a solution of eqn (1.1.1) on the interval I, where I is of the form $[t_0, T)$, $[t_0, T]$, or $[t_0, \infty)$, with $\tilde{t}_0 \leq t_0 < T$, if $x \colon [t_{-1}, t_0] \cup I \to \mathbb{R}^m$ is continuous, x is continuously differentiable for $t \in I$ and x satisfies eqn (1.1.1) for all $t \in I$.*

(b) *A function x is said to be a solution of the initial value problem (1.1.1) and (1.1.5) on the interval I, where I is of the form $[t_0, T)$, $[t_0, T]$, or $[t_0, \infty)$, if x is a solution of eqn (1.1.1) on the interval I and x satisfies (1.1.5).*

(c) *A function x is said to be a solution of eqn (1.1.1) if for some $t_0 \geq \tilde{t}_0$, x is a solution of eqn (1.1.1) on the interval $[t_0, \infty)$.*

(d) *A function x is said to be a solution of the initial value problem (1.1.1) and (1.1.5) if x is a solution of eqn (1.1.1) on the interval $[t_0, \infty)$ and x satisfies (1.1.5).*

Throughout this book, unless otherwise specified, for any m-dimensional vector $x = (x_1, \ldots, x_m)^T \in \mathbb{R}^m$, $\|x\|$ denotes any vector norm. For any $m \times m$ real matrix A, the associated matrix norm is then defined by $\|A\| = \max_{\|x\|=1} \|Ax\|$.

The following result, which is known as Gronwall's inequality, is needed in the proof of uniqueness theorems. For a proof see Driver (1977).

Lemma 1.1.1 (Gronwall's Inequality). *Let $I = [t_0, T)$ be an interval of real numbers and suppose that*

$$u(t) \leqslant c + \int_{t_0}^t v(s) u(s) \, ds \qquad for \ t \in I$$

where

$$c \in [0, \infty) \qquad and \qquad u, v \in C[I, \mathbb{R}^+].$$

Then

$$u(t) \leqslant c \exp\left(\int_{t_0}^t v(s) \, ds \right) \qquad for \ t \in I.$$

The following result is the basic global existence and uniqueness theorem for delay differential systems.

Theorem 1.1.1. *In addition to conditions* (1.1.2) *and* (1.1.3) *assume that there exists a function $p \in C[[\tilde{t}_0, \infty), \mathbb{R}^+]$ such that for all $t \geqslant \tilde{t}_0$ and for all $x_i, y_i \in \mathbb{R}^m$ for $i = 0, 1, \ldots, n$, the function f satisfies the following global Lipschitz condition:*

$$\| f(t, x_0, x_1, \ldots, x_n) - f(t, y_0, y_1, \ldots, y_n) \| \leqslant p(t) \sum_{i=0}^n \| x_i - y_i \|. \quad (1.1.6)$$

Let $t_0 \geqslant \tilde{t}_0$ and $\phi \in C[[t_{-1}, t_0], \mathbb{R}^m]$ be given. Then the initial value problem (1.1.1) *and* (1.1.5) *has exactly one solution in the interval $[t_0, \infty)$.*

Sketch of the Proof. Define the operator T on continuous functions $y: [t_{-1}, \infty) \to \mathbb{R}^m$ by

$$(Ty)(t) = \begin{cases} \phi(t), & t_{-1} \leqslant t < t_0 \\ \phi(t_0) + \displaystyle\int_{t_0}^t f(s, y(s), y(s - \tau_1(s)), \ldots, y(s - \tau_n(s))) \, ds, & t \geqslant t_0. \end{cases}$$

Clearly $(Ty)(t)$ is a continuous function on $[t_{-1}, \infty)$. We now define the following sequence of functions:

$$x_0(t) = \begin{cases} \phi(t), & t_{-1} \leqslant t < t_0 \\ \phi(t_0), & t \geqslant t_0 \end{cases}$$

and

$$x_{l+1} = Tx_l \qquad \text{for } l = 0, 1, 2, \ldots.$$

By using (1.1.6) one can show (as in Picard's method of successive approximations) that

$$\|x_{l+1}(t) - x_l(t)\| \leqslant M(t) \frac{(t - t_0)^l}{l!} \qquad \text{for } t \geqslant t_{-1} \text{ and } l = 0, 1, \ldots,$$

where $M(t) = \max_{t_0 \leqslant s \leqslant t} p(s)$. It follows that

$$x(t) = \lim_{l \to \infty} x_l(t) = \sum_{l=0}^{\infty} [x_{l+1}(t) - x_l(t)] + x_0(t)$$

exists for all $t \geqslant t_{-1}$, and x is a solution of the initial value problem (1.1.1) and (1.1.5) on $[t_0, \infty)$.

To show uniqueness, we assume that x and y are two solutions of (1.1.1) and (1.1.5) on the interval $[t_0, \infty)$. Then $x(t) = y(t)$ for $t_{-1} \leqslant t \leqslant t_0$. Set

$$u(t) = \max_{t_0 \leqslant s \leqslant t} \|x(s) - y(s)\|.$$

Clearly

$$u(t) \leqslant (n + 1) \int_{t_0}^{t} p(s)u(s) \, ds \qquad \text{for } t \geqslant t_0$$

and by Gronwall's inequality (with $c = 0$) it follows that $u(t) = 0$ for $t \geqslant t_0$. Hence $x(t) = y(t)$ for $t \geqslant t_0$ and the proof is complete.

Corollary 1.1.1. *Consider the linear non-autonomous delay system*

$$\dot{x}(t) + P_0(t)x(t) + \sum_{i=1}^{n} P_i(t)x(t - \tau_i(t)) = 0 \qquad (1.1.7)$$

where for some $\tilde{t}_0 \in \mathbb{R}$ and some positive integer m,

$$P_i \in C[[\tilde{t}_0, \infty], \mathbb{R}^{m \times m}] \qquad \text{for } i = 0, 1, \ldots, n,$$

$$\tau_i \in C[[\tilde{t}_0, \infty), \mathbb{R}^+] \qquad \text{and} \qquad \lim_{t \to \infty} [t - \tau_i(t)] = \infty \text{ for } i = 1, 2, \ldots, n.$$

Let $t_0 \geqslant \tilde{t}_0$ and $\phi \in C[[t_{-1}, t_0], \mathbb{R}^m]$ be given. Then the initial value problem (1.1.7) and (1.1.5) has exactly one solution in the interval $[t_0, \infty)$.

Corollary 1.1.2. *Consider the linear autonomous delay system*

$$\dot{x}(t) + P_0 x(t) + \sum_{i=1}^{n} P_i x(t - \tau_i) = 0, \qquad (1.1.8)$$

where for $i = 0, 1, \ldots, n$ the coefficients P_i are real $m \times m$ matrices and for $i = 1, 2, \ldots, n$ the delays τ_i are non-negative real numbers. Then for every $t_0 \in \mathbb{R}$ and for every $\phi \in C[[t_{-1}, t_0], \mathbb{R}^m]$, where $t_{-1} = t_0 - \max\{\tau_1, \tau_2, \ldots, \tau_n\}$, the initial value problem (1.1.8) and (1.1.5) has exactly one solution in the interval $[t_0, \infty)$.

When the function f in eqn (1.1.1) does not depend explicitly on $x(t)$ and when the delays $\tau_i(t)$ are all positive, one can establish a global existence theorem for eqn (1.1.1) without the assumption that the Lipschitz condition (1.1.6) holds. This is accomplished by utilizing the so-called *method of steps*. In fact we shall now utilize the method of steps to obtain a global existence theorem for a differential system of a more general form, namely, for the so-called *neutral delay differential system*

$$\frac{\mathrm{d}}{\mathrm{d}t}[x(t) + g(t, x(t - \tau_1(t)), \ldots, x(t - \tau_l(t)))]$$
$$+ f(t, x(t - \sigma_1(t)), \ldots, x(t - \sigma_n(t))) = 0 \quad (1.1.9)$$

where for some $\tilde{t}_0 \in \mathbb{R}$ and some positive integer m,

$$g \in C[[\tilde{t}_0, \infty) \times \mathbb{R}^m \times \cdots \times \mathbb{R}^m, \mathbb{R}^m], \quad (1.1.10)$$

$$f \in C[[\tilde{t}_0, \infty) \times \mathbb{R}^m \times \cdots \times \mathbb{R}^m, \mathbb{R}^m], \quad (1.1.11)$$

and for $i = 1, \ldots, l$ and $j = 1, \ldots, n$

$$\left. \begin{array}{l} \tau_i \in C[[\tilde{t}_0, \infty), (0, \infty)], \ \sigma_j \in C[[\tilde{t}_0, \infty), (0, \infty)] \text{ and} \\[2mm] \lim_{t \to \infty} [t - \tau_i(t)] = \infty = \lim_{t \to \infty} [t - \sigma_j(t)]. \end{array} \right\} \quad (1.1.12)$$

For a given initial point $t_0 \geq \tilde{t}_0$, we now define t_{-1} to be

$$t_{-1} = \min\left\{ \min_{1 \leq i \leq l}\left\{ \inf_{t \geq t_0}\{t - \tau_i(t)\} \right\}, \ \min_{1 \leq j \leq n}\left\{ \inf_{t \geq t_0}\{t - \sigma_j(t)\} \right\} \right\}. \quad (1.1.13)$$

With eqn (1.1.9) we also associate the initial condition

$$x(t) = \phi(t) \qquad \text{for } t_{-1} \leq t \leq t_0 \quad (1.1.14)$$

where $\phi: [t_{-1}, t_0] \to \mathbb{R}^m$ is a given initial function.

Definition 1.1.2.
   (a) *A function x is said to be a solution on the interval I, where I is of the form $[t_0, T)$, $[t_0, T]$, or $[t_0, \infty)$ with $\tilde{t}_0 \leq t_0 < T$, if $x: [t_{-1}, t_0] \cup I \to \mathbb{R}^m$ is continuous, $x(t) + g(t, x(t - \tau_1(t)), \ldots, x(t - \tau_l(t)))$ is continuously differentiable for $t \in I$ and x satisfies eqn (1.1.9) for all $t \in I$.*
   (b) *A function x is said to be a solution of the initial value problem (1.1.9)*

and (1.1.14) on the interval I, where I is of the form $[t_0, T)$, $[t_0, T]$, or $[t_0, \infty)$, if x is a solution of eqn (1.1.9) on the interval I and x satisfies (1.1.14).

(c) A function x is said to be a solution of eqn (1.1.9) if for some $t_0 \geq \tilde{t}_0$, x is a solution of eqn (1.1.9) on the interval $[t_0, \infty)$.

(d) A function x is said to be a solution of the initial value problem (1.1.9) and (1.1.14) if x is a solution of eqn (1.1.9) on the interval $[t_0, \infty)$ and x satisfies (1.1.14).

Remark 1.1.1. *Throughout this book, when we deal with oscillation theory, a solution of a delay differential equation will be understood in the sense of Definition 1.1.1(c). Similarly, a solution of a neutral delay differential equation will be understood in the sense of Definition 1.1.2(c). That is, they are solutions on $[t_0, \infty)$ for some $t_0 \geq \tilde{t}_0$.*

The following result is the basic global existence and uniqueness theorem for neutral delay differential systems.

Theorem 1.1.2. *Assume that (1.1.10), (1.1.11), and (1.1.12) are satisfied. Let $t_0 \geq \tilde{t}_0$ and $\phi \in C[[t_{-1}, t_0], \mathbb{R}^m]$ be given. Then the initial value problem (1.1.9) and (1.1.14) has exactly one solution in the interval $[t_0, \infty)$.*

Proof. Let $T > t_0$ be an arbitrary fixed number. It suffices to show that the initial value problem (1.1.9) and (1.1.14) has exactly one solution in the interval $[t_0, T]$. Let

$$\delta = \min\left\{ \min_{1 \leq i \leq l} \min_{t_0 \leq t \leq T} \tau_i(t), \ \min_{1 \leq j \leq n} \min_{t_0 \leq t \leq T} \sigma_j(t) \right\}. \tag{1.1.15}$$

Clearly, $\delta > 0$. Let $N = \left[\dfrac{T - t_0}{\delta}\right]$ be the greatest integer part of $\dfrac{T - t_0}{\delta}$. Then for all $k = 0, 1, \ldots, N$ and $t \in [t_0 + k\delta, t_0 + (k+1)\delta]$,

$$t - \tau_i(t) \leq t_0 + k\delta \quad \text{for } 1 \leq i \leq l \text{ and } t - \sigma_j(t) \leq t_0 + k\delta \text{ for } 1 \leq j \leq n.$$

Set

$$x_0(t) = \phi(t) \quad \text{for } t_{-1} \leq t \leq t_0.$$

Then for $t_0 \leq t \leq t_0 + \delta$ the initial value problem (1.1.9) and (1.1.14) yields

$$\frac{d}{dt}[x(t) + g(t, x_0(t - \tau_1(t)), \ldots, x_0(t - \tau_l(t)))]$$

$$+ f(t, x_0(t - \sigma_1(t)), \ldots, x_0(t - \sigma_n(t))) = 0$$

with

$$x(t_0) = x_0(t_0).$$

The unique solution of this initial value problem is

$$x_1(t) = -g(t, x_0(t - \tau_1(t)), \ldots, x_0(t - \tau_l(t))) + x_0(t_0)$$

$$+ g(t_0, x_0(t_0 - \tau_1(t_0)), \ldots, x_0(t_0 - \tau_l(t_0)))$$

$$- \int_{t_0}^{t} f(s, x_0(s - \sigma_1(s)), \ldots, x_0(s - \sigma_n(s))) \, ds.$$

Clearly the function $\tilde{x}_1(t)$ defined by

$$\tilde{x}_1(t) = \begin{cases} x_0(t) & \text{for } t_{-1} \leqslant t \leqslant t_0 \\ x_1(t) & \text{for } t_0 < t \leqslant t_0 + \delta \end{cases}$$

satisfies the initial value problem (1.1.9) and (1.1.14) on the interval $[t_0, t_0 + \delta]$. One can proceed in a similar way in the interval $[t_0 + \delta, t_0 + 2\delta]$ and so on as far as needed. The proof is complete.

Theorem 1.1.2 can be extended to include equations of the form

$$\frac{d}{dt} [x(t) + g(t, x(t - \tau_1(t)), \ldots, x(t - \tau_l(t)))]$$

$$+ f(t, x(t), x(t - \sigma_1(t)), \ldots, x(t - \sigma_n(t))) = 0 \quad (1.1.16)$$

where the functions g, τ_i, and σ_j satisfy the conditions (1.1.10) and (1.1.12) while the function f satisfies the following hypothesis:

(H) $f \in C[[\tilde{t}_0, \infty) \times \mathbb{R}^m \times \cdots \times \mathbb{R}^m, \mathbb{R}^m]$ and for every compact set $D \subset [\tilde{t}_0, \infty) \times \mathbb{R}^{(n+1)m}$ there exists a constant $K = K(D) > 0$ such that

$$\| f(t, x_0, x_1, \ldots, x_n) - f(t, y_0, x_1, \ldots, x_n) \| \leqslant K \| x_0 - y_0 \|$$

for all $(t, x_0, x_1, \ldots, x_n), (t, y_0, x_1, \ldots, x_n) \in D$.

Theorem 1.1.3. *Assume that conditions (1.1.10), (1.1.12) and hypothesis (H) are satisfied. Let $t_0 \geqslant \tilde{t}_0$ and $\phi \in C[[t_{-1}, t_0], \mathbb{R}^m]$ be given. Then the initial value problem (1.1.16) and (1.1.14) has exactly one solution in the interval $[t_0, \infty)$.*

Sketch of the Proof. The proof makes use of the method of steps. Let $T > t_0$ be an arbitrary fixed number and let δ be defined by (1.1.15). Set

$$x_0(t) = \phi(t) \quad \text{for } t_{-1} \leqslant t \leqslant t_0.$$

Then for $t_0 \leqslant t \leqslant t_0 + \delta$ the initial value problem (1.1.16) and (1.1.14) yields

$$x(t) = -g(t, x_0(t - \tau_1(t)), \ldots, x_0(t - \tau_l(t))) + x_0(t_0)$$
$$+ g(t_0, x_0(t_0 - \tau_1(t_0)), \ldots, x_0(t_0 - \tau_l(t_0)))$$
$$- \int_{t_0}^{t} f(s, x(s), x_0(s - \sigma_1(s)), \ldots, x(s - \sigma_n(s))) \, \mathrm{d}s.$$

In view of the hypothesis (H), this integral equation has exactly one solution $x_1(t)$ on $0 \leqslant t \leqslant \delta$. Clearly the function

$$\tilde{x}(t) = \begin{cases} x_0(t), & t_{-1} \leqslant t \leqslant t_0 \\ x_1(t), & t_0 \leqslant t \leqslant t_0 + \delta \end{cases}$$

is the unique solution of (1.1.16) and (1.1.14) on the interval $[t_0, t_0 + \delta]$. By repeating the same procedure in the interval $[t_0 + \delta, t_0 + 2\delta]$ and so on, the result follows.

Corollary 1.1.3. *Consider the linear non-autonomous delay system*

$$\frac{\mathrm{d}}{\mathrm{d}t} \left[x(t) + \sum_{i=1}^{l} P_i(t) x(t - \tau_i(t)) \right] + Q_0(t) x(t) + \sum_{j=1}^{n} Q_j(t) x(t - \sigma_j(t)) = 0$$
$$(1.1.17)$$

where for some $\tilde{t}_0 \in \mathbb{R}$ and some positive integer m,

$$P_i \in C[[\tilde{t}_0, \infty), \mathbb{R}^{m \times m}] \qquad \text{for } i = 1, 2, \ldots, l$$
$$Q_j \in C[[\tilde{t}_0, \infty), \mathbb{R}^{m \times m}] \qquad \text{for } j = 0, 1, 2, \ldots, n$$

and the delays τ_i and σ_j satisfy the condition (1.1.12). Let $t_0 \geqslant \tilde{t}_0$ and $\phi \in C[[t_{-1}, t_0], \mathbb{R}^m]$ be given. Then the initial value problem (1.1.17) and (1.1.14) has exactly one solution in the interval $[t_0, \infty)$.

Corollary 1.1.4. *Consider the linear autonomous delay system*

$$\frac{\mathrm{d}}{\mathrm{d}t} \left[x(t) + \sum_{i=1}^{l} P_i x(t - \tau_i) \right] + Q_0 x(t) + \sum_{j=1}^{n} Q_j x(t - \sigma_j) = 0 \quad (1.1.18)$$

where for $i = 1, 2, \ldots, l$ and for $j = 0, 1, \ldots, n$ the coefficients P_i and Q_j are real $m \times m$ matrices and for $i = 1, 2, \ldots, l$ and $j = 1, 2, \ldots, n$ the delays τ_i and σ_j are positive real numbers. Then for every $t_0 \in \mathbb{R}$ and for every $\phi \in C[[t_{-1}, t_0], \mathbb{R}^m]$, where $t_{-1} = t_0 - \max\{\tau_1, \tau_2, \ldots, \tau_l, \sigma_1, \sigma_2, \ldots, \sigma_n\}$, the initial value problem (1.1.18) and (1.1.14) has exactly one solution in the interval $[t_0, \infty)$.

The following result can be established by using the method of steps. Its proof is similar to that of Theorem 1.1.3 and will be omitted.

Theorem 1.1.4. *In addition to the hypotheses of Corollary 1.1.3, assume that for all $i = 1, 2, \ldots, l$ the functions $P_i(t)$ and $\tau_i(t)$ are absolutely continuous with locally bounded derivatives $\dot{P}_i(t)$ and $\dot{\tau}_i(t)$. Let $t_0 \geq \tilde{t}_0$ and $\phi \in C[[t_{-1}, t_0], \mathbb{R}^m]$ be given such that ϕ is absolutely continuous with locally bounded derivative $\dot{\phi}(t)$ on $[\tilde{t}_{-1}, t_0]$ where $\tilde{t}_{-1} = \min_{1 \leq i \leq l}\{\inf_{t \geq t_0}\{t - \tau_i(t)\}\}$. Then the unique solution $x(t)$ of the initial value problem (1.1.17) and (1.1.14) is absolutely continuous with locally bounded derivative on $[t_0, \infty)$ and satisfies the equation*

$$\dot{x}(t) + \sum_{i=1}^{l} P_i(t)[1 - \dot{\tau}_i(t)]\dot{x}(t - \tau_i(t)) + \sum_{i=1}^{l} \dot{P}_i(t)x(t - \tau_i(t))$$

$$+ Q_0(t)x(t) + \sum_{j=1}^{n} Q_j(t)x(t - \sigma_j(t)) = 0$$

for almost every $t \geq t_0$.

We close this section with an application on the generalized delay logistic equation

$$\dot{N}(t) = N(t)\left[\alpha - \beta N(t - \tau) + \gamma \dot{N}(t - \sigma) + \delta \frac{\dot{N}(t - \rho)}{N(t - \rho)}\right] \quad (1.1.19)$$

where

$$\alpha, \beta, \tau, \sigma, \rho \in (0, \infty) \quad \text{and} \quad \gamma, \delta \in \mathbb{R}. \quad (1.1.20)$$

With eqn (1.1.19) one associates an initial function of the form

$$N(t) = \phi(t), \quad -r \leq t \leq 0 \quad \text{with } r = \max\{\tau, \sigma, \rho\} \quad (1.1.21)$$

where the function ϕ satisfies the following condition:

$$\left.\begin{array}{l} \phi \in C[[-r, 0], \mathbb{R}^+], \ \phi(t) > 0 \text{ for } -\rho \leq t \leq 0 \text{ and } \phi(t) \\[4pt] \text{is absolutely continuous with locally bounded derivative} \\[4pt] \text{on } -\max\{\sigma, \rho\} \leq t \leq 0. \end{array}\right\} \quad (1.1.22)$$

Theorem 1.1.5. *Assume that (1.1.20) is satisfied and let ϕ be an initial function which satisfies condition (1.1.22). Then the initial value problem (1.1.19) and (1.1.21) has a unique solution on $[0, \infty)$. More precisely there exists a unique function $N \in C[[-r, \infty), \mathbb{R}^+]$ such that*

(1) $N(t)$ satisfies (1.1.21) and $N(t) > 0$ for $t \geqslant 0$;

(2) $N(t)$ is absolutely continuous with locally bounded derivative for $t \geqslant -\max\{\sigma, \rho\}$;

(3) $N(t)$ satisfies (1.1.19) for almost every $t \geqslant 0$.

Proof. The proof is an elementary application of the method of steps. Let $\delta = \min\{\tau, \sigma, \rho\}$. Then on the interval $0 \leqslant t \leqslant \delta$, the initial value problem (1.1.19) and (1.1.21) reduces to

$$\dot{N}(t) = f(t)N(t), \qquad N(0) = \phi(0) > 0 \tag{1.1.23}$$

where $f(t) = \alpha - \beta\phi(t - \tau) + \gamma\dot{\phi}(t - \sigma) + \delta\dfrac{\dot{\phi}(t - \rho)}{\phi(t - \rho)}$. Clearly f is locally bounded and locally integrable. The unique solution of (1.1.23) is

$$N(t) = \phi(0) \exp \int_{t_0}^{t} f(s)\, \mathrm{d}s, \qquad 0 \leqslant t \leqslant \delta$$

which is positive and absolutely continuous with locally bounded derivative. One can proceed in a similar fashion in the interval $[\delta, 2\delta]$ and so on. The proof is complete.

1.2 Exponential boundedness of solutions

Our aim in this section is to show that every solution of a linear autonomous delay system,

$$\frac{\mathrm{d}}{\mathrm{d}t}\left[x(t) + \sum_{i=1}^{l} P_i x(t - \tau_i) \right] + Q_0 x(t) + \sum_{j=1}^{n} Q_j x(t - \sigma_j) = 0 \tag{1.2.1}$$

is *exponentially bounded*. That is, if $x(t)$ is a solution of eqn (1.2.1) on $[0, \infty)$, then there exist positive constants M and α such that

$$\| x(t) \| \leqslant M\, \mathrm{e}^{\alpha t} \qquad \text{for } t \geqslant 0. \tag{1.2.2}$$

This result will be used in Sections 2.1, 5.1, and 6.4 where we employ Laplace transforms to establish necessary and sufficient conditions for the oscillation of all solutions of linear autonomous differential equations. Clearly if $x(t)$ satisfies (1.2.2), then the Laplace transforms $X(s)$ of $x(t)$ which is given by

$$X(s) = \int_{0}^{\infty} \mathrm{e}^{-st} x(t)\, \mathrm{d}t,$$

exists for $\operatorname{Re} s > \alpha$.

Theorem 1.2.1. *Assume that the coefficients P_i and Q_j of eqn (1.2.1) are real $m \times m$ matrices and the delays τ_i and σ_j are positive numbers. Let $x(t)$ be a solution of eqn (1.2.1) on $[0, \infty)$. Then there exist positive constants M and α such that (1.2.2) holds.*

Proof. Set $r = \max\{\tau_1, \ldots, \tau_l, \sigma_1, \ldots, \sigma_n\}$ and $\tau = \min\{\tau_1, \ldots, \tau_l\}$. By integrating both sides of eqn (1.2.1) from 0 to t we find

$$x(t) = x_0 - \sum_{i=1}^{l} P_i x(t - \tau_i) + \int_0^t \left[Q_0 x(s) + \sum_{j=1}^{n} Q_j x(s - \sigma_j) \right] ds, \qquad t \geq 0$$

where

$$x_0 = x(0) + \sum_{i=1}^{l} P_i x(-\tau_i).$$

Set

$$u(t) = \max_{-r \leq s \leq t} \|x(s)\| \qquad \text{for } t \geq -r, \qquad c = \|x_0\| + u(0),$$

$$A = \sum_{i=1}^{l} \|P_i\| \qquad \text{and} \qquad B = 1 + \sum_{j=0}^{n} \|Q_j\|.$$

Then $u(t)$ is a non-decreasing continuous function and

$$\|x(t)\| \leq \|x_0\| + \sum_{i=1}^{l} \|P_i\| \|x(t - \tau_i)\| + \int_0^t \left[\|Q_0\| \|x(s)\| \right.$$

$$\left. + \sum_{j=1}^{n} \|Q_j\| \|x(s - \sigma_j)\| \right] ds$$

$$\leq \|x_0\| + Au(t - \tau) + B \int_0^t u(s) \, ds, \qquad t \geq 0.$$

Hence

$$u(t) \leq c + Au(t - \tau) + B \int_0^t u(s) \, ds, \qquad t \geq 0. \qquad (1.2.3)$$

When $A = 0$, that is, in the case of a non-neutral differential equation, inequality (1.2.3) reduces to Gronwall's inequality:

$$u(t) \leq c + B \int_0^t u(s) \, ds.$$

Hence by Lemma 1.1.1,

$$u(t) \leq c \, e^{Bt}$$

which shows that solutions of linear autonomous delay differential equations are indeed exponentially bounded. When $A \neq 0$, inequality (1.2.3) may be thought of as being a generalized Gronwall's inequality. In the remaining part of the proof we show that (1.2.3) also implies that $u(t)$ is exponentially bounded.

Set $F(\lambda) = \lambda A\, e^{-\lambda\tau} + B - \lambda$. As $F(0)F(\infty) < 0$, there must exist an $\alpha > 0$ such that $F(\alpha) = 0$. Let ε be an arbitrary positive number and set

$$c_\varepsilon = \max\left\{(\varepsilon + c)(1 - A\, e^{-\alpha\tau})^{-1}, \ \max_{-r \leqslant s \leqslant 0} e^{-\alpha s}[u(s) + \varepsilon]\right\}.$$

We now claim that (1.2.3) implies that

$$u(t) < c_\varepsilon\, e^{\alpha t}, \qquad t \geqslant -r. \tag{1.2.4}$$

Clearly (1.2.4) is valid for $-r \leqslant t \leqslant 0$. Assume, for the sake of contradiction, that there exists a $t_1 > 0$ such that

$$u(t) < c_\varepsilon\, e^{\alpha t}, \qquad -r \leqslant t < t_1 \qquad \text{and} \qquad u(t_1) = c_\varepsilon\, e^{\alpha t_1}. \tag{1.2.5}$$

Then it follows from (1.2.3) that

$$u(t_1) < c + Ac_\varepsilon\, e^{\alpha(t_1 - \tau)} + B \int_0^{t_1} c_\varepsilon\, e^{\alpha s}\, ds = \left(c - \frac{B}{\alpha}c_\varepsilon\right) + c_\varepsilon\left(A\, e^{-\alpha\tau} + \frac{B}{\alpha}\right)e^{\alpha t_1}$$

$$= \left(c - \frac{B}{\alpha}c_\varepsilon\right) + c_\varepsilon\, e^{\alpha t_1}. \tag{1.2.6}$$

From the definition of α we see that $1 - A\, e^{-\alpha\tau} = \dfrac{B}{\alpha} > 0$ and so

$$c - \frac{B}{\alpha}c_\varepsilon \leqslant c - \frac{B}{\alpha}\frac{\varepsilon + c}{1 - A\, e^{-\alpha\tau}} < c - \frac{B}{\alpha}\frac{c}{1 - A\, e^{-\alpha\tau}}$$

$$= -\frac{c}{\alpha(1 - A\, e^{-\alpha\tau})}(B - \alpha + \alpha A\, e^{-\alpha\tau}) = 0.$$

Therefore (1.2.6) implies that $u(t_1) < c_\varepsilon\, e^{\alpha t_1}$ which contradicts (1.2.5) and completes the proof of the claim that (1.2.4) holds. Now as ε is arbitrary, (1.2.4) implies that $u(t) \leqslant M\, e^{\alpha t}$ for $t \geqslant 0$ where

$$M = \max\left\{c(1 - A\, e^{\alpha\tau})^{-1}, \ \max_{-r \leqslant s \leqslant 0} e^{-\alpha s}u(s)\right\}.$$

Since by the definition of $u(t)$ we know that $\|x(t)\| \leqslant u(t)$ for $t \geqslant 0$, the proof is complete.

1.3 The Laplace transform

In Sections 2.1, 5.1, and 6.3 we shall apply the Laplace transform to obtain necessary and sufficient conditions for the oscillation of linear autonomous delay differential equations in terms of their characteristic roots. Our aim in this section is to review some basic facts about Laplace transforms which will be needed in the above-mentioned sections. For further information on Laplace transforms see Widder (1971).

Let $x: [0, \infty) \to \mathbb{R}$ be a real-valued function. The Laplace transform of $x(t)$ is denoted by $L[x(t)]$ or $X(s)$ and is given by the improper integral

$$X(s) = \int_0^\infty e^{-st} x(t) \, dt. \tag{1.3.1}$$

By definition $X(s)$ is defined for all values of the complex variable s for which the improper integral in (1.3.1) converges in the sense that

$$\lim_{R \to \infty} \int_0^R e^{-st} x(t) \, dt$$

exists and is finite.

It can be shown—see for example Widder (1971)—that for a given function $x(t)$, the integral in (1.3.1) may behave in one of the following three ways:

(1) it converges for all complex numbers s;

(2) it diverges for all complex numbers s;

(3) there exists a real number σ_0 such that the integral in (1.3.1) converges for all s with $\text{Re } s > \sigma_0$ and diverges for all s with $\text{Re } s < \sigma_0$.

When (3) holds, the number σ_0 is called *the abscissa of convergence of $X(s)$*. When (1) holds, we say that the abscissa of convergence of $X(s)$ is $\sigma_0 = -\infty$. Finally when (2) holds, we say that the abscissa of convergence of $X(s)$ is $\sigma_0 = +\infty$. For example, the abscissae of convergence of the Laplace transforms of the functions e^{-t^2}, e^{3t}, and e^{t^2} are $-\infty$, 3, and $+\infty$ respectively.

The following lemma gives a sufficient condition for the existence of Laplace transform.

Lemma 1.3.1. *Let $x \in C[[0, \infty), \mathbb{R}]$ and suppose that there exist positive constants M and α such that*

$$|x(t)| \leq M e^{\alpha t} \quad \text{for } t \geq 0. \tag{1.3.2}$$

Then the abscissa of convergence σ_0 of the Laplace transform $X(s)$ of $x(t)$

satisfies

$$\sigma_0 \leqslant \alpha.$$

Furthermore, $X(s)$ exists and is an analytic function of s for $\mathrm{Re}\, s > \sigma_0$.

As we proved in Theorem 1.2.1, every solution of a linear autonomous differential equation is exponentially bounded, that is, satisfies (1.3.2). Therefore we now have licence to take the Laplace transform of such solutions and $X(s)$ exists for $\mathrm{Re}\, s > \alpha$.

The following lemma gives the Laplace transform of the shift function $x(t - \tau)$ and of the derivative $\dot{x}(t)$ in terms of the Laplace transform of $x(t)$. Its proof follows easily by an integration by parts.

Lemma 1.3.2.
  (a) *Let $x \in C[[-\tau, \infty), \mathbb{R}]$ and let $\sigma_0 < \infty$ be the abscissa of convergence of the Laplace transform $X(s)$ of $x(t)$. Then the Laplace transform of $x(t - \tau)$ has the same abscissa of convergence and*

$$L[x(t - \tau)] = \int_0^\infty e^{-st} x(t - \tau)\, \mathrm{d}t = e^{-s\tau} X(s) + e^{-s\tau} \int_{-\tau}^0 e^{-st} x(t)\, \mathrm{d}t. \quad (1.3.3)$$

for all s with $\mathrm{Re}\, s > \sigma_0$.
  (b) *Let $x \in C^1[[0, \infty), \mathbb{R}]$ and let $\sigma_0 < \infty$ be the abscissa of convergence of the Laplace transform $X(s)$ of $x(t)$. Then the Laplace transform of $\dot{x}(t)$ has the same abscissa of convergence and*

$$L[\dot{x}(t)] = \int_0^\infty e^{-st} \dot{x}(t)\, \mathrm{d}t = sX(s) - x(0)$$

for all s with $\mathrm{Re}\, s > \sigma_0$.

The following result from Widder (1971) will be needed in Sections 2.1, 5.1, and 6.3.

Theorem 1.3.1. *Let $x \in C[[0, \infty), \mathbb{R}^+]$ and assume that the abscissa of convergence σ_0 of the Laplace transform $X(s)$ of $x(t)$ is finite. Then $X(s)$ has a singularity at the point $s = \sigma_0$. More precisely, there exists a sequence*

$$s_n = \alpha_n + i\beta_n \qquad for\ n = 1, 2, \ldots$$

such that

$$\alpha_n \geqslant \alpha_0 \ \ for\ n \geqslant 1, \quad \lim_{n \to \infty} \alpha_n = \sigma_0, \quad \lim_{n \to \infty} \beta_n = 0 \quad and \quad \lim_{n \to \infty} |X(s_n)| = \infty.$$

1.4 The z-transform

The results in this section will be needed in Section 7.1. Let $\{a_n\}$ be a sequence of real numbers defined for $n = 0, 1, 2, \ldots$. The z-transform of this sequence is denoted by $Z(a_n)$ and is defined to be the series

$$Z(a_n) = \sum_{n=0}^{\infty} a_n/z^n. \tag{1.4.1}$$

The z-transform $Z(a_n)$ is assumed to be defined for all values of the complex variable z for which the series (1.4.1) converges. Clearly, if there exist positive numbers b and c such that

$$|a_n| \leqslant bc^n \qquad \text{for } n = 0, 1, 2, \ldots \tag{1.4.2}$$

then the series (1.4.1) converges for all $|z| > c$ and in this region it defines a complex analytic function of the variable z.

Consider the linear difference equation

$$a_{n+k} + P_1 a_{n+k-1} + \cdots + P_k a_n = 0, \qquad n = 0, 1, 2, \ldots \tag{1.4.3}$$

where k is a positive integer and the coefficients $P_1, P_2, \ldots, P_k$ are real numbers. It is interesting to note that every solution $\{a_n\}$ of eqn (1.4.3) satisfies (1.4.2) for some positive constants b and c. This follows immediately from the way that the solutions of eqn (1.4.3) can be expressed in terms of the roots of the characteristic equation

$$\lambda^k + P_1 \lambda^{k-1} + \cdots + P_k = 0. \tag{1.4.4}$$

See, for example, Finizio and Ladas (1982). A simple proof, free of characteristic roots, is given below in Lemma 1.4.2 for the general case where the coefficients $P_1, \ldots, P_k$ in eqn (1.4.3) are real $r \times r$ matrices.

Clearly the z-transform is a linear function. That is, if α and β are any constants and $\{a_n\}$ and $\{b_n\}$ are any sequences of complex numbers, then

$$Z(\alpha a_n + \beta b_n) = \alpha Z(a_n) + \beta Z(b_n).$$

The following lemma will be useful in Section 7.1.

Lemma 1.4.1. *Let* $k \in \{1, 2, \ldots\}$. *Then*

$$Z(a_{n+k}) = z^k Z(a_n) - \sum_{n=0}^{k-1} a_n z^{k-n}. \tag{1.4.5}$$

Proof. By the definition of the z-transform, we have

$$Z(a_{n+k}) = \sum_{n=0}^{\infty} \frac{a_{n+k}}{z^n} = z^k \sum_{n=0}^{\infty} \frac{a_{n+k}}{z^{n+k}} = z^k \sum_{n=k}^{\infty} \frac{a_n}{z^n} = z^k \left(\sum_{n=0}^{\infty} \frac{a_n}{z^n} - \sum_{n=0}^{k-1} a_n z^{-n} \right)$$

$$= z^k Z(a_n) - \sum_{n=0}^{k-1} a_n z^{k-n}$$

and the proof is complete.

By taking the z-transform of both sides of eqn (1.4.3) and by using the linearity of the z-transform and (1.4.5) we find (with $P_0 = 1$),

$$\sum_{i=0}^{k} P_i \left[z^{k-i} Z(a_n) - \sum_{j=0}^{k-i-1} a_j z^{k-i-j} \right] = 0$$

or

$$(z^k + P_1 z^{k-1} + \cdots + P_k) Z(a_n) = \sum_{i=0}^{k} P_i \sum_{j=0}^{k-i-1} a_j z^{k-i-j} \qquad (1.4.6)$$

where $P_0 = 1$.

Assume that the coefficients $P_1, \ldots, P_k$ in eqn (1.4.3) are real $r \times r$ matrices. Then a solution of eqn (1.4.3) is a sequence $\{a_n\}$ of vectors in $\mathbb{R}^r$. The z-transform of such a sequence is also defined by (1.4.1). As in the scalar case, if there exist positive numbers b and c such that

$$\|a_n\| \leqslant bc^n \qquad \text{for } n = 0, 1, 2, \ldots \qquad (1.4.7)$$

then the z-transform of $\{a_n\}$ converges for all $|z| > c$ and defines a complex analytic function of the variable z.

The next lemma provides a simple proof that every solution of a linear difference equation with constant coefficient matrices satisfies (1.4.7).

Lemma 1.4.2. *Assume that $P_1, P_2, \ldots, P_k$ are $r \times r$ matrices with real entries and let $\{a_n\}$ be any solution of eqn (1.4.3). Let $b = \max\{\|a_0\|, \ldots, \|a_k\|\}$ and let $c \in [1, \infty)$ be chosen in such a way that*

$$\|P_1\| c^{-1} + \cdots + \|P_k\| c^{-k} \leqslant 1. \qquad (1.4.8)$$

Then (1.4.7) is satisfied.

Proof. Clearly

$$\|a_n\| \leqslant b \leqslant bc^n \qquad \text{for } n = 0, 1, \ldots, k. \qquad (1.4.9)$$

Assume for the sake of contradiction that (1.4.7) is not true. Then there exists an $n_0 \geqslant k + 1$ such that

$$\|a_n\| \leqslant bc^n \qquad \text{for } n = 0, 1, \ldots, n_0 \qquad (1.4.10)$$

and

$$\|a_{n_0 + 1}\| > bc^{n_0 + 1}. \qquad (1.4.11)$$

For $n = (n_0 - k) + 1$, eqn (1.4.3) yields

$$a_{n_0 + 1} = -(P_1 a_{n_0} + \cdots + P_k a_{n_0 + 1 - k}).$$

Hence (1.4.10) and (1.4.8) imply

$$\|a_{n_0 + 1}\| \leqslant \|P_1\| bc^{n_0} + \cdots + \|P_k\| bc^{n_0 + 1 - k} = bc^{n_0 + 1}\left(\sum_{i=1}^{k} \|P_i\| c^{-i}\right) \leqslant bc^{n_0 + 1}$$

which contradicts (1.4.11) and completes the proof of the lemma.

For further information on z-transforms see Mickens (1987).

The following result, which is extracted from Hille (1963) is the discrete analogue of Theorem 1.3.1 and will be needed in Section 7.1.

Theorem 1.4.1. *Let $\{a_n\}$ be a sequence of non-negative real numbers and let*

$$\rho = \inf\left\{r \in \mathbb{R}: \sum_{n=0}^{\infty} \frac{a_n}{r^n} \text{ exists}\right\}$$

be finite. Then the z-transform $Z(a_n)$ of $\{a_n\}$ has a singularity at the point $z = \rho$. More precisely, there exists a sequence

$$z_n = x_n + iy_n \qquad \text{for } n = 1, 2, \ldots$$

such that

$$x_n \geqslant p \quad \text{for } n \geqslant 1, \quad \lim_{n \to \infty} x_n = \rho, \quad \lim_{n \to \infty} y_n = 0 \quad \text{and} \quad \lim_{n \to \infty} |Z(a_n)| = \infty.$$

1.5 Some basic lemmas

The results in this section will be needed in establishing the asymptotic behaviour of non-oscillatory solutions of certain neutral differential equations. These results may also have further applications in analysis.

Lemma 1.5.1. *Let $f, g: [t_0, \infty) \to \mathbb{R}$ be such that*

$$f(t) = g(t) + pg(t - c), \qquad t \geqslant t_0 + \max\{0, c\} \qquad (1.5.1)$$

where $p, c \in \mathbb{R}$ and $p \neq 1$. Assume that $\lim_{t \to \infty} f(t) \equiv l \in \mathbb{R}$ exists. Then the *following statements hold:*

1. *If* $\liminf_{t \to \infty} g(t) \equiv a \in \mathbb{R}$ *then* $l = (1 + p)a$.

2. *If* $\limsup_{t \to \infty} g(t) \equiv b \in \mathbb{R}$ *then* $l = (1 + p)b$.

Proof. We shall prove (1). The proof of (2) is similar and will be omitted. From (1.5.1) we see that

$$f(t + c) - f(t) = g(t + c) + (p - 1)g(t) - pg(t - c). \qquad (1.5.2)$$

Let $\{t_n\}$ be a sequence of points such that

$$\lim_{n \to \infty} t_n = \infty \qquad \text{and} \qquad \lim_{n \to \infty} g(t_n) = a. \qquad (1.5.3)$$

Case 1. Assume $p > 1$. By replacing t by $t_n + c$ in (1.5.2) and by taking limits we see that

$$\lim_{n \to \infty} [g(t_n + 2c) + (p - 1)g(t_n + c)] = pa. \qquad (1.5.4)$$

Set

$$a_1 = \liminf_{n \to \infty} g(t_n + 2c) \qquad \text{and} \qquad a_2 = \liminf_{n \to \infty} g(t_n + c).$$

Then clearly, $a_1 \geqslant a$, $a_2 \geqslant a$ and (1.5.4) implies that

$$a_1 + (p - 1)a_2 \leqslant pa. \qquad (1.5.5)$$

We now claim that

$$a_1 = a_2 = a. \qquad (1.5.6)$$

Otherwise

$$a_1 > a \qquad \text{or} \qquad a_2 > a$$

and so

$$a_1 + (p - 1)a_2 > a + (p - 1)a = pa$$

which contradicts (1.5.5). It follows from (1.5.6) and (1.5.3) that there exists a subsequence $\{t_{n_k}\}$ of $\{t_n\}$ such that

$$\lim_{k \to \infty} g(t_{n_k} + 2c) = \lim_{k \to \infty} g(t_{n_k} + c) = a.$$

By replacing t with $t_{n_k} + 2c$ in (1.5.1) and by taking limits as $k \to \infty$ we find

$$l = (1 + p)a$$

which completes the proof when $p > 1$.

Case 2: $0 \leqslant p < 1$. By replacing t by $t_n - c$ in (1.5.2) and by taking limits we find

$$\lim_{n \to \infty} \left[(1 - p)g(t_n - c) + pg(t_n - 2c)\right] = a.$$

The results now follows by an argument similar to that in Case 1.

Case 3: $p < 0$. By replacing t by t_n in (1.5.2) and by taking limits we obtain

$$\lim_{n \to \infty} \left[g(t_n + c) - pg(t_n - c)\right] = (1 - p)a,$$

from which the result follows as in Case 1. The proof of the lemma is complete.

Corollary 1.5.1. *Let $f, g: [t_0, \infty) \to \mathbb{R}$ satisfy*

$$f(t) = g(t) + pg(t - c), \qquad t \geqslant t_0 + \max\{0, c\}$$

where $p, c \in \mathbb{R}$ and $p \neq 1$. Assume that g is bounded on $[t_0, \infty)$ and that $\lim_{t \to \infty} f(t) \equiv l$ exists. Then the following statements hold:

1. *If $p = -1$ then $l = 0$.*

2. *If $p \neq \pm 1$ then $\lim_{t \to \infty} g(t)$ exists.*

Proof. As $g(t)$ is bounded, both

$$a \equiv \liminf_{t \to \infty} g(t) \qquad \text{and} \qquad b \equiv \limsup_{t \to \infty} g(t)$$

are real numbers and so both (1) and (2) of Lemma 1.5.1 hold. That is,

$$l = (1 + p)a = (1 + p)b.$$

Clearly if $p = -1$, $l = 0$ and if $p \neq \pm 1$, $a = b$ and so $\lim_{t \to \infty} g(t)$ exists.

Remark 1.5.1. *The following example shows that the conclusion of Corollary 1.5.1 may not be true when $p = 1$. Take*

$$f(t) \equiv 1, \qquad g(t) = \sin^2 t, \qquad c = \pi/2, \qquad \text{and} \qquad t_0 = 0.$$

Then $\lim_{t \to \infty} f(t) = 1$ and $g(t)$ is bounded but $\lim_{t \to \infty} g(t)$ does not exist.

Lemma 1.5.2. *Let $F, G, P: [t_0, \infty) \to \mathbb{R}$ and $c \in \mathbb{R}$ be such that*

$$F(t) = G(t) + P(t)G(t - c) \qquad \text{for } t \geqslant t_0 + \max\{0, c\}. \qquad (1.5.7)$$

Assume that there exist numbers $P_1, P_2, P_3, P_4 \in \mathbb{R}$ such that $P(t)$ is in one of the following ranges:

(1) $p_1 \leqslant P(t) \leqslant 0$ (1.5.8)

(2) $0 \leqslant P(t) \leqslant P_2 < 1$

(3) $1 < P_3 \leqslant P(t) \leqslant P_4$.

Suppose that $G(t) > 0$ for $t \geqslant t_0$, $\lim \inf_{t \to \infty} G(t) = 0$ and that $\lim_{t \to \infty} F(t) \equiv L \in \mathbb{R}$ exists. Then $L = 0$.

Proof. From (1.5.7) we see that

$$F(t + c) - F(t) = G(t + c) + [P(t + c) - 1]G(t) - P(t)G(t - c)$$

and so

$$\lim_{t \to \infty} \{G(t + c) + [P(t + c) - 1]G(t) - P(t)G(t - c)\} = 0. \quad (1.5.9)$$

Let $\{t_n\}$ be a sequence of points such that

$$\lim_{n \to \infty} t_n = \infty \quad \text{and} \quad \lim_{n \to \infty} G(t_n) = 0. \quad (1.5.10)$$

We shall prove the lemma when (1) holds. The cases where (2) or (3) holds are similar and will be omitted. By replacing t by t_n in (1.5.9) and by using (1.5.10) and the fact that $P(t)$ is bounded, we obtain

$$\lim_{n \to \infty} [G(t_n + c) - P(t_n)G(t_n - c)] = 0.$$

As $G(t_n + c) > 0$ and $P(t_n)G(t_n - c) \leqslant 0$, it follows that $\lim_{n \to \infty} [P(t_n)G(t_n - c)] = 0$ and so $L = \lim_{n \to \infty} F(t_n) = \lim_{n \to \infty} [G(t_n) + P(t_n)G(t_n - c)] = 0$. The proof is complete.

Lemma 1.5.3. *Let $p \in [0, 1)$, $\tau \in (0, \infty)$, $t_0 \in \mathbb{R}$, $x \in C[[t_0 - \tau, \infty), \mathbb{R}^+]$ and assume that for every $\varepsilon > 0$ there exists a $t_\varepsilon \geqslant t_0$ such that*

$$x(t) \leqslant (p + \varepsilon)x(t - \tau) + \varepsilon \qquad \text{for } t \geqslant t_\varepsilon. \quad (1.5.11)$$

Then

$$\lim_{t \to \infty} x(t) = 0. \quad (1.5.12)$$

Proof. First we claim that $x(t)$ is a bounded function. Otherwise there exists a sequence of points $\{t_n\}$ such that

$$\lim_{n \to \infty} t_n = \infty, \quad x(t_n) = \max_{t_0 - \tau \leqslant s \leqslant t_n} x(s) \text{ for } n = 1, 2, \ldots \quad \text{and} \quad \lim_{n \to \infty} x(t_n) = \infty.$$

$$(1.5.13)$$

Choose $\varepsilon \in (0, 1 - p)$. Then for n sufficiently large, (1.5.11) yields

$$x(t_n) \leqslant (p + \varepsilon)x(t_n - \tau) + \varepsilon \leqslant (p + \varepsilon)x(t_n) + \varepsilon$$

or

$$0 \leqslant (1 - p - \varepsilon)x(t_n) \leqslant \varepsilon. \tag{1.5.14}$$

Clearly (1.5.14) contradicts (1.5.13) and so $x(t)$ is a bounded function. Set

$$M = \limsup_{t \to \infty} x(t) < \infty$$

and choose $\varepsilon > 0$ so small that

$$M > (p + \varepsilon)M + \varepsilon.$$

On the other hand (1.5.11) implies that

$$M \leqslant (p + \varepsilon)M + \varepsilon$$

which is a contradiction, and the proof is complete.

Lemma 1.5.4. *Let $a \in (-\infty, 0)$, $\tau \in (0, \infty)$, $t_0 \in \mathbb{R}$ and suppose that a function $x \in C[[t_0 - \tau, \infty), \mathbb{R}]$ satisfies the inequality*

$$x(t) \leqslant a + \max_{t-\tau \leqslant s \leqslant t} x(s) \qquad \text{for } t \geqslant t_0. \tag{1.5.15}$$

Then x cannot be a non-negative function.

Proof. Assume, for the sake of contradiction, that $x(t) \geqslant 0$ for $t \geqslant t_0$. First we claim that $x(t)$ is a bounded function. Otherwise there exists a $t_1 \geqslant t_0$ such that

$$x(t_1) = \max_{t_0 - \tau \leqslant s \leqslant t_1} x(s).$$

Then (1.5.15) yields

$$x(t_1) \leqslant a + \max_{t_1 - \tau \leqslant s \leqslant t_1} x(s) \leqslant a + x(t_1) < x(t_1)$$

which is a contradiction. Hence

$$M = \limsup_{t \to \infty} x(t) < \infty.$$

It now follows from (1.5.15) that

$$M \leqslant a + M < M$$

which is impossible. The proof is complete.

Lemma 1.5.5. *Let $F, G, P \in C[[t_0, \infty), \mathbb{R}]$ and $c \in (0, \infty)$ be such that*

$$F(t) = G(t) + P(t)G(t - c), \qquad t \geqslant t_0 + c.$$

Assume that

$$F(t) > 0 \quad and \quad G(t) > 0 \quad for\ t \geqslant t_0, \qquad \lim_{t \to \infty} F(t) = 0,$$

and suppose that there exists a $P \in (-1, 0]$ such that

$$-1 < P \leqslant P(t) \leqslant 0.$$

Then

$$\lim_{t \to \infty} G(t) = 0.$$

Proof. First we claim that $G(t)$ must be a bounded function. Otherwise there exists a sequence $\{t_n\}$ such that

$$\lim_{n \to \infty} t_n = \infty, \qquad G(t_n) = \max\{G(s): s \leqslant t_n\} \qquad for\ n = 1, 2, \ldots$$

and

$$\lim_{n \to \infty} G(t_n) = \infty.$$

Then

$$F(t_n) = G(t_n) + P(t_n)G(t_n - c) \geqslant G(t_n)(1 + P) \to \infty, \qquad as\ n \to \infty,$$

which is a contradiction. Thus $G(t)$ is indeed bounded. Let $l = \limsup_{t \to \infty} G(t)$ and let $\{\tilde{t}_n\}$ be a sequence of points such that $\lim_{n \to \infty} \tilde{t}_n = \infty$ and $\lim_{n \to \infty} G(\tilde{t}_n) = l$. Then

$$PG(\tilde{t}_n - c) \leqslant P(\tilde{t}_n)G(\tilde{t}_n - c) = F(\tilde{t}_n) - G(\tilde{t}_n)$$

or

$$G(\tilde{t}_n - c) \geqslant \frac{F(\tilde{t}_n) - G(\tilde{t}_n)}{P} \to \frac{-l}{P} \qquad as\ n \to \infty.$$

Hence

$$l \geqslant -l/P$$

or $l(1 + P) \leqslant 0$ which implies that $l = 0$, and the proof is complete.

1.6 Some basic lemmas on differential inequalities

The results in this section will be useful in establishing certain properties of non-oscillatory solutions of differential equations.

Lemma 1.6.1. *Let p and τ be positive constants. Let $x(t)$ be an eventually positive solution of the delay differential inequality*

$$\dot{x}(t) + px(t - \tau) \leqslant 0 \tag{1.6.1}$$

and let $y(t)$ be an eventually positive solution of the advance differential inequality

$$\dot{y}(t) - py(t + \tau) \geqslant 0. \tag{1.6.2}$$

Then for t sufficiently large,

$$x(t - \tau) \leqslant Bx(t) \tag{1.6.3}$$

and

$$y(t + \tau) \leqslant By(t) \tag{1.6.4}$$

where

$$B = 4/(p\tau)^2.$$

Proof. We shall prove (1.6.3). The proof of (1.6.4) is similar and will be omitted. Assume that t_0 is such that $x(t) > 0$ for $t \geqslant t_0 - \tau$ and $x(t)$ satisfies (1.6.1) for $t \geqslant t_0$. For given $s \geqslant t_0 + \tau$ we integrate both sides of (1.6.1) from s to $s + \tau/2$, and by using the fact that x is decreasing for $t \geqslant t_0$ we find that

$$x(s + \tau/2) - x(s) + (p\tau/2)x(s - \tau/2) \leqslant 0$$

or

$$(p\tau/2)x(s - \tau/2) \leqslant x(s) \qquad \text{for } s \geqslant t_0 + \tau. \tag{1.6.5}$$

For any $t \geqslant t_0 + 3\tau/2$ we apply (1.6.5) for $s = t - \tau/2$ and for $s = t$, and find

$$(p\tau/2)x(t - \tau) \leqslant x(t - \tau/2) \qquad \text{and} \qquad (p\tau/2)x(t - \tau/2) \leqslant x(t).$$

The assertion (1.6.3) now follows by combining these inequalities.

Lemma 1.6.2. *Let $v(t)$ be a positive and continuously differentiable function on some interval $[t_0, \infty)$. Assume that there exist positive numbers A and α such that for t sufficiently large,*

$$\dot{v}(t) \leqslant 0 \qquad \text{and} \qquad v(t - \alpha) < Av(t) \tag{1.6.6}$$

or

$$\dot{v}(t) \geqslant 0 \qquad and \qquad v(t + \alpha) < Av(t). \qquad (1.6.7)$$

Set

$$\Lambda = \{\lambda \geqslant 0 : \dot{v}(t) + \lambda v(t) \leqslant 0 \text{ for } t \text{ sufficiently large}\} \text{ if } (1.6.6) \text{ holds},$$

or

$$\Lambda = \{\lambda \geqslant 0 : -\dot{v}(t) + \lambda v(t) \leqslant 0 \text{ for } t \text{ sufficiently large}\} \text{ if } (1.6.7) \text{ holds}.$$

Then $A > 1$ *and*

$$\lambda_0 = \frac{\ln A}{\alpha} \notin \Lambda.$$

Proof. We shall prove the lemma when (1.6.6) holds. The case when (1.6.7) holds is similar and will be omitted. Assume that (1.6.6) holds and, for the sake of contradiction, assume that $\lambda_0 \in \Lambda$. Then eventually,

$$\frac{d}{dt}[e^{\lambda_0 t}v(t)] = e^{\lambda_0 t}[\dot{v}(t) + \lambda_0 v(t)] \leqslant 0,$$

which implies that the function $e^{\lambda_0 t}v(t)$ is eventually decreasing. Hence for t sufficiently large,

$$e^{\lambda_0(t-\alpha)}v(t - \alpha) \geqslant e^{\lambda_0 t}v(t)$$

or

$$v(t - \alpha) \geqslant e^{\lambda_0 \alpha}v(t) = e^{\ln A}v(t) = Av(t)$$

which contradicts (1.6.6) and completes the proof of the lemma.

Lemma 1.6.3. *Suppose that* $P \in C[[T, \infty), (0, \infty)]$, $\tau \in C[[T, \infty), [0, \infty)]$,

$$\liminf_{t \to \infty} (t - \tau(t)) = \infty \qquad (1.6.8)$$

and

$$\liminf_{t \to \infty} \int_{t-\tau(t)}^{t} P(s)\, ds > 0. \qquad (1.6.9)$$

Set

$$T_{-1} = \inf_{t \geqslant T} \{t - \tau(t)\}.$$

Assume that $\alpha \in C[[T_{-1}, \infty), (-\infty, 0]]$ satisfies the inequality

$$\alpha(t) + P(t) \exp\left[-\int_{t-\tau(t)}^{t} \alpha(s) \, ds\right] \leqslant 0, \qquad t \geqslant T. \tag{1.6.10}$$

Then

$$\liminf_{t \to \infty} \left[-\int_{t-\tau(t)}^{t} \alpha(s) \, ds\right] < \infty. \tag{1.6.11}$$

Proof. Set

$$x(t) = \exp\left(\int_{T-1}^{t} \alpha(s) \, ds\right) \qquad \text{for } t \geqslant T_{-1}.$$

Then

$$\frac{x(t - \tau(t))}{x(t)} = \exp\left(-\int_{t-\tau(t)}^{t} \alpha(s) \, ds\right) \tag{1.6.12}$$

and so (1.6.10) implies that

$$\dot{x}(t) + P(t)x(t - \tau(t)) \leqslant 0, \qquad t \geqslant T. \tag{1.6.13}$$

As $x(t) > 0$ for $t \geqslant T$, and because of (1.6.8), there exists a $T_1 \geqslant T$ such that

$$x(t - \tau(t)) > 0 \qquad \text{for } t \geqslant T_1.$$

Also, (1.6.13) implies that

$$\dot{x}(t) < 0 \qquad \text{for } t \geqslant T_1. \tag{1.6.14}$$

Define

$$h(t) = \max_{T_1 \leqslant s \leqslant t} \{s - \tau(s)\}. \tag{1.6.15}$$

Then, in view of (1.6.14),

$$x(t - \tau(t)) \geqslant x(h(t)) \qquad \text{for } t \geqslant T_1,$$

and so (1.6.13) yields

$$\dot{x}(t) + P(t)x(h(t)) \leqslant 0, \qquad t \geqslant T_1. \tag{1.6.16}$$

We now claim that

$$\liminf_{t \to \infty} \int_{h(t)}^{t} P(s) \, ds > 0. \tag{1.6.17}$$

Otherwise there exists a sequence $\{\rho_n\}$ such that $\lim_{n \to \infty} \rho_n = \infty$ and

$$\lim_{n \to \infty} \int_{h(\rho_n)}^{\rho_n} P(s) \, ds = 0.$$

On the other hand, for each $n = 1, 2, \ldots$ there exists $\xi_n \in [T_1, \rho_n]$ such that, because of (1.6.15) and (1.6.8),

$$h(\rho_n) = \xi_n - \tau(\xi_n) \qquad \text{and} \qquad \lim_{n \to +\infty} \xi_n = \infty.$$

Then

$$\int_{h(\rho_n)}^{\rho_n} P(s) \, ds = \int_{\xi_n - \tau(\xi_n)}^{\xi_n} P(s) \, ds + \int_{\xi_n}^{\rho_n} P(s) \, ds \geq \int_{\xi_n - \tau(\xi_n)}^{\xi_n} P(s) \, ds,$$

which implies that

$$\lim_{n \to \infty} \int_{\xi_n - \tau(\xi_n)}^{\xi_n} P(s) \, ds = 0.$$

This contradicts (1.6.9) and so (1.6.17) has been established.

Let $\{t_n\}$ and $\{t_n^*\}$ be two increasing sequences of points tending to ∞, as $n \to \infty$, and such that for some $\gamma > 0$,

$$\int_{h(t_n)}^{t_n^*} P(s) \, ds \geq \gamma \qquad \text{and} \qquad \int_{t_n^*}^{t_n} P(s) \, ds \geq \gamma \qquad \text{for } n = 1, 2, \ldots.$$

This is possible because of (1.6.17). By integrating (1.6.16) from $h(t_n)$ to t_n^* we find

$$x(t_n^*) - x(h(t_n)) + \int_{h(t_n)}^{t_n^*} P(s) x(h(s)) \, ds \leq 0$$

or

$$\gamma x(h(t_n^*)) \leq x(h(t_n)). \tag{1.6.18}$$

Similarly, by integrating (1.6.16) from t_n^* to t_n we find

$$\gamma x(h(t_n)) \leq x(t_n^*). \tag{1.6.19}$$

It follows from (1.6.18) and (1.6.19) that

$$\gamma^2 x(h(t_n^*)) \leq x(t_n^*).$$

Let $\sigma_n \leq t_n^*$ be such that

$$h(t_n^*) = \sigma_n - \tau(\sigma_n).$$

Then

$$\gamma^2 x(\sigma_n - \tau(\sigma_n)) \leqslant x(t_n^*) \leqslant x(\sigma_n)$$

and by the definition of x we see that

$$\exp\left(-\int_{\sigma_n - \tau(\sigma_n)}^{\sigma_n} \alpha(s)\,\mathrm{d}s\right) \leqslant 1/\gamma^2,$$

from which (1.6.11) is an immediate consequence. The proof is complete.

1.7 Some useful theorems from analysis

In this section we state for the convenience of readers several results from analysis which are needed throughout this book. Among them are Schauder's fixed point theorem, the Banach contraction principle, the Knaster–Tarski fixed point theorem and the Lebesgue dominated convergence theorem. These theorems are employed, for example, when we want to establish that a certain delay differential equation has a positive (or non-oscillatory) solution.

First we introduce some terminology. A vector space X on which we have defined a norm $\|\cdot\|$ is called a *normed vector space*. A sequence $\{x_n\}$ of vector in a normed vector space X is said to *converge* to the vector $x \in X$ if and only if the sequence of non-negative numbers $\|x_n - x\|$ converges to zero. In such a case we write

$$\lim_{n \to \infty} x_n = x$$

and we call x the *limit* of $\{x_n\}$.

We say that $\{x_n\}$ is a *Cauchy sequence* of vectors in X if for every $\varepsilon > 0$ there exists an $N = N(\varepsilon)$ such that for all $n, m \geqslant N(\varepsilon)$ we have $\|x_n - x_m\| < \varepsilon$. Clearly a convergent sequence is a Cauchy sequence, but the converse may not be true. A space X where every Cauchy sequence of elements of X converges to an element of X is called a *complete* space. A *Banach space* is a complete normed vector space. That is, a Banach space if a vector space X with a norm in which every Cauchy sequence of vectors converges to a vector in X.

Example 1.7.1. An important example of a Banach space for our investigations is the space X of all real, bounded, and continuous functions f on some interval $[T, \infty)$ with the sup norm

$$\|f\| = \sup_{T \leqslant x < \infty} |f(x)|.$$

Let M be a subset of a Banach space. We say that a point $x \in X$ is a *limit point* of M if there exists a sequence of vectors in M which converges to x. We say that M is *closed* if M contains all of its limit points. The union of M and its limit points is called the *closure* of M and will be denoted by $\bar{M}$. A set M is called *convex* if for every $x, y \in M$ and for every $\lambda \in [0, 1]$,

$$\lambda x + (1 - \lambda) y \in M.$$

We say that a subset M of a Banach space X is *compact* if every sequence of vectors in M contains a subsequence which converges to a vector in M. We say that M is *relatively compact* if every sequence of vectors in M contains a subsequence which converges to a vector in X. That is, M is relatively compact if $\bar{M}$ is compact.

By the Ascoli–Arzela theorem, a subset M of the space $C[[a, b], \mathbb{R}]$ with the sup-norm is relatively compact provided that M is equicontinuous and uniformly bounded. We say that a family M of continuous functions from $[a, b]$ to $\mathbb{R}$ is *equicontinuous* if for every $\varepsilon > 0$ there exists a $\delta = \delta(\varepsilon) > 0$ such that for all $t_1, t_2 \in [a, b]$ with $|t_1 - t_2| < \delta$ and for all $f \in M$ we have $|f(t_1) - f(t_2)| < \varepsilon$. The family M is called *uniformly bounded* if there exists a positive number B such that

$$|f(t)| \leqslant B \qquad \text{for all } t \in [a, b] \text{ and for all } f \in M.$$

The Ascoli–Arzela theorem states that any sequence $\{f_n\}$ of function from $[a, b]$ into $\mathbb{R}^m$, whose terms are equicontinuous and uniformly bounded, contains a subsequence $\{f_{n_k}\}$ which converges uniformly on $[a, b]$.

Clearly a family M of differentiable functions from $[a, b]$ into $\mathbb{R}$, with uniformly bounded derivatives, is equicontinuous. Indeed, if $|f'(t)| \leqslant K$ for every $f \in M$, then by the mean value theorem,

$$|f(t_1) - f(t_2)| \leqslant K |t_1 - t_2| \tag{1.7.1}$$

and equicontinuity follows with $\delta = \varepsilon/K$. If the interval $[a, b]$ is finite, then (1.7.1) implies that the family M is also uniformly bounded. Indeed, by taking $t_1 = t$ and $t_2 = b$, (1.7.1) implies that

$$|f(t)| - |f(b)| \leqslant |f(t) - f(b)| \leqslant K |t - b|$$

and so

$$|f(t)| \leqslant |f(b)| + k(b - a).$$

Let M be a subset of a Banach space and let $T: M \to M$ be a function (or a mapping) of M into itself. We say that T is *continuous at the point* $x_0 \in M$ if for every $\varepsilon > 0$ there exists a $\delta = \delta(\varepsilon, x_0) > 0$ such that for all $x \in M$ with $\|x - x_0\| < \delta$ we have $\|Tx - Tx_0\| < \varepsilon$. We say that $T: M \to M$ is a *continuous function* if T is continuous at every point $x \in M$.

We are now ready to state Schauder's fixed-point theorem (Schauder 1930).

Theorem 1.7.1 (Schauder's fixed-point theorem). *Let M be a closed, convex and non-empty subset of a Banach space X. Let*

$$T: M \to M$$

be a continuous function such that TM is a relatively compact subset of X. Then T has at least one fixed point in M. That is, there exists an $x \in M$ such that $Tx = x$.

Before we state the Banach contraction mappping principle, we introduce some terminology. Let X be any set. A *distance* in X is a function $d: X \times X \to X$ having the following properties for all $x, y, z \in X$:

(1) $d(x, y) \geq 0$ and $d(x, y) = 0$ if and only if $x = y$;

(2) $d(x, y) = d(y, x)$;

(3) $d(x, z) \leq d(x, y) + d(y, z)$ (triangle inequality).

A *metric space* is a set X together with a given distance in X. Let $\{x_n\}$ be a sequence of points in a metric space X. We say that $\{x_n\}$ converges to $x \in X$ if the sequence of non-negative numbers $d(x_n, x)$ converges to zero. A sequence $\{x_n\}$ is called a *Cauchy sequence* if for every $\varepsilon > 0$ there exists an $N = N(\varepsilon)$ such that for all $n, m \geq N(\varepsilon)$ we have $d(x_n, x_m) < \varepsilon$. A *complete metric space* is a metric space X in which every Cauchy sequence converges to a point in X.

Example 1.7.2. An important example of a complete metric space for our investigations is the following. Let $\mu, \eta \in (0, \infty)$ and $T \in \mathbb{R}$. Define the set of functions

$$X = \{x \in C[[T, \infty), \mathbb{R}]: |x(t)| \leq \mu \text{ for } t \geq T\}$$

and for $x_1, x_2 \in X$ define the distance

$$d(x_1, x_2) = \sup_{t \geq T} |x_1(t) - x_2(t)| \, e^{-\eta t}.$$

Then (X, d) is a complete metric space.

Let (X, d) be a metric space and let $T: X \to X$. We say that T is a *contraction mapping* on X if there exists a number $r \in [0, 1)$ such that

$$d(Tx, Ty) \leq rd(x, y) \qquad \text{for every } x, y \in X.$$

For a proof of the following theorem, called the *Banach contraction principle*, see Simmons (1963) or Coppel (1965).

Theorem 1.7.2 (the Banach contraction principle). *A contraction mapping on a complete metric space has exactly one fixed point.*

A non-empty and closed subset K of a Banach space X is called a *cone* if it possesses the following properties:

(1) if $\alpha \in \mathbb{R}^+$ and $x \in K$, then $\alpha x \in K$;

(2) if $x, y \in K$ then $x + y \in K$;

(3) if $x \in K - \{0\}$ then $-x \notin K$.

We say that a Banach space X is *partially ordered* if X contains a cone K with non-empty interior. The ordering $\leqslant$ in X is then defined as follows:

$$x \leqslant y \qquad \text{if and only if } y - x \in K.$$

Let M be a subset of a partially ordered Banach space X. Set

$$\tilde{M} = \{x \in X : y \leqslant x \text{ for every } y \in M\}.$$

We say that the point $x_0 \in X$ is the *infimum* of M if $x_0 \in \tilde{M}$ and for every $x \in \tilde{M}$, $x_0 \leqslant x$. The *supremum* of M is defined in a similar way.

We are now ready to state the Knaster–Tarski fixed-point theorem (Tarski 1955).

Theorem 1.7.3 (the Knaster–Tarski fixed-point theorem). *Let X be a partially ordered Banach space with ordering $\leqslant$. Let M be a subset of X with the following properties: the infimum of M belongs to M and every non-empty subset of M has a supremum which belongs to M. Let $T: M \to M$ be an increasing mapping, that is, $x \leqslant y$ implies $Tx \leqslant Ty$. Then T has a fixed point in M.*

Finally we present the statement of the Lebesgue dominated convergence theorem which we shall use on several occasions throughout this monograph. See Kolmogorov and Fomin (1970) or Rudin (1966).

Theorem 1.7.4 (Lebesgue's dominated convergence theorem). *Let $\{f_n\}$ be a sequence of functions such that*

$$\lim_{n \to \infty} f_n(x) = f(x) \qquad \text{a.e. in } A$$

and such that for every $n = 1, 2, \ldots$

$$|f_n(x)| \leqslant g(x) \qquad a.e.\ in\ A$$

where g is integrable on A. Then

$$\lim_{n \to \infty} \int_A f_n(x)\, \mathrm{d}\mu = \int_A f(x)\, \mathrm{d}\mu.$$

1.8 Notes

Lemmas 1.5.3 and 1.5.4 are new. For the remaining results in Section 1.5 see Chuanxi *et al.* (1989) and Chuanxi and Ladas (1989*b*).

Lemma 1.6.1 is extracted from Ladas *et al.* (1983*b*). Lemma 1.6.2 is from Grammatikopoulos *et al.* (1988*b*). Lemma 1.6.3 is from Györi (1984) but the proof here is new.

It is of great interest, for its own sake and for applications to neutral equations with several delays, to extend Lemma 1.5.1 to functions f and g such that

$$f(t) = \sum_{i=1}^{n} p_i g(t - c_i)$$

where $p_i, c_i \in \mathbb{R}$ for $i = 1, 2, \ldots, n$. A similar extension of Lemma 1.5.2 is also of immense interest.

2

OSCILLATIONS OF LINEAR SCALAR DELAY EQUATIONS

In this chapter we present some of the most basic results in the theory of oscillations of linear delay differential equations with constant and variable coefficients and with constant delays. We begin with a detailed description of the concept of *oscillation*.

Consider the linear delay differential equation

$$\dot{x}(t) + \sum_{i=1}^{n} P_i(t)x(t - \tau_i) = 0 \qquad (2.0.1)$$

where

$$P_i \in C[[t_0, \infty), \mathbb{R}] \qquad \text{and} \qquad \tau_i \in \mathbb{R}^+ \qquad \text{for } i = 1, 2, \ldots, n. \quad (2.0.2)$$

Let $\tau = \max\{\tau_1, \tau_2, \ldots, \tau_n\}$. We recall from Section 1.1 that by a *solution* of eqn (2.0.1) we mean a function $x \in C[[t_1 - \tau, \infty), \mathbb{R}]$, for some $t_1 \geq t_0$, such that x is continuously differentiable on $[t_1, \infty)$ and x satisfies eqn (2.0.1) for $t \geq t_1$. Such a solution is called *a solution on* $[t_1, \infty)$.

Let an *initial point* t_1 and an *initial function* $\phi \in C[[t_1 - \tau, t_1], \mathbb{R}]$ be given. Then, as we proved in Section 1.1, eqn (2.0.1) has a unique solution x on $[t_1, \infty)$ such that

$$x(t) = \phi(t) \qquad \text{for } t_1 - \tau \leq t \leq t_1. \qquad (2.0.3)$$

Let x be a continuous function defined on some infinite interval $[a, \infty)$. The function x is said to *oscillate* or to be *oscillatory* if x has arbitrarily large zeros. That is, for every $b > a$ there exists a point $c > b$ such that $x(c) = 0$. Otherwise x is called *non-oscillatory*. That is, x is non-oscillatory if there exists a $b > a$ such that $x(t) \neq 0$ for $t > b$. As x is continuous, if it is non-oscillatory it must be eventually positive or eventually negative. That is, there exists a $T \in \mathbb{R}$ such that $x(t)$ is positive for $t \geq T$ or is negative for $t \geq T$.

For a linear differential equation the opposite of a solution is also a solution of the same equation and so, if a linear differential equation has a non-oscillatory solution, then it also has an eventually positive (as well as eventually negative) solution.

Assume that (2.0.2) holds. When we say in this book that 'every solution of eqn (2.0.1) oscillates' we mean that for every initial point $t_1 \geq t_0$ and for every initial function $\phi \in C[[t_1 - \tau, t_1], \mathbb{R}]$ the unique solution x of the initial value problem (2.0.1) and (2.0.3) oscillates, that is, x has arbitrarily large zeros. When we want to prove that eqn (2.0.1) has a non-oscillatory solution it suffices to prove that eqn (2.0.1) has an eventually positive solution x. That is, there exist points $t_2 \geqslant t_1 \geqslant t_0$ such that x is a solution on $[t_1, \infty)$ and $x(t) > 0$ for $t \geqslant t_2$.

For the special case of a linear autonomous delay equation where the coefficients are constants, when we say that every solution oscillates we mean the following: for every $t_1 \in \mathbb{R}$ and for every initial function $\phi \in C[[t_1 - \tau, t_1], \mathbb{R}]$, where τ is the maximum delay, the unique solution of the equation which satisfies the condition (2.0.3) oscillates.

For equations with both delay and advanced arguments and for equations with just advanced arguments, as a rule, we do not have existence and uniqueness theorems for solutions on infinite intervals of the form $[a, \infty)$. For such equations, when we say that all solutions oscillate we mean that every continuous function which satisfies the equation on some infinite interval $[a, \infty)$ has arbitrarily large zeros.

2.1 Necessary and sufficient conditions for oscillations. The autonomous case

Consider the linear autonomous delay differential equation

$$\dot{x}(t) + \sum_{i=1}^{n} p_i x(t - \tau_i) = 0 \tag{2.1.1}$$

where the coefficients p_i are real numbers and the delays τ_i are non-negative real numbers. With eqn (2.1.1) one associates its characteristic equation

$$\lambda + \sum_{i=1}^{n} p_i e^{-\lambda \tau_i} = 0. \tag{2.1.2}$$

Our first aim in this section is to establish the following fundamental result for the oscillation of all solutions of eqn (2.1.1).

Theorem 2.1.1. *Assume that*

$$p_i \in \mathbb{R} \quad \text{and} \quad \tau_i \in \mathbb{R}^+ \quad \text{for } i = 1, 2, \ldots, n. \tag{2.1.3}$$

Then the following statements are equivalent.

(a) *Every solution of eqn (2.1.1) oscillates.*

(b) *The characteristic equation (2.1.2) has no real roots.*

Proof. The proof that (a) $\Rightarrow$ (b) is elementary. This is because if the characteristic equation (2.1.2) has a real root λ_0 then $e^{\lambda_0 t}$ is a non-oscillatory solution of eqn (2.1.1).

The proof that (b) $\Rightarrow$ (a) makes use of Laplace transforms and Theorem 1.3.1. Assume, for the sake of contradiction, that (b) holds and that eqn (2.1.1) has an eventually positive solution $x(t)$. As eqn (2.1.1) is autonomous, we may (and do) assume that

$$x(t) > 0 \qquad \text{for } t \geqslant -\tau \text{ where } \tau = \max_{1 \leqslant i \leqslant n} \tau_i.$$

Clearly $\tau > 0$, for otherwise eqn (2.1.2) has a real root. By Theorem 1.2.1 we know that there exist constants M and μ such that

$$|x(t)| \leqslant M \, e^{\mu t}, \qquad t \geqslant -\tau.$$

Thus the Laplace transform

$$X(s) = \int_0^\infty e^{-st} x(t) \, dt \tag{2.1.4}$$

exists for Re $s > \mu$. Let σ_0 be the abscissa of convergence of $X(s)$, that is (see Section 1.3), $\sigma_0 = \inf\{\sigma \in \mathbb{R} : X(\sigma) \text{ exists}\}$. Then for any $i = 1, \ldots, n$, the Laplace transform of $x(t - \tau_i)$ exists and has abscissa of convergence σ_0. Furthermore by Lemma 1.3.2,

$$\int_0^\infty e^{-st} \dot{x}(t) \, dt = sX(s) - x(0), \qquad \text{Re } s > \sigma_0$$

and for $i = 1, 2, \ldots, n$

$$\int_0^\infty e^{-st} x(t - \tau_i) \, dt = e^{-s\tau_i} X(s) + e^{-s\tau_i} \int_{-\tau_i}^0 e^{-st} x(t) \, dt, \qquad \text{Re } s > \sigma_0.$$

Therefore by taking Laplace transforms of both sides of (2.1.1) we obtain

$$F(s)X(s) = \Phi(s), \qquad \text{Re } s > \sigma_0 \tag{2.1.5}$$

where

$$F(s) = s + \sum_{i=1}^n p_i \, e^{-s\tau_i} \tag{2.1.6}$$

and

$$\Phi(s) = x(0) - \sum_{i=1}^n p_i \, e^{-s\tau_i} \int_{-\tau_i}^0 e^{-st} x(t) \, dt. \tag{2.1.7}$$

Clearly, $F(s)$ and $\Phi(s)$ are entire functions. Also by hypothesis, $F(s) \neq 0$ for all real s. It follows from (2.1.5) that

$$X(s) = \frac{\Phi(s)}{F(s)}, \qquad \text{Re } s > \sigma_0. \tag{2.1.8}$$

We now claim that $\sigma_0 = -\infty$. Otherwise $\sigma_0 > -\infty$ and by Theorem 1.3.1 the point $s = \sigma_0$ must be a singularity of the quotient $\Phi(s)/F(s)$. But this quotient has no singularity on the real axis (the numerator and denominator are entire functions and by hypothesis the denominator has no real zeros). Thus $\sigma_0 = -\infty$ and (2.1.8) becomes

$$X(s) = \frac{\Phi(s)}{F(s)} \qquad \text{for all } s \in \mathbb{R}. \tag{2.1.9}$$

One can now see that as $s \to -\infty$, through real values, (2.1.9) leads to a contradiction because $X(s)$ and $F(s)$ are always positive while $\Phi(s)$ becomes eventually negative. The positivity of $X(s)$ follows from (2.1.4) and the fact that $x(t) > 0$ for $t \geqslant 0$. The positivity of $F(s)$ follows from (2.1.6) and the facts that $F(\infty) = \infty$ and that the characteristic equation has no real roots. Without loss of generality we may assume that the delays in eqn (2.1.1) are distinct and that the coefficients p_i are different from zero. Let τ_{i_0} be the maximum delay in eqn (2.1.1). Then the corresponding coefficient $p_{i_0} > 0$, for otherwise $\lim_{s \to \infty} F(s) = -\infty$ and the dominant term in (2.1.7), as $s \to -\infty$, is $p_{i_0} e^{-s\tau_{i_0}}$. Clearly $\lim_{s \to -\infty} \Phi(s) = -\infty$. The proof is complete.

The proof of Theorem 2.1.1 makes use of the fact that the solutions of linear delay differential equations are exponentially bounded. See Theorem 1.2.1, which we used in the proof of Theorem 2.1.1 when we assumed that the solution $x(t)$ had a Laplace transform. For non-delay differential equations, that is, when $\tau_i \in \mathbb{R}$, it is not known whether their solutions are exponentially bounded. If we could prove it, then Theorem 2.1.1 and its proof would also be valid for such equations. However, we still believe that the result is true and we pose it as an open problem in Section 2.8.

Although we cannot prove this open problem in general, we can establish it in the following special case.

Theorem 2.1.2. *Consider the differential equation*

$$\dot{x}(t) + px(t - \tau) = 0 \tag{2.1.10}$$

where

$$p, \tau \in \mathbb{R}. \tag{2.1.11}$$

Then the following statements are equivalent.

(a) *Every solution of eqn (2.1.10) oscillates.*

(b) *The characteristic equation*

$$\lambda + p\,e^{-\lambda\tau} = 0 \tag{2.1.12}$$

has no real roots.

Proof. (a) $\Rightarrow$ (b). The proof is obvious.

(b) $\Rightarrow$ (a). Set $F(\lambda) = \lambda + p\,e^{-\lambda\tau}$. As $F(\lambda)$ has no real roots, it follows that $\tau \neq 0$.

Case 1: $\tau < 0$. As $F(-\infty) = -\infty$, it follows that $F(0) = p < 0$. Then $F(\infty) = -\infty$ and because eqn (2.1.12) has no real roots it follows that there exists a positive constant m such that

$$\lambda + p\,e^{-\lambda\tau} \leqslant -m \qquad \text{for } \lambda \in \mathbb{R}. \tag{2.1.13}$$

Assume, for the sake of contradiction, that eqn (2.1.10) has an eventually positive solution $x(t)$. Then eventually $\dot{x}(t) = -px(t-\tau) > 0$. Define the set

$$\Lambda = \{\lambda \geqslant 0 : -\dot{x}(t) + \lambda x(t) \leqslant 0 \text{ for } t \text{ sufficiently large}\}.$$

Clearly $0 \in \Lambda$ and Λ is a subinterval of $\mathbb{R}^+$. We show that Λ has the following contradictory properties:

(P_1) There exist positive numbers λ_1 and λ_2 such that

$$\lambda_1 \in \Lambda \qquad \text{and} \qquad \lambda_2 \notin \Lambda.$$

(P_2) $\lambda \in \Lambda \Rightarrow \lambda + m \in \Lambda$ where m is as defined in (2.1.13).

Observe that $x(t)$ is increasing and so $\dot{x}(t) + px(t) \geqslant 0$, which implies that $-p \in \Lambda$. From Lemma 1.6.2 and eqn (2.1.10) it follows that

$$\lambda_2 = \ln[4/(\tau p)^2]/-\tau \notin \Lambda.$$

Next, we turn to the proof of (P_2). Let $\lambda \in \Lambda$ and set $\phi(t) = e^{-\lambda t}x(t)$. Then $\dot{\phi}(t) = -e^{-\lambda t}[-\dot{x}(t) + \lambda x(t)] \geqslant 0$, which shows that $\phi(t)$ is increasing. Now

$$-\dot{x}(t) + (\lambda + m)x(t) = px(t-\tau) + (\lambda + m)x(t)$$

$$= p\phi(t-\tau)\,e^{\lambda(t-\tau)} + (\lambda + m)\phi(t)\,e^{\lambda t}$$

$$\leqslant \phi(t)\,e^{\lambda t}(p\,e^{-\lambda\tau} + \lambda + m) \leqslant \phi(t)\,e^{\lambda t}(-m + m)$$

$$= 0.$$

This proves (P_2) and so the proof of the theorem is complete in this case.

Case 2: $\tau > 0$. The proof in this case follows from Theorem 2.1.1. However, we also present a proof which does not make use of Theorem 1.2.1. As $F(\infty) = \infty$, it follows that $F(0) = p > 0$. Then $F(-\infty) = \infty$ and because of hypothesis (b), there exists a positive constant m such that

$$\lambda + p\,e^{-\lambda\tau} \geqslant m \qquad \text{for } \lambda \in \mathbb{R}. \tag{2.1.13'}$$

Assume for the sake of contradiction that eqn (2.1.10) has an eventually positive solution $x(t)$. Define the set

$$\Lambda = \{\lambda \geq 0 : \dot{x}(t) + \lambda x(t) \leq 0 \text{ for } t \text{ sufficiently large}\}.$$

As in Case 1, Λ is a non-empty subinterval of $\mathbb{R}^+$ and it suffices to show that it has the contradictory properties (P_1) and (P_2) where m is as defined in (2.1.13′).

Observe that $\dot{x}(t) + px(t) \leq 0$, which implies that $\lambda_1 = p \in \Lambda$. By applying Lemmas 1.6.1 and 1.6.2 to (2.1.10) we obtain

$$\lambda_2 = \ln[4/(\tau p)^2]/\tau \notin \Lambda.$$

Let $\lambda \in \Lambda$ and set $\phi(t) = e^{\lambda t}x(t)$. Then $\dot{\phi}(t) = e^{\lambda t}[\dot{x}(t) + \lambda x(t)] \leq 0$, which implies $\phi(t)$ is non-increasing. Now

$$\dot{x}(t) + (\lambda + m)x(t) = -px(t - \tau) + (\lambda + m)x(t)$$
$$= -p\,e^{-\lambda(t-\tau)}\phi(t - \tau) + (\lambda + m)\,e^{-\lambda t}\phi(t)$$
$$\leq e^{-\lambda t}\phi(t)[-p\,e^{\lambda \tau} + \lambda + m] \leq e^{-\lambda t}\phi(t)[-m + m] = 0,$$

which shows that $\lambda + m \in \Lambda$. The proof of the theorem is complete.

2.2 Explicit conditions for oscillations

Consider the delay differential equation

$$\dot{x}(t) + \sum_{i=1}^{n} p_i x(t - \tau_i) = 0 \tag{2.2.1}$$

where

$$p_i \in \mathbb{R} \quad \text{and} \quad \tau_i \in \mathbb{R}^+ \quad \text{for } i = 1, 2, \ldots, n.$$

Our aim in this section is to obtain sufficient conditions, given explicitly in terms of the coefficients and the delays, for the oscillation of all solutions of eqn (2.2.1). Among other conditions, we show that if

$$p_i, \tau_i \in \mathbb{R}^+ \quad \text{for } i = 1, 2, \ldots, n \quad \text{and} \quad \sum_{i=1}^{n} p_i \tau_i > \frac{1}{e}$$

then every solution of eqn (2.2.1) oscillates.

The main tool for the proofs of the results in this section is Theorem 2.1.1, which states that 'every solution of eqn (2.2.1) oscillates if and only if the characteristic equation

$$\lambda + \sum_{i=1}^{n} p_i\,e^{-\lambda \tau_i} = 0 \tag{2.2.2}$$

has no real roots'.

Theorem 2.2.1. *Assume that*

$$p_i, \tau_i \geqslant 0 \quad for\ i = 1, 2, \ldots, n.$$

Then each of the following two conditions is sufficient for the oscillation of all solutions of eqn (2.2.1).

(a) $\displaystyle\sum_{i=1}^{n} p_i \tau_i > \frac{1}{e};$
$\hspace{6cm}$ (2.2.3)

(b) $\displaystyle\left(\prod_{i=1}^{n} p_i\right)^{1/n}\left(\sum_{i=1}^{n} \tau_i\right) > \frac{1}{e}.$
$\hspace{5cm}$ (2.2.4)

Proof.

(a) By making use of the inequality

$$e^x \geqslant ex \quad for\ x \geqslant 0$$
$\hspace{8cm}$ (2.2.5)

we see that for $\lambda < 0$,

$$\lambda + \sum_{i=1}^{n} p_i\, e^{-\lambda \tau_i} \geqslant \lambda + \sum_{i=1}^{n} p_i(-\lambda \tau_i)\, e = -\lambda\, e\left(-\frac{1}{e} + \sum_{i=1}^{n} p_i \tau_i\right) > 0,$$

which shows that eqn (2.2.2) has no negative roots. As eqn (2.2.2) has no roots in $\mathbb{R}^+$ either, the result follows as a consequence of Theorem 2.1.1.

(b) By employing the arithmetic mean–geometric mean inequality

$$\left(\prod_{i=1}^{n} p_i\right)^{1/n} \leqslant \frac{1}{n}\sum_{i=1}^{n} p_i$$

and then by (2.2.5), we see that for $\lambda < 0$,

$$\lambda + \sum_{i=1}^{n} p_i\, e^{-\lambda \tau_i} \geqslant \lambda + n\left(\prod_{i=1}^{n} p_i\, e^{-\lambda \tau_i}\right)^{1/n} = \lambda + n\left(\prod_{i=1}^{n} p_i^{1/n}\right)\exp\left(-\frac{\lambda}{n}\sum_{i=1}^{n} \tau_i\right)$$

$$\geqslant \lambda + n\left(\prod_{i=1}^{n} p_i\right)^{1/n}\left(-e\frac{\lambda}{n}\sum_{i=1}^{n} \tau_i\right)$$

$$= -\lambda\, e\left[-\frac{1}{e} + \left(\prod_{i=1}^{n} p_i\right)^{1/n}\left(\sum_{i=1}^{n} \tau_i\right)\right] > 0$$

which shows that eqn (2.2.2) has no negative roots. As eqn (2.2.2) has no roots in $\mathbb{R}^+$, the result follows by applying Theorem 2.1.1. The proof is complete.

Theorem 2.2.2. *Assume that*

$$p_i, \tau_i \geqslant 0 \quad for\ i = 1, 2, \ldots, n.$$

Then

$$\left(\sum_{i=1}^{n} p_i\right)\left(\max_{1\leqslant i\leqslant n} \tau_i\right) \leqslant \frac{1}{e} \qquad (2.2.6)$$

is a sufficient condition for the existence of a non-oscillatory solution of eqn (2.2.1), while

$$\left(\sum_{i=1}^{n} p_i\right)\left(\min_{1\leqslant i\leqslant n} \tau_i\right) > \frac{1}{e} \qquad (2.2.7)$$

is a sufficient condition for all solutions of eqn (2.2.1) to be oscillatory.

Proof. Assume that (2.2.6) holds. It suffices to show that eqn (2.2.2) has a real root. Set

$$\tau = \max_{1\leqslant i\leqslant n} \tau_i \qquad \text{and} \qquad F(\lambda) = \lambda + \sum_{i=1}^{n} p_i\, e^{-\lambda\tau_i}.$$

Clearly if $\tau = 0$, eqn (2.2.2) has the real root $\lambda = -\sum_{i=1}^{n} p_i$. On the other hand if $\tau > 0$, we have

$$F(0)F\left(-\frac{1}{\tau}\right) = \left(\sum_{i=1}^{n} p_i\right)\left(-\frac{1}{\tau} + \sum_{i=1}^{n} p_i\, e^{\tau_i/\tau}\right) \leqslant \left(\sum_{i=1}^{n} p_i\right)\left(-\frac{1}{\tau} + \sum_{i=1}^{n} p_i\, e\right)$$

$$= \frac{e}{\tau}\left(\sum_{i=1}^{n} p_i\right)\left[-\frac{1}{e} + \left(\sum_{i=1}^{n} p_i\right)\tau\right] \leqslant 0$$

which shows that eqn (2.2.2) has a real root in $[-1/\tau, 0]$. The fact that (2.2.7) is a sufficient condition for all solutions of eqn (2.2.1) to oscillate follows immediately as a corollary of condition (a) of Theorem 2.2.1. The proof is complete.

Theorem 2.2.3. *Consider the differential equation*

$$\dot{x}(t) + px(t - \tau) = 0 \qquad (2.2.8)$$

where $p, \tau \in \mathbb{R}$. Then the following statements are equivalent.
(a) Every solution of eqn (2.2.8) oscillates.
(b) $p\tau > 1/e$. $\qquad\qquad\qquad\qquad\qquad\qquad\qquad\qquad$ (2.2.9)

Proof. In view of Theorem 2.1.2 it suffices to show that (b) is equivalent to the statement:
(c) The characteristic equation

$$F(\lambda) \equiv \lambda + p\, e^{-\lambda\tau} = 0 \qquad (2.2.10)$$

has no real roots.

First we should make the observation that if either (b) or (c) is satisfied then

$$\text{either} \quad p > 0 \quad \text{and} \quad \tau > 0$$

$$\text{or} \quad p < 0 \quad \text{and} \quad \tau < 0.$$

Now we can easily compute the critical points of $F(\lambda)$ and evaluate its extreme values. The proof that (b) $\Rightarrow$ (c) is an immediate consequence of the following facts. If $p > 0$ and $\tau > 0$ then

$$F(-\infty) = F(\infty) = \infty \quad \text{and} \quad \min_{\lambda \in \mathbb{R}} F(\lambda) = \ln(p\tau\, e)/\tau > 0.$$

On the other hand, if $p < 0$ and $\tau < 0$ then

$$F(-\infty) = F(\infty) = -\infty \quad \text{and} \quad \max_{\lambda \in \mathbb{R}} F(\lambda) = \ln(p\tau\, e)/\tau < 0.$$

The proof is complete.

Corollary 2.2.1. *Consider the differential equation*

$$\dot{x}(t) + px(t) + qx(t - \tau) = 0 \tag{2.2.11}$$

where

$$p, q, \tau \in \mathbb{R}.$$

Then

$$q\tau\, e^{p\tau} > 1/e$$

is a necessary and sufficient condition for the oscillation of all solutions of eqn (2.2.11).

Proof. The transformation

$$x(t) = e^{-pt} y(t),$$

which is oscillation-invariant, reduces eqn (2.2.11) to

$$\dot{y}(t) + q\, e^{pt} y(t - \tau) = 0.$$

The result now follows from Theorem 2.2.3.

Finally we present an explicit condition for the oscillation of all solutions of the delay differential equation (DDE) with positive and negative coefficients

$$\dot{x}(t) + px(t - \tau) - qx(t - \sigma) = 0 \tag{2.2.12}$$

where

$$p, q, \tau, \sigma \in \mathbb{R}^+. \tag{2.2.13}$$

Theorem 2.2.4. *Consider the DDE (2.2.12) and assume that (2.2.13) holds. Then*

$$p > q \quad and \quad \tau \geqslant \sigma \qquad (2.2.14)$$

is a necessary condition for all solutions of eqn (2.2.12) to oscillate, while

$$p > q, \quad \tau \geqslant \sigma, \quad q(\tau - \sigma) \leqslant 1 \quad and \quad (p - q)\tau > \frac{1}{e}[1 - q(\tau - \sigma)]$$

$$(2.2.15)$$

is a sufficient condition for all solutions of eqn (2.2.12) to oscillate.

Proof. The characteristic equation of eqn (2.2.12) is

$$F(\lambda) = \lambda + p\,e^{-\lambda\tau} - q\,e^{-\lambda\sigma} = 0. \qquad (2.2.16)$$

Assume that every solution of eqn (2.2.12) oscillates. Then eqn (2.2.16) has no real roots. As $F(\infty) = \infty$, it follows that $F(0) = p - q > 0$, that is, $p > q$. Also $\tau \geqslant \sigma$, for otherwise, $\tau < \sigma$ (and $q > 0$), which implies that $F(-\infty) = -\infty$. Next, assume that (2.2.15) holds and, for the sake of contradiction, assume that eqn (2.2.16) has a real root λ_0. Then, in view of eqn (2.2.16),

$$\lambda_0\left(1 - q\int_\sigma^\tau e^{-\lambda_0 s}\,ds\right) = \lambda_0 + q(e^{-\lambda_0\tau} - e^{-\lambda_0\sigma}) = -(p - q)\,e^{-\lambda_0\tau}. \quad (2.2.17)$$

For $\lambda \geqslant 0$,

$$1 - q\int_\sigma^\tau e^{-\lambda s}\,ds \geqslant 1 - q\int_\sigma^\tau ds = 1 - q(\tau - \sigma) \geqslant 0$$

and so (2.2.17) implies that $\lambda_0 < 0$. Then

$$0 < 1 - q\int_\sigma^\tau e^{-\lambda_0 s}\,ds < 1 - q\int_\sigma^\tau ds = 1 - q(\tau - \sigma),$$

and (2.2.17) yields

$$\lambda_0[1 - q(\tau - \sigma)] + (p - q)\,e^{-\lambda_0\tau} < 0.$$

That is,

$$1 - q(\tau - \sigma) > 0 \quad and \quad \lambda_0 + \frac{(p - q)}{1 - q(\tau - \sigma)}e^{-\lambda_0\tau} < 0.$$

Thus the equation

$$\lambda + \frac{(p - q)}{1 - q(\tau - \sigma)}e^{-\lambda\tau} = 0$$

has a real root in $(\lambda_0, 0)$ which by Theorem 2.2.3 implies that

$$(p - q)\tau \leqslant \frac{1}{e}[1 - q(\tau - \sigma)].$$

This contradicts the last inequality in (2.1.15) and completes the proof of the theorem.

2.3 Sufficient conditions for oscillations and for non-oscillations. The non-autonomous case

Consider the linear non-autonomous delay differential equation

$$\dot{x}(t) + P(t)x(t - \tau) = 0, \qquad t \geqslant t_0. \tag{2.3.1}$$

Our aim in this section is to establish the following results.

Theorem 2.3.1. *Assume that*

$$P \in C[[t_0, \infty), \mathbb{R}^+], \qquad \tau > 0 \tag{2.3.2}$$

and

$$\liminf_{t \to +\infty} \int_{t-\tau}^{t} P(s)\, ds > \frac{1}{e}. \tag{2.3.3}$$

Then every solution of eqn (2.3.1) *oscillates.*

Theorem 2.3.2. *Assume that*

$$P \in C[[t_0 - \tau, \infty), \mathbb{R}^+], \qquad \tau > 0 \tag{2.3.2'}$$

and

$$\int_{t-\tau}^{t} P(s)\, ds \leqslant \frac{1}{e} \qquad \text{for } t \geqslant t_0. \tag{2.3.4}$$

Then eqn (2.3.1) *has a positive solution.*

The case where the delay τ is also variable will be discussed in Sections 3.3 and 3.4 by different methods.

It should be noticed that the above results are sharp in the sense that the lower bound $1/e$ cannot be improved. Moreover, when $P(t)$ is identically equal to a constant p, then (2.3.3) reduces to $p\tau\, e > 1$, which, as we showed in Theorem 2.2.3, is a necessary and sufficient condition for the oscillation of all solutions of

$$\dot{x}(t) + px(t - \tau) = 0. \tag{2.3.5}$$

Similarly, (2.3.4) reduces to $p\tau e \leqslant 1$, which is necessary and sufficient for the existence of a positive solution of (2.3.5).

Proof of Theorem 2.3.1. Assume, for the sake of contradiction, that eqn (2.3.1) has an eventually positive solution $x(t)$. Then there exists a $t^* \geqslant t_0 + \tau$ such that, for $t \geqslant t^*$,

$$x(t) > 0, \qquad x(t - \tau) > 0, \qquad \dot{x}(t) \leqslant 0 \qquad \text{and} \qquad x(t - \tau) \geqslant x(t).$$

Also, from (2.3.3) it follows that there exists a constant $c > 0$ and a $t_1 \geqslant t^*$ such that

$$\int_{t-\tau}^{t} p(s)\, ds \geqslant c > 1/e, \qquad t \geqslant t_1. \tag{2.3.6}$$

Then

$$\dot{x}(t) + p(t)x(t) \leqslant 0, \qquad t \geqslant t_1$$

or

$$\frac{\dot{x}(t)}{x(t)} + p(t) \leqslant 0, \qquad t \geqslant t_1.$$

By integrating both sides from $t - \tau$ to t and by using (2.3.6), we find

$$\ln \frac{x(t)}{x(t - \tau)} + c \leqslant 0, \qquad t \geqslant t_1 + \tau$$

or

$$e^c x(t) \leqslant x(t - \tau), \qquad t \geqslant t_1 + \tau.$$

One can easily show that

$$e^c \geqslant ec \qquad \text{for all } c \in \mathbb{R}$$

and so

$$(ec)x(t) \leqslant x(t - \tau), \qquad t \geqslant t_1 + \tau.$$

Repeating the above procedure, it follows by induction that for any positive integer k,

$$(ec)^k x(t) \leqslant x(t - \tau), \qquad t \geqslant t_1 + k\tau. \tag{2.3.7}$$

Choose k such that

$$(2/c)^2 < (ec)^k \tag{2.3.8}$$

which is possible because by (2.3.6), $ec > 1$. Now fix a $\tilde{t} \geqslant t_1 + k\tau$. Then because of (2.3.6), there exists a $\xi \in (\tilde{t}, \tilde{t} + \tau)$ such that

$$\int_{\tilde{t}}^{\xi} p(s)\,ds \geqslant \frac{c}{2} \quad \text{and} \quad \int_{\xi}^{\tilde{t}+\tau} p(s)\,ds \geqslant \frac{c}{2}. \tag{2.3.9}$$

By integrating eqn (2.3.1) over the intervals $[\tilde{t}, \xi]$ and $[\xi, \tilde{t} + \tau]$, we find

$$x(\xi) - x(\tilde{t}) + \int_{\tilde{t}}^{\xi} p(s)x(s - \tau)\,ds = 0 \tag{2.3.10}$$

and

$$x(\tilde{t} + \tau) - x(\xi) + \int_{\xi}^{\tilde{t}+\tau} p(s)x(s - \tau)\,ds = 0. \tag{2.3.11}$$

By omitting the first terms in (2.3.10) and (2.3.11) and by using the decreasing nature of $x(t)$ and (2.3.9), we find

$$-x(\tilde{t}) + x(\xi - \tau)\frac{c}{2} < 0 \quad \text{and} \quad -x(\xi) + x(\tilde{t})\frac{c}{2} < 0$$

or

$$x(\xi) > \frac{c}{2}x(\tilde{t}) > \left(\frac{c}{2}\right)^2 x(\xi - \tau).$$

This and (2.3.7) imply that

$$(ec)^k \leqslant \frac{x(\xi - \tau)}{x(\xi)} < \left(\frac{2}{c}\right)^2,$$

which contradicts (2.3.8) and completes the proof of the theorem.

Proof of Theorem 2.3.2. We employ Schauder's fixed-point theorem (see Theorem 1.7.1). Let us denote by X the Banach space of all real, bounded and continuous functions on $[t_0 - \tau, \infty)$ with the sup norm (see Example 1.7.1). Let M be the subset of X consisting of those functions $x(t)$ in X which satisfy the following properties:

(1) $x(t)$ is non-increasing for $t \geqslant t_0$ and

$$x(t) \equiv 1 \quad \text{for } t_0 - \tau \leqslant t \leqslant t_0;$$

(2) $\exp\left(-e\int_{t_0}^{t} P(s)\,ds\right) \leqslant x(t) \leqslant 1 \quad \text{for } t \geqslant t_0;$

(3) $x(t - \tau) \leqslant ex(t) \quad \text{for } t \geqslant t_0.$

For example, $x(t) = 1$ for $t \geqslant t_0 - \tau$ is such a function.

Consider the mapping T on M defined as follows:

$$(Tx)(t) = \begin{cases} 1, & t_0 - \tau \leqslant t \leqslant t_0 \\ \exp\left(-\int_{t_0}^{t} \dfrac{P(s)x(s-\tau)}{x(s)}\,ds\right), & t \geqslant t_0. \end{cases} \quad (2.3.12)$$

We claim that T maps M into M. Indeed, $(Tx)(t)$ is a non-increasing, continuous function and in view of (3) for $t \geqslant t_0$,

$$(Tx)(t) = \exp\left(-\int_{t_0}^{t} \frac{P(s)x(s-\tau)}{x(s)}\,ds\right) \geqslant \exp\left(-e\int_{t_0}^{t} P(s)\,ds\right).$$

Also for $t \geqslant t_0$, by using (2.3.4) and (3) we see that

$$(Tx)(t-\tau) = \exp\left(-\int_{t_0}^{t-\tau} \frac{P(s)x(s-\tau)}{x(s)}\,ds\right)$$

$$= (Tx)(t)\exp\left(\int_{t-\tau}^{t} \frac{P(s)x(s-\tau)}{x(s)}\,ds\right) \leqslant (Tx)(t)\,e^{e(1/e)} = e(Tx)(t).$$

Thus $T\colon M \to M$. Now clearly M is a closed and convex subset of X and $T\colon M \to M$ is a continuous function. Also TM is a relatively compact subset of X because TM is equicontinuous and uniformly bounded. Hence all the hypotheses of Schauder's fixed-point theorem are satisfied, so T has a fixed point x. That is, there exists a function $x \in M$ such that $Tx = x$. It follows from (2.3.12) that

$$x(t) = 1 \qquad \text{for } t_0 - \tau \leqslant t \leqslant t_0$$

and

$$x(t) = \exp\left(-\int_{t_0}^{t} \frac{P(s)x(s-\tau)}{x(s)}\,ds\right), \qquad t \geqslant t_0.$$

Hence $x(t)$ is positive and

$$\dot{x}(t) = -\frac{P(t)x(t-\tau)}{x(t)}\exp\left(-\int_{t_0}^{t} \frac{P(s)x(s-\tau)}{x(s)}\,ds\right)$$

$$= -\frac{P(t)x(t-\tau)}{x(t)}\,x(t) = -p(t)x(t-\tau).$$

That is, $x(t)$ is a positive solution of eqn (2.3.1). The proof is complete.

The proof of Theorem 2.3.1 applied verbatim to the delay differential inequalities

$$\dot{x}(t) + P(t)x(t-\tau) \leqslant 0, \qquad t \geqslant t_0 \qquad (2.3.13)$$

and

$$\dot{x}(t) + P(t)x(t - \tau) \geqslant 0, \qquad t \geqslant t_0 \qquad (2.3.14)$$

leads to the following result.

Theorem 2.3.3. *Assume that conditions (2.3.2) and (2.3.3) are satisfied. Then (2.3.13) cannot have an eventually positive solution and (2.3.14) cannot have an eventually negative solution.*

Consider the advance delay differential equations and inequalities

$$\dot{y}(t) - P(t)y(t + \tau) = 0 \qquad (2.3.15)$$

$$\dot{y}(t) - P(t)y(t + \tau) \geqslant 0 \qquad (2.3.16)$$

$$\dot{y}(t) - P(t)y(t + \tau) \leqslant 0. \qquad (2.3.17)$$

A proof similar to that of Theorem 2.3.1 applied to (2.3.15)–(2.3.17) leads to the following result.

Theorem 2.3.4. *Assume that (2.3.2) and (2.3.3) are satisfied. Then the following statements hold:*

(1) *inequality (2.3.16) has no eventually positive solution;*

(2) *inequality (2.3.17) has no eventually negative solution;*

(3) *every solution of eqn (2.3.15) oscillates.*

2.4 Oscillations of linear delay equations with asymptotically constant coefficients

Our aim in this section is to establish a sufficient, as well as a necessary and sufficient, condition for the oscillation of all solutions of the linear delay differential equation with asymptotically constant coefficients:

$$\dot{x}(t) + \sum_{j=1}^{n} Q_j(t)x(t - \tau_j) = 0, \qquad t \geqslant t_0 \qquad (2.4.1)$$

where for $j = 1, 2, \ldots, n$,

$$Q_j \in C[[t_0, \infty), \mathbb{R}^+], \qquad \tau_j \geqslant 0 \qquad \text{and} \qquad \lim_{t \to \infty} Q_j(t) \equiv q_j. \qquad (2.4.2)$$

Theorem 2.4.1.

(a) *Assume that (2.4.2) holds and that every solution of the limiting equation*

$$\dot{z}(t) + \sum_{j=1}^{n} q_j z(t - \tau_j) = 0 \tag{2.4.3}$$

oscillates. Then every solution of eqn (2.4.1) also oscillates.

(b) *In addition to condition (2.4.2), assume that for t sufficiently large,*

$$Q_j(t) \leq q_j \qquad for \ j = 1, 2, \ldots, n. \tag{2.4.4}$$

Then every solution of eqn (2.4.1) oscillates if and only if every solution of eqn (2.4.3) oscillates.

Proof of part (a) *of Theorem* 2.4.1. By Theorem 2.1.1 the characteristic equation of eqn (2.4.3),

$$F(\lambda) \equiv \lambda + \sum_{j=1}^{n} q_j e^{-\lambda \tau_j} = 0 \tag{2.4.5}$$

has no real roots. As $F(\infty) = \infty$, it follows that

$$F(\lambda) > 0 \qquad for \ \lambda \in \mathbb{R}.$$

In particular,

$$F(0) = \sum_{j=1}^{n} q_j > 0.$$

Therefore at least one of the coefficients q_j is positive. Also, for some positive q_{j0} the corresponding delay τ_{j0} must be positive. [Otherwise, $\lambda = -\sum_{j=1}^{n} q_j$ would be a root of eqn (2.4.5).] Then $F(-\infty) = \infty$ and so $m = \min_{\lambda \in \mathbb{R}} F(\lambda)$ exists and is positive. Thus

$$\lambda + \sum_{j=1}^{n} q_j e^{-\lambda \tau_j} \geq m, \qquad \lambda \in \mathbb{R}$$

or equivalently,

$$\sum_{j=1}^{n} q_j e^{\lambda \tau_j} \geq \lambda + m, \qquad \lambda \in \mathbb{R}. \tag{2.4.6}$$

Now assume for the sake of contradiction that eqn (2.4.1) has an eventually positive solution $x(t)$. Then

$$\dot{x}(t) + Q_{j0}(t)x(t - \tau_{j0}) \leq 0 \tag{2.4.7}$$

where the index j_0 is chosen in such a way that $q_{j0} > 0$ and $\tau_{j0} > 0$. Clearly for t sufficiently large,

$$\dot{x}(t) + \tfrac{1}{2}q_{j_0}x(t - \tau_{j_0}) \leqslant 0. \tag{2.4.8}$$

Let

$$\Lambda = \{\lambda \geqslant 0 : \dot{x}(t) + \lambda x(t) \leqslant 0, \text{ eventually for } t \text{ large}\}.$$

Clearly, Λ is a non-empty subinterval of $\mathbb{R}^+$. The proof that every solution of eqn (2.4.1) oscillates will be accomplished by showing that Λ has the following contradictory properties:

(P_1) Λ is bounded above.

(P_2) $\lambda \in \Lambda \Rightarrow (\lambda + m/2) \in \Lambda$, where m is the positive constant which satisfies (2.4.6).

The proof of (P_1) follows from (2.4.8) and Lemma 1.6.2. In order to establish (P_2), let $\lambda \in \Lambda$ and set

$$\psi(t) = e^{\lambda t}x(t).$$

Then $\dot{\psi}(t) = e^{\lambda t}[\dot{x}(t) + \lambda x(t)] \leqslant 0$, which shows that $\psi(t)$ is decreasing. Now, choose $\varepsilon > 0$ such that for all j's with $q_j > 0$, and for t sufficiently large, $Q_j(t) \geqslant q_j - \varepsilon > 0$ and $\varepsilon \sum_{j=1}^{n} e^{\lambda \tau_j} \leqslant m/2$. Then $-Q_j(t) \leqslant -(q_j - \varepsilon)$ for all $j = 0, 1, \ldots, n$ and so

$$\dot{x}(t) + (\lambda + m/2)x(t) = -\sum_{j=1}^{n} Q_j(t)x(t - \tau_j) + (\lambda + m/2)x(t)$$

$$\leqslant e^{-\lambda t}\left[-\sum_{j=1}^{n} Q_j(t)\,e^{\lambda \tau_j}\psi(t - \tau_j) + (\lambda + m/2)\psi(t)\right]$$

$$\leqslant e^{-\lambda t}\psi(t)\left[-\sum_{j=1}^{n} Q_j(t)\,e^{\lambda \tau_j} + \lambda + m/2\right]$$

$$\leqslant e^{-\lambda t}\psi(t)\left[-\sum_{j=1}^{n} (q_j - \varepsilon)\,e^{\lambda \tau_j} + \lambda + m/2\right]$$

$$\leqslant e^{-\lambda t}\psi(t)(-\lambda - m + m/2 + \lambda + m/2) = 0$$

which proves (P_2) and completes the proof of (a).

Proof of part (b) *of Theorem* 2.4.1. It is clear from the proof of part (a) that if every solution of eqn (2.4.3) oscillates, so does every solution of eqn (2.4.1). To prove that the converse is also true it suffices to show that if eqn (2.4.3) has a real (and thus negative) root μ, then eqn (2.4.1) has a non-oscillatory solution $x(t)$. In fact we prove a little more, namely, that for any given $x_0 > 0$, eqn (2.4.1) has a positive solution $x(t)$ such that, if $\tau = \max_{1 \leqslant j \leqslant n} \tau_j$, then

$x_0 e^{\mu(t-t_0)} \leqslant x(t) \leqslant x_0$ for $t \geqslant t_0$ and $x(t) = x_0 e^{\mu(t-t_0)}$ for $t_0 - \tau \leqslant t \leqslant t_0$.

To this end, let us denote by X the Banach space of all real, bounded and continuous functions on $[t_0 - \tau, \infty)$ with the sup norm (see Example 1.7.1). Let M be the subset of X consisting of those functions $x(t)$ in X which satisfy the following properties:

(1) $x(t)$ is non-increasing for $t \geqslant t_0$ and

$$x(t) \equiv x_0 e^{\mu(t-t_0)} \qquad \text{for } t_0 - \tau \leqslant t \leqslant t_0;$$

(2) $x_0 e^{\mu(t-t_0)} \leqslant x(t) \leqslant x_0$ for $t \geqslant t_0$.

(3) $x(t - \tau_j) \leqslant e^{-\mu\tau_j} x(t)$ for $t \geqslant t_0$ and $j = 1, 2, \ldots, n$.

Consider the mapping T on M defined as follows:

$$(Tx)(t) = \begin{cases} x_0 e^{\mu(t-t_0)}, & t_0 - \tau \leqslant t \leqslant t_0 \\ x_0 \exp\left(-\sum_{j=1}^{n} \int_{t_0}^{t} \frac{Q_j(s) x(s - \tau_j)}{x(s)} \, ds \right), & t \geqslant t_0. \end{cases}$$

We employ Schauder's fixed-point theorem (see Theorem 1.7.1). Clearly, $(Tx)(t)$ is a non-increasing continuous function and $(Tx)(t) \leqslant x_0$ for $t \geqslant t_0$. Also, for $t \geqslant t_0$, and in view of (2.4.4),

$$(Tx)(t) \geqslant x_0 \exp\left(-\sum_{j=1}^{n} q_j \int_{t_0}^{t} \frac{x(s - \tau_j)}{x(s)} \, ds \right) \geqslant x_0 \exp\left(-\sum_{j=1}^{n} q_j e^{-\mu\tau_j} \int_{t_0}^{t} ds \right)$$

$$= x_0 e^{\mu(t-t_0)}$$

where in the last step we used the fact that μ is a root of the characteristic equation (2.4.5).

Again, using (2.4.4) and the hypothesis that μ is a root of (2.4.5) we see that for every $k = 1, 2, \ldots, n$ and for $t \geqslant t_0$,

$$(Tx)(t - \tau_k) = x_0 \exp\left(-\sum_{j=1}^{n} \int_{t_0}^{t-\tau_k} \frac{Q_j(s) x(s - \tau_j)}{x(s)} \, ds \right)$$

$$= (Tx)(t) \exp\left(\sum_{j=1}^{n} \int_{t-\tau_k}^{t} \frac{Q_j(s) x(s - \tau_j)}{x(s)} \, ds \right)$$

$$\leqslant (Tx)(t) \exp\left(\sum_{j=1}^{n} q_j \int_{t-\tau_k}^{t} e^{-\mu\tau_j} \, ds \right)$$

$$= (Tx)(t) \exp\left(\tau_k \sum_{j=1}^{n} q_j e^{-\mu\tau_j} \right) = (Tx)(t) e^{-\mu\tau_k}.$$

Thus we have established that T maps M into M. Clearly M is non-empty (because $e^{\mu(t-t_0)} \in M$), closed and convex, and $T: M \to M$ is continuous.

Finally we show that TM is relatively compact in X by showing that $(d/dt)[(Tx)(t)]$ is uniformly bounded (see Section 1.7). In fact,

$$\frac{d}{dt}[(Tx)(t)] = -\left[\sum_{j=1}^{n} \frac{Q_j(t)x(t-\tau_j)}{x(t)}\right](Tx)(t)$$

and so

$$\left|\frac{d}{dt}[(Tx)(t)]\right| = \left[\sum_{j=1}^{n} \frac{Q_j(t)x(t-\tau_j)}{x(t)}\right](Tx)(t) \leqslant \left[\sum_{j=1}^{n} q_j \frac{x(t-\tau_j)}{x(t)}\right]x_0$$

$$\leqslant x_0 \sum_{j=1}^{n} q_j\, e^{-\mu\tau_j} = -\mu x_0.$$

Thus all the hypotheses of Schauder's fixed-point theorem are satisfied and the mapping T has a fixed point $x \in M$. Clearly $x(t)$ is a positive solution of eqn (2.4.1). The proof of the theorem is complete.

Remark 2.4.1. *Consider the differential inequality*

$$\dot{x}(t) + \sum_{j=1}^{n} Q_j(t)x(t-\tau_j) \leqslant 0, \qquad t \geqslant t_0 \qquad (2.4.9)$$

where the coefficients and the delays satisfy (2.4.2). *Then by applying the proof of part* (a) *of Theorem 2.4.1 verbatim, we see that if every solution of* (2.4.3) *oscillates, then* (2.4.9) *cannot have an eventually positive solution. That is, equivalently, if* (2.4.9) *has an eventually positive solution then so does eqn* (2.4.3). *In particular, we have the following result.*

Corollary 2.4.1. *Assume that*

$$q_j \in (0, \infty) \qquad and \qquad \tau_j \in [0, \infty) \qquad for\ j = 1, 2, \ldots, n \qquad (2.4.10)$$

and suppose that the differential inequality

$$\dot{x}(t) + \sum_{j=1}^{n} q_j x(t-\tau_j) \leqslant 0$$

has an eventually positive solution. Then the equation

$$\dot{y}(t) + \sum_{j=1}^{n} q_j y(t-\tau_j) = 0$$

also has an eventually positive solution.

A similar result holds for advanced delay differential inequalities.

Corollary 2.4.2. *Assume that* (2.4.10) *holds and suppose that the differential inequality*

$$\dot{x}(t) - \sum_{j=1}^{n} q_j x(t + \tau_j) \geqslant 0$$

has an eventually positive solution. Then the equation

$$\dot{y}(t) - \sum_{j=1}^{n} q_j y(t + \tau_j) = 0$$

also has an eventually positive solution.

2.5 Nicholson's blowflies

The delay differential equation

$$\dot{N}(t) = -\delta N(t) + P N(t - \tau) e^{-aN(t-\tau)}, \qquad t \geqslant 0 \qquad (2.5.1)$$

was used by Gurney *et al.* (1980) to describe the dynamics of Nicholson's blowflies. Here P is the maximum per capita daily egg production rate, $1/a$ is the size at which the population reproduces at its maximum rate, δ is the per capita daily adult death rate, τ is the generation time and $N(t)$ is the size of the population at time t. We consider only those solutions of eqn (2.5.1) which correspond to initial conditions of the form

$$N(t) = \phi(t), \qquad -\tau \leqslant t \leqslant 0 \qquad \text{with } \phi \in C[[-\tau, 0], \mathbb{R}^+] \text{ and } \phi(0) > 0.$$

$$(2.5.2)$$

Clearly (2.5.1) with (2.5.2) has a unique solution $N(t)$ which exists and is positive for all $t \geqslant 0$.

For $P > \delta$, the positive equilibrium N^* of (2.5.1) is given by

$$N^* = \frac{1}{a} \ln(P/\delta). \qquad (2.5.3)$$

Set $N(t) = N^* + (1/a)x(t)$. Then $x(t)$ satisfies the delay differential equation

$$\dot{x}(t) + \delta x(t) + a\delta N^*[1 - e^{-x(t-\tau)}] - \delta x(t - \tau) e^{-x(t-\tau)} = 0 \quad (2.5.4)$$

and $N(t)$ oscillates about N^* if and only if $x(t)$ oscillates about zero.

The main result in this section is the following:

Theorem 2.5.1.
(a) *Assume that* $\delta, P, \tau, a \in (0, \infty)$ *are such that*

$$P \geqslant \delta e \qquad (2.5.5)$$

and

$$\delta\tau \, e^{\delta\tau}[\ln(P/\delta) - 1] > 1/e. \qquad (2.5.6)$$

Then the solution $N(t)$ of (2.5.1) and (2.5.2) oscillates about the equilibrium N^*.
  (b) *Assume that $\delta, P, \tau, a \in (0, \infty)$ are such that*

$$P > \delta \, e^2. \qquad (2.5.5')$$

Then the solution $N(t)$ of (2.5.1) and (2.5.2) oscillates about N^* if and only if (2.5.6) holds.

Proof.
  (a) Assume that conditions (2.5.5) and (2.5.6) hold and, for the sake of contradiction, assume that the solution of (2.5.1) and (2.5.2) does not oscillate about N^*. Then eqn (2.5.4) has a non-oscillatory solution $x(t)$. We assume that $x(t)$ is eventually positive. The case where $x(t)$ is eventually negative is similar and will be omitted. As eqn (2.5.4) is autonomous we may (and do) assume that $x(t) > 0$ for $t \geq -\tau$. First we prove that

$$\lim_{t \to \infty} x(t) = 0. \qquad (2.5.7)$$

To this end, we set

$$u(t) = x(t) + \delta \int_{t-\tau}^{t} x(s) \, ds + a\delta N^* \int_{0}^{t} [1 - e^{-x(s-\tau)}] \, ds, \qquad t \geq 0. \quad (2.5.8)$$

Then $u(t) > 0$ for $t \geq 0$ and

$$\dot{u}(t) = -\delta x(t - \tau)[1 - e^{-x(t-\tau)}] \leq 0, \qquad t \geq 0.$$

Hence $u(t)$ decreases and so from (2.5.8) we see that $x(t)$ is bounded and

$$\int_{0}^{\infty} [1 - e^{-x(s-\tau)}] \, ds < \infty. \qquad (2.5.9)$$

Let $M > 0$ be such that

$$0 < x(t) \leq M \qquad \text{for } t \geq 0.$$

Clearly,

$$x \leq e^{M}(1 - e^{-x}) \qquad \text{for } 0 \leq x \leq M$$

and so by using (2.5.9) we find that

$$\int_{0}^{\infty} x(s) \, ds \leq e^{M} \int_{0}^{\infty} [1 - e^{-x(s-\tau)}] \, ds < \infty. \qquad (2.5.10)$$

Also

$$\int_0^\infty x(s - \tau) e^{-x(s-\tau)} ds \leqslant \int_0^\infty x(s - \tau) ds < \infty. \qquad (2.5.11)$$

In view of (2.5.9), (2.5.10), and (2.5.11) we see from eqn (2.5.4) that $\dot{x} \in L^1[0, \infty)$ and so the $\lim_{t \to \infty} x(t)$ exists. As x is itself in $L^1[0, \infty)$, it follows that (2.5.7) holds.

Now we rewrite eqn (2.5.4) in the form

$$\dot{x}(t) + \delta x(t) + Q(t)x(t - \tau) = 0, \qquad t \geqslant 0 \qquad (2.5.12)$$

where

$$Q(t) = \frac{a\delta N^*[1 - e^{-x(t-\tau)}]}{x(t - \tau)} - \delta e^{-x(t-\tau)}$$

and observe that because of (2.5.7),

$$\lim_{t \to \infty} Q(t) = a\delta N^* - \delta = \delta[\ln(P/\delta) - 1] > 0.$$

The limiting equation of (2.5.12) is

$$\dot{z}(t) + \delta z(t) + \delta[\ln(P/\delta) - 1]z(t - \tau) = 0 \qquad (2.5.13)$$

and because of (2.5.6) and Corollary 2.2.1 every solution of (2.5.3) oscillates. By Theorem 2.4.1(a), every solution of (2.5.12) or equivalently every solution of (2.5.4) oscillates. This contradicts the assumption that $x(t) > 0$ and completes the proof of (a).

(b) Clearly (2.5.5′) and (2.5.6) imply that $N(t)$ oscillates about N^*. Conversely, assume that (2.5.5′) holds and that $N(t)$ oscillates about N^*. We must prove that (2.5.6) holds or equivalently that every solution of (2.5.13) oscillates. This will be an immediate consequence of Theorem 2.4.1(b) provided that we can show that for t sufficiently large,

$$Q(t) \leqslant \delta[\ln(P/\delta) - 1]. \qquad (2.5.14)$$

To this end, observe that for t sufficiently large, $u \equiv x(t - \tau)$ is sufficiently small. But for u sufficiently small one can see that

$$\frac{a\delta N^*(1 - e^{-u})}{u} - \delta e^{-u} \leqslant \delta[\ln(P/\delta) - 1].$$

The proof is complete.

2.6 Oscillations in non-autonomous equations with positive and negative coefficients

Consider the linear delay differential equation with positive and negative coefficients

$$\dot{y}(t) + P(t)y(t - \tau) - Q(t)y(t - \sigma) = 0 \qquad (2.6.1)$$

where

$$P, Q \in C[[t_0, \infty), \mathbb{R}^+] \qquad \text{and} \qquad \tau, \sigma \in [0, \infty). \qquad (2.6.2)$$

Our aim in this section is to obtain a sufficient condition for the oscillation of all solutions of eqn (2.6.1). The following lemma will be useful in the proof of the main result.

Lemma 2.6.1. *Assume that (2.6.2) holds,*

$$\tau \geqslant \sigma, \qquad (2.6.3)$$

$$P(t) \geqslant Q(t + \sigma - \tau) \qquad \text{and} \qquad P(t) \not\equiv Q(t + \sigma - \tau) \qquad \text{for } t \geqslant t_0 + \tau - \sigma \qquad (2.6.4)$$

and

$$\int_{t-\tau}^{t-\sigma} Q(s)\, ds \leqslant 1 \qquad \text{for } t \geqslant t_0 + \tau. \qquad (2.6.5)$$

Let y(t) be an eventually positive solution of eqn (2.6.1) and set

$$z(t) = y(t) - \int_{t-\tau}^{t-\sigma} Q(s + \sigma)y(s)\, ds \qquad \text{for } t \geqslant t_0 + \tau - \sigma. \qquad (2.6.6)$$

Then eventually z(t) is a non-increasing and positive function.

Proof. Assume that $t_1 \geqslant t_0 + \tau$ is such that $y(t) > 0$ for $t \geqslant t_1 - \tau$. Then

$$\dot{z}(t) = -[P(t) - Q(t + \sigma - \tau)]y(t - \tau) \leqslant 0 \qquad (2.6.7)$$

for $t \geqslant t_1 + \tau$. Moreover, since $P(t) \not\equiv Q(t + \sigma - \tau)$ for $t \geqslant t_0 + \tau - \sigma$ and $y(t)$ is eventually positive, (2.6.7) yields that there exists a sequence $\{s_n\}_{n=1}^{\infty}$ such that $\dot{z}(s_n) < 0$ for all $n \geqslant 1$ and $s_n \to +\infty$, as $n \to +\infty$. We now show that $z(t)$ is eventually positive. Otherwise there exists a $t_2 \geqslant t_1$ such that $z(t_2) \leqslant 0$. Because $\dot{z}(t) \leqslant 0$ for all $t \geqslant t_1 + \tau$ and $\dot{z}(t) \not\equiv 0$ on $[t_1 + \tau, \infty)$, there exists a $t_3 \geqslant t_2$ such that $z(t_3) < 0$ and $z(t) \leqslant z(t_3)$ for all $t \geqslant t_3$. Thus from

(2.6.6) it follows that for $t \geqslant t_3$,

$$y(t) = z(t) + \int_{t-\tau}^{t-\sigma} Q(s + \sigma) y(s) \, ds$$

$$\leqslant z(t_3) + \int_{t-\tau}^{t-\sigma} Q(s + \sigma) \, ds \left(\max_{t-\tau \leqslant s \leqslant t-\sigma} y(s) \right).$$

By using (2.6.5), this yields

$$y(t) \leqslant z(t_2) + \max_{t-\tau \leqslant s \leqslant t} y(s) \qquad \text{for all } t \geqslant t_3,$$

where $z(t_2) < 0$. Thus by Lemma 1.5.4 we see that $y(t)$ cannot be a non-negative function on $[t_3, \infty)$. This is a contradiction and the proof of the lemma is complete.

The next result provides a sufficient condition for the oscillation of all solutions of eqn (2.6.1).

Theorem 2.6.1. *Assume that* (2.6.2)–(2.6.5) *hold and that*

$$\liminf_{t \to \infty} \int_{t-\tau}^{t} [P(s) - Q(s + \sigma - \tau)] \, ds > 1/e. \qquad (2.6.8)$$

Then every solution of eqn (2.6.1) *oscillates.*

Proof. Assume, for the sake of contradiction, that eqn (2.6.1) has an eventually positive solution $y(t)$. By Lemma 2.6.1 it follows that the function $z(t)$, which is defined by (2.6.6), is an eventually positive function. Also by (2.6.7) and the fact that eventually $0 < z(t) \leqslant y(t)$, we see that eventually,

$$\dot{z}(t) + [P(t) - Q(t + \sigma - \tau)] z(t - \tau) \leqslant 0. \qquad (2.6.9)$$

But in view of (2.6.8) it follows from Theorem 2.3.3 that the inequality (2.6.9) cannot have an eventually positive solution. This contradicts the fact that $z(t)$ is eventually positive and completes the proof.

2.7 Notes

To the best of our knowledge, Tramov (1975) should be credited with the statement of Theorem 2.1.1 (for equations with positive coefficients). The same result was independently discovered and proved by Ladas *et al.* (1983b, 1984a) and later on by Hunt and Yorke (personal communication).

Theorem 2.1.1 was established for distributed delays by using the exponential series representation of the solutions in Arino *et al.* (1984). The Laplace transform method was first employed by Arino and Györi (1990) to establish necessary and sufficient conditions for the oscillation of linear autonomous equations via the characteristic equation. The proof of Theorem 2.1.1 is extracted from Györi et al. (1989) and the proof of Theorem 2.1.2 is from Ladas et al. (1983b).

The results of Section 2.2 are extracted from Ladas and Sficas (1984) and Hunt and Yorke (1984). The proofs which are given here, however, are new.

The results of Section 2.3 are derived from Ladas (1979) and Koplatadze and Chanturia (1982). The condition (2.3.3) for the oscillation of all solutions of eqn (2.3.1) is due to Ladas (1979), while the condition (2.2.9) for the oscillation of all solutions of eqn (2.2.8) is due to Myskis (1972).

For the results in Section 2.4, see Kulenovic et al. (1987b).

Section 2.5 is from Kulenovic and Ladas (1987b) and the results in Section 2.6 are adapted from Chuanxi and Ladas (1990a).

For some important results in oscillation theory which have not been included in this book for lack of space, see Ladde et al. (1987) and the references cited therein, Dahiya (1982), Dahiya et al. (1984), Domshlak and Aliev (1988), Kitamura and Kusano (1980), Kusano (1984), Mahfoud (1979), Naito (1984), Olah (1982), Onose (1986), Philos (1980), Philos et al. (1982), Seifert (1986, 1990), Shevelo and Varekh (1979), Shevelo et al. (1982), H. Smith (1976), Stavroulakis (1982), Sugie (1988), Takano (1983, 1985), Zhang and Gopalsamy (1990), Buchanan (1974), Elbert (1976), Fink (1983), Grace and Lalli (1982), Ivanov and Shevelo (1981), Kreith and Ladas (1985), Ladas et al. (1972b), Sing (1979), and the book of Swanson (1968).

2.8 Open problems

2.8.1 (Conjecture)   Assume that

$$p_i \in \mathbb{R} \quad \text{and} \quad \tau_i \in \mathbb{R} \quad \text{for } i = 1, 2, \dots, n.$$

Then the following statements are equivalent.

(a) Every solution of eqn (2.1.1) oscillates.

(b) The characteristic equation (2.1.2) has no real roots.

See Theorem 2.1.2, Farrell et al. (1990), Arino and Györi (1990), and Györi et al. (1989).

2.8.2   Extend Theorems 2.2.1 and 2.2.2 to equations with both positive and negative coefficients. See Arino et al. (1987).

2.8.3 Extend Theorems 2.3.1 and 2.3.3 to equations with oscillating coefficients. See Ladas *et al.* (1984c).

2.8.4 Extend Theorem 2.4.1 to equations with oscillating and asymptotically constant coefficients.

2.8.5 (Conjecture) Assume that

$$p_1, p_2, \tau_1, \tau_2 \in (0, \infty), \qquad \tau_1 \neq \tau_2,$$

$$1 + p_1 - p_2 > 0$$

and that the equation

$$1 + p_1 e^{-\lambda \tau_1} - p_2 e^{-\lambda \tau_2} = 0$$

has no positive roots. Then the equation

$$y(t) + p_1 y(t - \tau_1) - p_2 y(t - \tau_2) = \text{constant} \qquad \text{for } t \geq 0$$

has no continuous, eventually positive, unbounded solutions.

3

GENERALIZED CHARACTERISTIC EQUATION AND EXISTENCE OF POSITIVE SOLUTIONS

As we know from stability theory, the location of the roots of the characteristic equation

$$\lambda + \sum_{i=1}^{n} p_i\, e^{-\lambda \tau_i} = 0 \tag{3.0.1}$$

of the autonomous delay differential equation

$$\dot{x}(t) + \sum_{i=1}^{n} p_i x(t - \tau_i) = 0 \tag{3.0.2}$$

determines the asymptotic stability of the trivial solution of eqn (3.0.2). As we saw in Theorem 2.1.1, the oscillatory character of eqn (3.0.2.) is also determined by the location of the characteristic roots.

Our aim in this chapter is to introduce a generalized characteristic equation for non-autonomous delay differential equations and then utilize it to obtain sufficient conditions for the oscillation of all solutions. Furthermore, we utilize the generalized characteristic equation to obtain some powerful comparison results and to demonstrate the existence of positive solutions for a wide class of linear delay equations with variable coefficients and variable delays.

Of particular importance for oscillation theory is Corollary 3.2.2, which states that under appropriate hypotheses if a differential inequality has a positive solution, so does the corresponding equation.

3.1 A generalized characteristic equation

Consider the non-autonomous linear delay differential equation

$$\dot{x}(t) + \sum_{i=1}^{n} p_i(t)x(t - \tau_i(t)) = 0, \qquad t_0 \leqslant t < T \tag{3.1.1}$$

where $t_0 < T \leqslant \infty$ and for $i = 1, 2, \ldots, n$,

$$p_i \in C[[t_0, T), \mathbb{R}] \qquad \text{and} \qquad \tau_i \in C[[t_0, T), \mathbb{R}^+]. \tag{3.1.2}$$

Let

$$t_{-1} = \min_{1 \leqslant i \leqslant n} \left\{ \inf_{t_0 \leqslant t < T} \{t - \tau_i(t)\} \right\}. \tag{3.1.3}$$

With eqn ((3.1.1) one associates an initial condition of the form

$$x(t) = \phi(t), \qquad t_{-1} \leqslant t \leqslant t_0 \qquad \text{with } \phi \in C[[t_{-1}, t_0], \mathbb{R}]. \quad (3.1.4)$$

The unique solution of the IVP (initial value problem) (3.1.1.) and (3.1.4.) is denoted by $x(\phi)(t)$ and exists throughout the interval $t_0 \leqslant t < T$.

For every $t \in [t_0, T)$ and every $i = 1, 2, \ldots, n$ we define the functions

$$h_i(t) = \min\{t_0, t - \tau_i(t)\} \text{ and } H_i(t) = \max\{t_0, t - \tau_i(t)\}. \quad (3.1.5)$$

In this section we introduce the concept of generalized *characteristic equation* associated with eqn (3.1.1), namely, the integral equation

$$\alpha(t) + \sum_{i=1}^{n} p_i(t) \frac{\phi(h_i(t))}{\phi(t_0)} \exp\left(-\int_{H_i(t)}^{t} \alpha(s) \, ds\right) = 0, \qquad t_0 \leqslant t < T \quad (3.1.6)$$

and investigate how it relates to the existence of positive solutions of eqn (3.1.1).

When the coefficients and delays of eqn (3.1.1) are constants $p_i(t) \equiv p_i$ and $\tau_i(t) = \tau_i$ for $i = 1, 2, \ldots, n$, eqn (3.1.1) reduces to the linear autonomous delay equation

$$\dot{x}(t) + \sum_{i=1}^{n} p_i x(t - \tau_i) = 0 \quad (3.1.7)$$

whose characteristic equation is

$$\lambda + \sum_{i=1}^{n} p_i e^{-\lambda \tau_i} = 0. \quad (3.1.8)$$

We should note that eqn (3.1.8) is obtained from eqn (3.1.7) by looking for solutions of the form $x(t) = e^{\lambda t}$.

In the case of eqn (3.1.7) and when t is sufficiently large, eqn (3.1.6) takes the form

$$\alpha(t) + \sum_{i=1}^{n} p_i \exp\left(-\int_{t-\tau_i}^{t} \alpha(s) \, ds\right) = 0. \quad (3.1.9)$$

Now if we look for a solution of the form $\alpha(t) = \lambda$, eqn (3.1.9) gives precisely eqn (3.1.8).

The following result is analogous to Theorem 2.1.1, which can be restated in an equivalent way to say that 'eqn (3.1.7) has a positive solution if and only if its characteristic equation (3.1.8) has a real root'.

Theorem 3.1.1. *Assume that (3.1.2) holds and that $\phi \in C[[t_{-1}, t_0], \mathbb{R}]$ with $\phi(t_0) > 0$. Then the following statements are equivalent:*

(a) *The solution of the IVP (3.1.1) and (3.1.4) is positive on $t_0 \leqslant t < T$.*

(b) *The generalized characteristic equation (3.1.6) has a continuous solution on $t_0 \leqslant t < T$.*

(c) *There exists functions $\beta, \gamma \in C[[t_0, T), \mathbb{R}]$ such that*

$$\beta(t) \leqslant \gamma(t), \qquad t_0 \leqslant t < T \tag{3.1.10}$$

and such that for any function $\delta \in C[[t_0, T), \mathbb{R}]$ between β and γ,

$$\beta(t) \leqslant \delta(t) \leqslant \gamma(t), \qquad t_0 \leqslant t < T \tag{3.1.11}$$

the following inequality holds:

$$\beta(t) \leqslant (S\delta)(t) \leqslant \gamma(t), \qquad t_0 \leqslant t < T \tag{3.1.12}$$

where by definition

$$(S\delta)(t) \equiv - \sum_{i=1}^{n} p_i(t) \frac{\phi(h_i(t))}{\phi(t_0)} \exp\left(-\int_{H_i(t)}^{t} \delta(s) \, ds\right).$$

Proof.

(a) $\Rightarrow$ (b). Let $x(t) = x(\phi)(t)$ be the solution of the IVP (3.1.1) and (3.1.4). By hypothesis, $x(t) > 0$ for $t_0 \leqslant t < T$. Now we claim that the continuous function $\alpha(t)$ defined by

$$\alpha(t) = \frac{\dot{x}(t)}{x(t)}, \qquad t_0 \leqslant t < T \tag{3.1.13}$$

is a solution of (3.1.6) on $t_0 \leqslant t < T$. Indeed, from (3.1.13) we see that

$$x(t) = \phi(t_0) \exp\left(\int_{t_0}^{t} \alpha(s) \, ds\right), \qquad t_0 \leqslant t < T \tag{3.1.14}$$

and so

$$\frac{x(H_i(t))}{x(t)} = \exp\left(-\int_{H_i(t)}^{t} \alpha(s) \, ds\right), \qquad t_0 \leqslant t < T \quad \text{and } i = 1, 2, \ldots, n. \tag{3.1.15}$$

By dividing both sides of (3.1.1) by $x(t)$ and using (3.1.13) and (3.1.15) we see that

$$\alpha(t) + \sum_{i=1}^{n} p_i(t) \frac{x(t - \tau_i(t))}{x(H_i(t))} \exp\left(-\int_{H_i(t)}^{t} \alpha(s) \, ds\right) = 0, \qquad t_0 \leqslant t < T.$$

To show that $\alpha(t)$ is a solution of (3.1.6) it remains to prove that for $t_0 \leqslant t < T$ and for every $i = 1, 2, \ldots, n$,

$$\frac{x(t - \tau_i(t))}{x(H_i(t))} = \frac{\phi(h_i(t))}{\phi(t_0)}.$$

To this end observe that if $t - \tau_i(t) \geqslant t_0$, then $h_i(t) = t_0$ and $H_i(t) = t - \tau_i(t)$, and so:

$$\frac{x(t - \tau_i(t))}{x(H_i(t))} = \frac{x(H_i(t))}{x(H_i(t))} = 1 = \frac{\phi(t_0)}{\phi(t_0)} = \frac{\phi(h_i(t))}{\phi(t_0)}.$$

On the other hand, if $t - \tau_i(t) < t_0$, then $h_i(t) = t - \tau_i(t)$ and $H_i(t) = t_0$, and so again:

$$\frac{x(t - \tau_i(t))}{x(H_i(t))} = \frac{\phi(h_i(t))}{x(t_0)} = \frac{\phi(h_i(t))}{\phi(t_0)}.$$

The proof that (a) $\Rightarrow$ (b) is complete.

(b) $\Rightarrow$ (c). If $\alpha(t)$ is a solution of (3.1.6), then take $\beta(t) \equiv \gamma(t) \equiv \alpha(t)$, $t_0 \leqslant t < T$ and the proof is obvious because from (3.1.6), $\alpha(t) = (S\alpha)(t)$.

(c) $\Rightarrow$ (a). Our strategy here is first to show that, under hypothesis (c), eqn (3.1.6) has a continuous solution $\alpha(t)$ on $t_0 \leqslant t < T$. Second, we show that the function $x(t)$ defined by

$$x(t) = \begin{cases} \phi(t), & t_{-1} \leqslant t < t_0 \\ \phi(t_0) \exp\left(\int_{t_0}^{t} \alpha(s) \, ds \right), & t_0 \leqslant t < T \end{cases} \tag{3.1.16}$$

is a positive solution of the IVP (3.1.1) and (3.1.4).

The continuous solution of eqn (3.1.6) will be constructed (by the method of successive approximations) as the limit of a sequence of functions $\{\alpha_k(t)\}$ defined as follows. Take any function $\alpha_0 \in C[[t_0, T), \mathbb{R}]$ between β and γ:

$$\beta(t) \leqslant \alpha_0(t) \leqslant \gamma(t), \qquad t_0 \leqslant t < T \tag{3.1.17}$$

and set

$$\alpha_{k+1}(t) = (S\alpha_k)(t), \qquad t_0 \leqslant t < T \quad \text{and } k = 0, 1, \dots. \tag{3.1.18}$$

By using the hypothesis (3.1.12) and by induction, it follows that for all $k = 1, 2, \dots,$

$$\beta(t) \leqslant \alpha_k(t) \leqslant \gamma(t), \qquad t_0 \leqslant t < T \tag{3.1.19}$$

and clearly $\alpha_k \in C[[t_0, T), \mathbb{R}]$.

Next we prove that the sequence $\{\alpha_k(t)\}$ converges uniformly on any compact subinterval $[t_0, T_1]$ of $[t_0, T)$. Set

$$L = \max_{t_0 \leqslant t \leqslant T_1} \{\max\{|\beta(t)|, |\gamma(t)|\}\}$$

$$M = \max_{t_0 \leqslant t \leqslant T_1} \sum_{i=1}^{n} \left| p_i(t) \frac{\phi(h_i(t))}{\phi(t_0)} \right|$$

$$N = M \, e^{L(T_1 - t_0)}.$$

Then from (3.1.19),

$$\max_{t_0 \leqslant t \leqslant T_1} |\alpha_k(t)| \leqslant L.$$

By using the mean value theorem, we have for all $i = 0, 1, 2, \ldots, n$, $k = 0, 1, 2, \ldots$ and $t_0 \leqslant t \leqslant T_1$,

$$\exp\left(-\int_{H_i(t)}^{t} \alpha_k(s) \, ds\right) - \exp\left(-\int_{H_i(t)}^{t} \alpha_{k-1}(s) \, ds\right)$$

$$= e^{-\mu_{k,i}(t)} \int_{H_i(t)}^{t} [\alpha_{k-1}(s) - \alpha_k(s)] \, ds$$

where $\mu_{k,i}(t)$ is between $\int_{H_i(t)}^{t} \alpha_k(s) \, ds$ and $\int_{H_i(t)}^{t} \alpha_{k-1}(s) \, ds$. Hence, as $H_i(t) \geqslant t_0$, we obtain $|\mu_{k,1}(t)| \leqslant L(T_1 - t_0)$ and

$$\left| \exp\left(-\int_{H_i(t)}^{t} \alpha_k(s) \, ds\right) - \exp\left(-\int_{H_i(t)}^{t} \alpha_{k-1}(s) \, ds\right) \right|$$

$$\leqslant e^{L(T_1 - t_0)} \int_{t_0}^{t} |\alpha_k(s) - \alpha_{k-1}(s)| \, ds.$$

Thus for all $k = 0, 1, 2, \ldots$ and $t_0 \leqslant t \leqslant T_1$,

$$|\alpha_{k+1}(t) - \alpha_k(t)| \leqslant N \int_{t_0}^{t} |\alpha_k(s) - \alpha_{k-1}(s)| \, ds. \qquad (3.1.20)$$

One can now show, by induction (as in Picard's method of successive approximations: Finizio and Ladas 1982, Appendix D) that for $k = 0, 1, 2, \ldots$ and $t_0 \leqslant t \leqslant T_1$,

$$|\alpha_{k+1}(t) - \alpha_k(t)| \leqslant 2L \frac{[N(t - t_0)]^k}{k!}.$$

It follows by the Weierstrass M-test that the series $\sum_{k=0}^{\infty} [\alpha_{k+1}(t) - \alpha_k(t)]$ converges uniformly on $t_0 \leqslant t \leqslant T_1$. Therefore the sequence

$$\alpha_k(t) = \alpha_0(t) + \sum_{j=0}^{k-1} [\alpha_{j+1}(t) - \alpha_j(t)], \qquad t_0 \leqslant t \leqslant T_1$$

for $k = 1, 2, \ldots$ also converges uniformly and so the limit function

$$\alpha(t) \equiv \lim_{k \to \infty} \alpha_k(t) \qquad (3.1.21)$$

is continuous on $t_0 \leqslant t \leqslant T_1$. Because of the uniform convergence,

$$\alpha(t) = \lim_{k \to \infty} \alpha_{k+1}(t) = \lim_{k \to \infty} \left[-\sum_{i=1}^{n} p_i(t) \frac{\phi(h_i(t))}{\phi(t_0)} \exp\left(-\int_{H_i(t)}^{t} \alpha_k(s)\, ds \right) \right]$$

$$= -\sum_{i=1}^{n} p_i(t) \frac{\phi(h_i(t))}{\phi(t_0)} \exp\left(-\int_{H_i(t)}^{t} \alpha(s)\, ds \right), \qquad t_0 \leqslant t \leqslant T_1$$

which shows that $\alpha(t)$ is a solution of (3.1.6) on $[t_0, T_1]$. As T_1 is an arbitrary fixed point in $[t_0, T)$, it follows that $\alpha(t)$ as defined by (3.1.21) is a solution of (3.1.6) on $t_0 \leqslant t < T$.

Finally, it suffices to show that $x(t)$ as defined by (3.1.16) is a solution of the IVP (3.1.1) and (3.1.4). [As $x(t)$ is clearly positive on $t_0 \leqslant t \leqslant T$ and in view of the fact that the IVP (3.1.1) and (3.1.4) has a unique solution, the proof would then be complete.] To this end observe that [by using arguments similar to those in the proof that (a) $\Rightarrow$ (b)]

$$\dot{x}(t) = \alpha(t)x(t) = -x(t) \sum_{i=1}^{n} p_i(t) \frac{\phi(h_i(t))}{\phi(t_0)} \exp\left(-\int_{H_i(t)}^{t} \alpha(s)\, ds \right)$$

$$= -x(t) \sum_{i=1}^{n} p_i(t) \frac{x(t - \tau_i(t))}{x(H_i(t))} \frac{x(H_i(t))}{x(t)} = -\sum_{i=1}^{n} p_i(t) x(t - \tau_i(t)), \quad t_0 \leqslant t < T$$

which completes the proof of the theorem.

The proof of the following result is essentially contained in the proof of Theorem 3.1.1.

Corollary 3.1.1. *Assume that (3.1.2) holds and that $\phi(t_0) > 0$. Then the solution $x(\phi)(t)$ of the IVP (3.1.1) and (3.1.4) is positive on $t_0 \leqslant t < T$ if and only if*

$$x(\phi)(t) = \phi(t_0) \exp\left(\int_{t_0}^{t} \alpha(s)\, ds \right), \qquad t_0 \leqslant t < T \qquad (3.1.22)$$

where $\alpha(t)$ is a continuous solution of (3.1.6). Furthermore, if $\beta, \gamma \in C[[t_0, T), \mathbb{R}]$ such that (3.1.10) holds and such that for any $\delta \in C[[t_0, T), \mathbb{R}]$, (3.1.11) implies (3.1.12), then the solutions $\alpha(t)$ and $x(\phi)(t)$ of (3.1.6) and (3.1.1) respectively satisfy the following inequalities:

$$\beta(t) \leqslant \alpha(t) \leqslant \gamma(t), \qquad t_0 \leqslant t < T \qquad (3.1.23)$$

and

$$\phi(t_0) \exp\left(\int_{t_0}^{t} \beta(s)\, ds \right) \leqslant x(\phi)(t) \leqslant \phi(t_0) \exp\left(\int_{t_0}^{t} \gamma(s)\, ds \right), \quad t_0 \leqslant t < T.$$

$$(3.1.24)$$

Remark 3.1.1. *For every* $\phi \in C[[t_0, T), \mathbb{R}]$ *with* $\phi(t_0) > 0$, *eqn (3.1.6) has at most one continuous solution on* $t_0 \leqslant t < T$. *This is because to two different solutions of (3.1.6) there would correspond two different (positive) solutions of the IVP (3.1.1) and (3.1.4), which is impossible. And if the IVP (3.1.1) and (3.1.4) has a solution* $x(\phi)(t)$ *which is positive on* $t_0 \leqslant t < T$, *then (3.1.6) has exactly one continuous solution on* $t_0 \leqslant t < T$.

Remark 3.1.2. *Assume* $T = \infty$ *and that* $\lim_{t \to \infty} (t - \tau_i(t)) = \infty$ *for* $i = 1, 2, \ldots, n$. *Then for t sufficiently large,* $h_i(t) = t_0$ *and* $H_i(t) = t - \tau_i(t)$ *for* $i = 1, 2, \ldots, n$, *so eqn (3.1.6) reduces to*

$$\alpha(t) + \sum_{i=1}^{n} p_i(t) \exp\left(-\int_{t-\tau_i(t)}^{t} \alpha(s) \, ds \right) = 0.$$

The following result is a useful corollary of Theorem 3.1.1.

Corollary 3.1.2. *Assume that the delay differential equation*

$$\dot{x}(t) + \sum_{i=1}^{n} p_i(t) x(t - \tau_i(t)) = 0, \qquad t \geqslant t_0$$

where for $i = 1, 2, \ldots, n$

$$p_i \in C[[t_0, \infty), \mathbb{R}], \qquad \tau_i \in C[[t_0, \infty), \mathbb{R}^+] \qquad and \qquad \lim_{t \to \infty} (t - \tau_i(t)) = \infty$$

has an eventually positive solution. Then there exists a $\tilde{t}_0 \geqslant t_0$ *and a function* $\alpha \in C[[\tilde{t}_{-1}, \infty), \mathbb{R}]$ *where*

$$\tilde{t}_{-1} = \min_{1 \leqslant i \leqslant n} \left\{ \inf_{t \geqslant \tilde{t}_0} \{t - \tau_i(t)\} \right\}$$

such that

$$\alpha(t) + \sum_{i=1}^{n} p_i(t) \exp\left(-\int_{t-\tau_i(t)}^{t} \alpha(s) \, ds \right) = 0, \qquad t \geqslant \tilde{t}_0. \qquad (3.1.25)$$

3.2 Differential inequalities and comparison results for positive solutions

In this section we present some comparison results for the positive solutions of the delay differential equation

$$\dot{y}(t) + \sum_{i=1}^{n} q_i(t) y(t - \tau_i(t)) = 0, \qquad t_0 \leqslant t < T \tag{3.2.1}$$

and the delay differential inequalities

$$\dot{x}(t) + \sum_{i=1}^{n} p_i(t) x(t - \tau_i(t)) \leqslant 0, \qquad t_0 \leqslant t < T \tag{3.2.2}$$

and

$$\dot{z}(t) + \sum_{i=1}^{n} r_i(t) z(t - \tau_i(t)) \geqslant 0, \qquad t_0 \leqslant t < T \tag{3.2.3}$$

where

$$p_i, q_i, r_i, \tau_i \in C[[t_0, T), \mathbb{R}^+] \qquad \text{for } i = 1, 2, \dots, n \tag{3.2.4}$$

and $t_0 < T \leqslant \infty$.

The initial interval associated with the above equations and inequalities is $t_{-1} \leqslant t \leqslant t_0$ where

$$t_{-1} = \min_{1 \leqslant i \leqslant n} \left\{ \inf_{t_0 \leqslant t < T} \{ t - \tau_i(t) \} \right\}. \tag{3.2.5}$$

The following theorem is the main result in this section.

Theorem 3.2.1. *Suppose that (3.2.4) holds and that*

$$p_i(t) \geqslant q_i(t) \geqslant r_i(t). \qquad t_0 \leqslant t < T \qquad \text{and} \qquad i = 1, 2, \dots, n. \tag{3.2.6}$$

Assume that $x(t)$, $y(t)$, and $z(t)$ are continuous solutions of (3.2.2), (3.2.1) and (3.2.3) respectively, such that

$$x(t) > 0, \qquad t_0 \leqslant t < T \tag{3.2.7}$$

$$z(t_0) \geqslant y(t_0) \geqslant x(t_0) \tag{3.2.8}$$

$$\frac{x(t)}{x(t_0)} \geqslant \frac{y(t)}{y(t_0)} \geqslant \frac{z(t)}{z(t_0)} \geqslant 0, \qquad t_{-1} \leqslant t < t_0. \tag{3.2.9}$$

Then

$$z(t) \geqslant y(t) \geqslant x(t), \qquad t_0 \leqslant t < T. \tag{3.2.10}$$

Proof. In view of (3.2.7) and (3.2.8), the functions $y(t)$ and $z(t)$ are positive in some right neighbourhood $[t_0, T_1)$ of t_0. We claim that $T_1 = T$. Otherwise, if $[t_0, T_1)$ is the maximal subinterval of $[t_0, T)$ where both $y(t)$ and $z(t)$ are positive, then $T_1 < T$ and

$$y(T_1) z(T_1) = 0. \tag{3.2.11}$$

Set

$$\alpha_0(t) = \frac{\dot{x}(t)}{x(t)}, \qquad \beta_0(t) = \frac{\dot{y}(t)}{y(t)} \qquad \text{and} \qquad \gamma_0(t) = \frac{\dot{z}(t)}{z(t)}$$

for all $t_0 \leqslant t < T_1$. Then, as in the proof of Theorem 3.1.1. and using the notation (3.1.5), we find, for $t_0 \leqslant t < T_1$,

$$\alpha_0(t) + \sum_{i=1}^{n} p_i(t) \frac{x(h_i(t))}{x(t_0)} \exp\left(-\int_{H_i(t)}^{t} \alpha_0(s) \, ds \right) \leqslant 0 \qquad (3.2.12)$$

$$\beta_0(t) + \sum_{i=1}^{n} q_i(t) \frac{y(h_i(t))}{y(t_0)} \exp\left(-\int_{H_i(t)}^{t} \beta_0(s) \, ds \right) = 0 \qquad (3.2.13)$$

$$\gamma_0(t) + \sum_{i=1}^{n} r_i(t) \frac{z(h_i(t))}{z(t_0)} \exp\left(-\int_{H_i(t)}^{t} \gamma_0(s) \, ds \right) \geqslant 0. \qquad (3.2.14)$$

We can now show that

$$\alpha_0(t) \leqslant \beta_0(t) \leqslant \gamma_0(t) \qquad \text{for all } t_0 \leqslant t < T_1. \qquad (3.2.15)$$

We show that $\alpha_0(t) \leqslant \beta_0(t)$, while the proof that $\beta_0(t) \leqslant \gamma_0(t)$ is similar and will be omitted. Let $\delta \in C[[t_0, T_1), \mathbb{R}]$ be an arbitrary function between $\alpha_0(t)$ and 0, $\alpha_0(t) \leqslant \delta(t) \leqslant 0, t_0 \leqslant t < T_1$. Then from (3.2.12), (3.2.6) and (3.2.13) we see that for $t_o \leqslant t < T_1$,

$$\alpha_0(t) \leqslant - \sum_{i=1}^{n} p_i(t) \frac{x(h_i(t))}{x(t_0)} \exp\left(-\int_{H_i(t)}^{t} \alpha_0(s) \, ds \right)$$

$$\leqslant - \sum_{i=1}^{n} q_i(t) \frac{y(h_i(t))}{y(t_0)} \exp\left(-\int_{H_i(t)}^{t} \delta(s) \, ds \right) \leqslant 0.$$

That is, the statement (c) of Theorem 3.1.1 is true [with $\beta(t) = \alpha_0(t)$ and $\gamma(t) \equiv 0$], so by the same theorem (and Corollary 3.1.1. and Remark 3.1.1), the equation

$$\alpha(t) + \sum_{i=1}^{n} q_i(t) \frac{y(h_i(t))}{y(t_0)} \exp\left(-\int_{H_i(t)}^{t} \alpha(s) \, ds \right) = 0 \qquad (3.2.16)$$

has exactly one solution on $t_0 \leqslant t < T_1$ and, moreover, the solution of (3.2.16) is between $\alpha_0(t)$ and 0 for $t_0 \leqslant t < T_1$. But, by (3.2.13), β_0 is a solution of (3.2.16) on $t_0 \leqslant t < T_1$. Hence $\alpha_0(t) \leqslant \alpha(t) \equiv \beta_0(t) \leqslant 0$ for $t_0 \leqslant t < T_1$ and (3.2.15) has been established. Clearly, on $t_0 \leqslant t < T_1$, by using the definition of α_0, β_0, and γ_0, we find

$$x(t) = x(t_0) \exp\left(\int_{t_0}^{t} \alpha_0(s) \, ds \right), \qquad y(t) = y(t_0) \exp\left(\int_{t_0}^{t} \beta_0(s) \, ds \right),$$

and

$$z(t) = z(t_0) \exp\left(\int_{t_0}^{t} \gamma_0(s)\, ds\right).$$

Hence (3.2.8) and (3.2.15) imply that

$$z(t) \geqslant y(t) \geqslant x(t), \qquad t_0 \leqslant t < T_1. \tag{3.2.17}$$

As $x(T_1) > 0$ and in view of the continuity of the functions $x(t)$, $y(t)$, and $z(t)$, it follows from (3.2.17) that $z(T_1) \geqslant y(T_1) > 0$. This contradicts (3.2.11) and proves that $T_1 = T$. Thus (3.2.17) holds with T_1 replaced by T. The proof is complete.

The following corollary of Theorem 3.2.1 compares two positive solutions $y(\phi)(t)$ and $y(\psi)(t)$ of eqn (3.2.1) by means of their initial functions ϕ and ψ respectively.

Corollary 3.2.1. *Assume that coefficients and the delays of eqn (3.2.1) are non-negative continuous functions in the interval $t_0 \leqslant t < T$. Let $\phi, \psi \in C[[t_{-1}, t_0], \mathbb{R}]$ be such that*

$$\psi(t_0) \geqslant \phi(t_0) > 0, \qquad \frac{\phi(t)}{\phi(t_0)} \geqslant \frac{\psi(t)}{\psi(t_0)} \geqslant 0, \qquad t_{-1} \leqslant t \leqslant t_0$$

and

$$y(\phi)(t) > 0, \qquad t_0 \leqslant t < T.$$

Then

$$y(\psi)(t) \geqslant y(\phi)(t), \qquad t_0 \leqslant t < T.$$

The following result is an immediate corollary of Theorem 3.2.1.

Corollary 3.2.2. *Assume that*

$$p_i, \tau_i \in C[[t_0, \infty), \mathbb{R}^+] \qquad for\ i = 1, 2, \dots, n.$$

Then the differential inequality

$$\dot{x}(t) + \sum_{i=1}^{n} p_i(t)x(t - \tau_i(t)) \leqslant 0, \qquad t \geqslant t_0$$

has an eventually positive solution if and only if the equation

$$\dot{y}(t) + \sum_{i=1}^{n} p_i(t)y(t - \tau_i(t)) = 0, \qquad t \geqslant t_0$$

has an eventually positive solution.

By utilizing some of the results that we have established so far in this book we obtain the following powerful theorem about autonomous differential equations and inequalities.

Theorem 3.2.2. *Assume that*

$$p_i \in (0, \infty) \quad and \quad \tau_i \in [0, \infty) \quad for \ i = 1, 2, \ldots, n.$$

Then the following statements are equivalent:

(a) *The delay differential equation*

$$\dot{x}(t) + \sum_{i=1}^{n} p_i x(t - \tau_i) = 0, \quad t \geq 0 \qquad (3.2.18)$$

has a positive solution.

(b) *The characteristic equation*

$$\lambda + \sum_{i=1}^{n} p_i e^{-\lambda \tau_i} = 0 \qquad (3.2.19)$$

has a real root.

(c) *The delay differential inequality*

$$\dot{y}(t) + \sum_{i=1}^{n} p_i y(t - \tau_i) \leq 0, \quad t \geq 0 \qquad (3.2.20)$$

has a positive solution.

(d) *There exists $\varepsilon_0 \in (0, 1)$ such that for every $\varepsilon \in [0, \varepsilon_0]$ the delay differential equation*

$$\dot{z}(t) + \sum_{i=1}^{n} (1 - \varepsilon) p_i z(t - \tau_i) = 0, \quad t \geq 0 \qquad (3.2.21)$$

has a positive solution.

Proof. The proof that (a) $\Leftrightarrow$ (b) follows from Theorem 2.1.1 and the autonomous nature of eqn (3.2.18). The proof that (a) $\Rightarrow$ (c) is obvious. On the other hand, (c) $\Rightarrow$ (a) follows from Theorem 3.2.1 by taking the solution of (3.2.18) with initial function equal to $y(t)$ in the initial interval. The proof will be complete if we show that (b) $\Leftrightarrow$ (d). As (d) is equivalent (by Theorem 2.1.1) to the statement that the characteristic equation of (3.2.21)

$$\mu + \sum_{i=1}^{n} (1 - \varepsilon) p_i e^{-\mu \tau_i} = 0 \qquad (3.2.22)$$

has a real root, it suffices to prove that the following statements are equivalent:

(b') Eqn (3.2.19) has no real root.

(d') There exists $\varepsilon_0 \in (0, 1)$ such that for every $\varepsilon \in [0, \varepsilon_0]$, eqn (3.2.22) has no real root.

Clearly (d') $\Rightarrow$ (b'). To prove (b') $\Rightarrow$ (d'), set

$$F(\lambda) = \lambda + \sum_{i=1}^{n} p_i e^{-\lambda \tau_i}$$

and observe that $F(\infty) = \infty$. As $F(\lambda) = 0$ has no real root, it follows that $F(\lambda) > 0$ for $\lambda \in \mathbb{R}$. Also $F(-\infty) = \infty$, for otherwise all the delays are equal to zero and (3.2.19) has no real root. Therefore, $\min_{\lambda \in \mathbb{R}} F(\lambda)$ exists and is a positive number m. Hence,

$$F(\lambda) \geqslant m \qquad \text{for } \lambda \in \mathbb{R}. \tag{3.2.23}$$

Set

$$G(\mu) = \mu + \frac{1}{2} \sum_{i=1}^{n} p_i e^{-\mu \tau_i}, \qquad \mu \in \mathbb{R}$$

and observe that $G(\infty) = G(-\infty) = \infty$. Thus there exists $\alpha > 0$ such that $G(\mu) > 0$ for $|\mu| > \alpha$. Now choose $\varepsilon_0 \in (0, \frac{1}{2})$ such that

$$\varepsilon_0 \sum_{i=1}^{n} p_i e^{\alpha \tau_i} \leqslant \frac{m}{2}.$$

We claim that for every $\varepsilon \in [0, \varepsilon_0]$, eqn (3.2.22) has no real roots. Indeed, if $\mu \in \mathbb{R}$ with $|\mu| > \alpha$, then

$$\mu + \sum_{i=1}^{n} (1 - \varepsilon) p_i e^{-\mu \tau_i} \geqslant \mu + \frac{1}{2} \sum_{i=1}^{n} p_i e^{-\mu \tau_i} > 0,$$

On the other hand, if $-\alpha \leqslant \mu \leqslant \alpha$ then

$$\mu + \sum_{i=1}^{n} (1 - \varepsilon) p_i e^{-\mu \tau_i} \geqslant \left(\mu + \sum_{i=1}^{n} p_i e^{-\mu \tau_i} \right) - \varepsilon \sum_{i=1}^{n} p_i e^{-\mu \tau_i}$$

$$\geqslant F(\mu) - \varepsilon_0 \sum_{i=1}^{n} p_i e^{\alpha \tau_i} \geqslant m - \frac{m}{2} = \frac{m}{2} > 0.$$

The proof of the theorem is complete.

One can establish comparison results for differential equations and inequalities with advanced arguments by a simple adaptation of the methods in the delay case. For example, the following result is the 'dual' of Corollary 3.2.2.

Corollary 3.2.3. *Assume that*

$$p_i, \tau_i \in C[[t_0, \infty), \mathbb{R}^+] \qquad \text{for } i = 1, 2, \ldots, n.$$

Then the differential inequality

$$\dot{x}(t) - \sum_{i=1}^{n} p_i(t)x(t + \tau_i(t)) \geqslant 0, \qquad t \geqslant t_0$$

has an eventually positive solution if and only if the equation

$$\dot{y}(t) - \sum_{i=1}^{n} p_i(t)y(t + \tau_i(t)) = 0$$

has an eventually positive solution.

3.3 Existence of positive solutions

Consider the linear non-autonomous delay differential equation with variable coefficients and variable delays

$$\dot{x}(t) + \sum_{i=1}^{n} p_i(t)x(t - \tau_i(t)) = 0, \qquad t_o \leqslant t < T \tag{3.3.1}$$

where $t_0 \leqslant T \leqslant \infty$ and for $i = 1, 2, \ldots, n$,

$$p_i \in C[[t_0, T), \mathbb{R}] \qquad \text{and} \qquad \tau_i \in C[[t_0, T), \mathbb{R}^+]. \tag{3.3.2}$$

Our aim in this section is to show that under appropriate hypotheses eqn (3.3.1) has a huge number of positive solutions on $t_0 \leqslant t < T$.

Let $t_{-1} = \min_{1 \leqslant i \leqslant n} \{\inf_{t_0 \leqslant t < T} \{(t - \tau_i(t))\}\}$

$$p_i^+(t) = \max\{0, p_i(t)\}] \qquad \text{for } t_0 \leqslant t < T \text{ and } i = 1, 2, \ldots, n$$

$$g(t) = \min_{1 \leqslant i \leqslant n} \{\max\{t_0, t - \tau_i(t)\}\} \qquad \text{for } t_0 \leqslant t < T$$

$$\Phi = \{\phi \in C[[t_{-1}, t_0], R^+] \mid \phi(t_0) > 0 \qquad \text{and}$$

$$\phi(t) \leqslant \phi(t_0) \qquad \text{for } t_{-1} \leqslant t \leqslant t_0\}. \tag{3.3.3}$$

Then the following result is true.

Theorem 3.3.1. *Assume that (3.3.2) holds and suppose that*

$$\sum_{i=1}^{n} \int_{g(t)}^{t} p_i^+(s) \, ds \leqslant \frac{1}{e} \qquad \text{for } t_0 \leqslant t < T. \tag{3.3.4}$$

Then for every $\phi \in \Phi$, the solution $x(\phi)(t)$ of eqn (3.3.1) through (t_0, ϕ) remains positive on $t_0 \leqslant t < T$.

Proof. Consider the equation

$$\dot{y}(t) + \sum_{i=1}^{n} p_i^+(t) y(t - \tau_i(t)) = 0, \qquad t_0 \leqslant t < T \qquad (3.3.5)$$

and its generalized characteristic equation

$$\alpha(t) + \sum_{i=1}^{n} p_i^+(t) \frac{\phi(h_i(t))}{\phi(t_0)} \exp\left(-\int_{H_i(t)}^{t} \alpha(s) \, ds\right) = 0, \qquad t_0 \leqslant t < T \qquad (3.3.6)$$

where $h_i(t)$ and $H_i(t)$ are given by (3.1.5). We now claim that statement (c) of Theorem 3.1.1 is true with $\beta(t) = -e \sum_{i=1}^{n} p_i^+(t)$ and $\gamma(t) = 0$ for $t_0 \leqslant t < T$. Indeed, for any function $\delta \in C[[t_0, T), \mathbb{R}]$ between β and γ we have for $t_0 \leqslant t < T$,

$$-\int_{H_i(t)}^{t} \delta(s) \, ds \leqslant e \int_{H_i(t)}^{t} \sum_{i=1}^{n} p_i^+(s) \, ds \leqslant e \int_{g(t)}^{t} \sum_{i=1}^{n} p_i^+(s) \, ds \leqslant 1.$$

Hence, (3.1.12) holds because, for $t_0 \leqslant t < T$,

$$\gamma(t) = 0 \geqslant - \sum_{i=1}^{n} p_i^+(t) \frac{\phi(h_i(t))}{\phi(t_0)} \exp\left(-\int_{H_i(t)}^{t} \delta(s) \, ds\right) \geqslant - \sum_{i=1}^{n} p_i^+(t) e = \beta(t).$$

Thus by Theorem 3.1.1, the solution $y(\phi)(t)$ of eqn (3.3.5) is positive on $t_0 \leqslant t < T$.

Now by an elementary application of Theorem 3.2.1 to the solutions $x(\phi)(t)$ and $y(\phi)(t)$ of eqns (3.3.1) and (3.3.5) we conclude that $x(\phi)(t) \geqslant y(\phi)(t) > 0$, $t_0 \leqslant t < T$. The proof is complete.

Corollary 3.3.1. *Assume that $\tau \in C[\mathbb{R}^+, \mathbb{R}^+]$ is such that*

$$\tau(t) \leqslant 1/e \qquad \text{for } t \geqslant 0. \qquad (3.3.7)$$

Then the delay differential equation

$$\dot{x}(t) + x(t - \tau(t)) = 0, \qquad t \geqslant 0 \qquad (3.3.8)$$

has a positive solution on $[t_{-1}, \infty)$ where $t_{-1} = \min_{t \geqslant 0}\{t - \tau(t)\}$.

Proof. Here $t_0 = 0$, $T = \infty$ and $g(t) = \max\{0, t - \tau(t)\}$. Hence $\int_{g(t)}^{t} ds = t - g(t) \leqslant 1/e$ for all $t \geqslant 0$ and the result is a consequence of Theorem 3.3.1.

In the case of constant delays, the condition (3.3.7) is necessary and sufficient for the existence of a positive solution for eqn (3.3.8). However, the following example shows that in the case of variable delays, condition (3.3.7) is not sharp.

Example 3.3.1. Consider the equation (3.3.8) with the delay $\tau(t)$ defined as follows:

$$\tau(t) = \begin{cases} 0, & 0 \leqslant t \leqslant \frac{1}{2} \\ t - \frac{1}{2}, & \frac{1}{2} < t \leqslant 1 \\ \frac{3}{2} - t, & 1 < t \leqslant \frac{3}{2} \\ 0, & \frac{3}{2} < t < \infty. \end{cases}$$

Clearly, $\tau(1) = \frac{1}{2} > 1/e$ and so (3.3.7) is not satisfied. In spite of that, we show that the solution of eqn (3.3.8) with initial function

$$x(t) = e^{-t} \qquad \text{for } t \leqslant 0 \tag{3.3.9}$$

is positive for $t \geqslant 0$. Indeed, by straightforward computations we find

$$x(t) = \begin{cases} e^{-t}, & 0 \leqslant t \leqslant \frac{1}{2} \\ (\frac{3}{2} - t) \, e^{-1/2}, & \frac{1}{2} < t \leqslant 1 \\ x(\frac{3}{2}) \, e^{-(t-3/2)}, & t \geqslant \frac{3}{2}. \end{cases}$$

The proof will be complete if we show that $x(t) > 0$ for $1 \leqslant t \leqslant \frac{3}{2}$. Assume, for the sake of contradiction, that $x(t)$ is not positive in the interval $[1, \frac{3}{2}]$. Then there must exist a point $t_1 \in (1, \frac{3}{2}]$ such that $x(t) > 0$ for $0 \leqslant t < t_1$ and $x(t_1) = 0$. It also follows from eqn (3.3.8) that $\dot{x}(t) < 0$ for $0 \leqslant t \leqslant t_1$. By integrating eqn (3.3.8) from 1 to t_1 we get

$$x(t_1) - x(1) + \int_1^{t_1} x(2s - \tfrac{3}{2}) \, ds = 0.$$

Hence

$$\tfrac{1}{2} e^{-1/2} = x(1) = \int_1^{t_1} x(2s - \tfrac{3}{2}) \, ds \leqslant \int_1^{3/2} x(2s - \tfrac{3}{2}) \, ds < \int_1^{3/2} x(\tfrac{1}{2}) \, ds = \tfrac{1}{2} e^{-1/2}$$

and this contradiction establishes our claim that the solution of (3.3.8) and (3.3.9) is positive for all t.

In the next theorems we give other sufficient conditions under which eqn (3.3.1) has positive solutions on $[t_0, T)$.

Theorem 3.3.2. *Assume that (3.3.2) holds and that there exists a positive number μ such that*

$$\sum_{i=1}^{n} |p_i(t)| \, e^{\mu \tau_i(t)} \leqslant \mu \qquad \text{for } t \geqslant t_0. \tag{3.3.10}$$

Then for every $\phi \in \Phi$, the solution $x(\phi)(t)$ of eqn (3.3.1) through (t_0, ϕ) remains positive on $t_0 \leqslant t < T$.

Proof. Let $\phi \in \Phi$ be a fixed initial function and consider the operator $(S\delta)(t)$ for $\delta \in C[[t_0, T), \mathbb{R}]$ defined by

$$(S\delta)(t) = -\sum_{i=1}^{n} p_i(t) \frac{\phi(h_i(t))}{\phi(t_0)} \exp\left(-\int_{H_i(t)}^{t} \delta(s) \, ds\right), \qquad t \in [t_0, T) \quad (3.3.11)$$

where $h_i(t)$ and $H_i(t)$ are given by (3.1.5). We now claim that statement (c) of Theorem 3.1.1 is true with $\beta(t) = -\mu$ and $\gamma(t) = \mu$ for $t_0 \leqslant t < T$. Indeed, for any function $\delta \in C[[t_0, T), \mathbb{R}]$ between β and γ we have

$$-\mu(t - H_i(t)) \leqslant \int_{H_i(t)}^{t} \delta(s) \, ds \leqslant \mu(t - H_i(t)), \qquad t_0 \leqslant t < T$$

and clearly (3.1.5) yields

$$-\mu\tau_i(t) \leqslant \int_{H_i(t)}^{t} \delta(s) \, ds \leqslant \mu\tau_i(t).$$

We now claim that (3.1.12) holds. Indeed, from (3.3.10) we have

$$\beta(t) = -\mu \leqslant -\sum_{i=1}^{n} |p_i(t)| \, e^{\mu\tau_i(t)} \leqslant (S\delta) \leqslant \sum_{i=1}^{n} |p_i(t)| \, e^{\mu\tau_i(t)} \leqslant \mu = \gamma(t)$$

for $t \in [t_0, T)$. Therefore by Theorem 3.1.1, the solution $x(\phi)(t)$ of eqn (3.3.1) is positive on $[t_0, T)$. The proof is complete.

Theorem 3.3.3. *Assume that (3.3.2) holds,*

$$0 \leqslant \tau_1(t) \leqslant \tau_2(t) \leqslant \cdots \leqslant \tau_n(t), \qquad (3.3.12)$$

and

$$\sum_{i=1}^{m} p_i(t) \leqslant 0 \qquad for \; m = 1, 2, \ldots, n \qquad and \qquad t \in [t_0, T). \qquad (3.3.13)$$

Then eqn (3.3.1) has a positive and increasing solution on $[t_0, T)$.

Proof. Let $\phi(t) = 1$ for $t_{-1} \leqslant t \leqslant t_0$ and consider the operator $(S\delta)(t)$ defined by (3.3.11) for $\delta \in C[[t_0, T), \mathbb{R}]$. We now claim that statement (c) of Theorem 3.1.1 is true with $\beta(t) = 0$ and $\gamma(t) = \sum_{i=1}^{n} |p_i(t)|$. Indeed,

$$(S\delta)(t) \leqslant \sum_{i=1}^{n} |p_i(t)| \exp\left(-\int_{H_i(t)}^{t} \delta(s) \, ds\right) \leqslant \sum_{i=1}^{n} |p_i(t)| = \gamma(t)$$

and because $H_1(t) \geqslant H_2(t) \geqslant \cdots \geqslant H_n(t)$, the relation (3.3.13) yields

$$(S\delta)(t) \geqslant -[p_1(t) + p_2(t)] \exp\left(-\int_{H_2(t)}^t \delta(s)\,ds\right)$$

$$-\sum_{i=3}^n p_i(t) \exp\left(-\int_{H_i(t)}^t \delta(s)\,ds\right)$$

$$\geqslant \left[-\sum_{i=1}^n p_i(t)\right] \exp\left(-\int_{H_n(t)}^t \delta(s)\,ds\right) \geqslant 0 = \beta(t), \qquad \text{for } t \in [t_0, T).$$

Therefore by Theorem 3.1.1, the solution $x(t) = x(\phi)(t)$ of eqn (3.3.1) is positive on $[t_0, T)$. Moreover, by Corollary 3.1.1 we know that

$$x(t) = \exp\left(\int_{t_0}^t \alpha(s)\,ds\right), \qquad t_0 \leqslant t < T.$$

where $\alpha(t)$ is a continuous solution of (3.1.6) between $\beta(t) = 0$ and $\gamma(t)$ for all $t \in [t_0, T)$. Hence $x(t)$ is an increasing solution of eqn (3.1.1) and the proof is complete.

Corollary 3.3.2. *Consider the delay differential equation*

$$\dot{x}(t) - q(t)x(t - \tau) + \sum_{i=1}^n p_i(t)x(t - \tau_i) = 0 \tag{3.3.14}$$

where for $i = 1, 2, \ldots, n$,

$$q, p_i \in C[[t_0, \infty), \mathbb{R}], \qquad \tau, \tau_i \in [0, \infty),$$

$$\sum_{i=1}^n |p_i(t)| \leqslant q(t) \qquad and \qquad \tau \leqslant \min\{\tau_1, \tau_2, \ldots, \tau_n\}.$$

Then for every $t_1 \geqslant t_0$, eqn (3.3.14) has a positive solution on $[t_1, \infty)$.

3.4 Sufficient conditions for the oscillation of non-autonomous equations

Our aim in this section is to establish several sufficient conditions for the oscillation of all solutions of the linear non-autonomous delay differential equation

$$\dot{x}(t) + \sum_{i=1}^n p_i(t)x(t - \tau_i(t)) = 0, \qquad t \geqslant 0 \tag{3.4.1}$$

where

$$p_i, \tau_i \in C[[0, \infty), \mathbb{R}^+] \qquad \text{for } i = 1, 2, \ldots, n. \tag{3.4.2}$$

The main tools for the proofs will be Lemma 1.6.3 and Corollary 3.1.2. The first result is an extension of Theorem 2.3.1 to the case of several variable delays.

Theorem 3.4.1. *Assume that (3.4.2) holds. Set*

$$\tau(t) = \max_{1 \leqslant i \leqslant n} \{\tau_i(t)\}, \qquad t \geqslant 0 \tag{3.4.3}$$

and suppose that

$$\lim_{t \to \infty} (t - \tau(t)) = \infty. \tag{3.4.4}$$

Then

$$\liminf_{t \to \infty} \int_{t-\tau(t)}^{t} \sum_{i=1}^{n} p_i(s) \, ds > 1/e \tag{3.4.5}$$

is a sufficient condition for the oscillation of all solutions of eqn (3.4.1).

Proof. Assume, for the sake of contradiction, that eqn (3.4.1) has an eventually positive solution $x(t)$. Then there exists $T \geqslant 0$ such that $x(t) > 0$ for $t \geqslant T_{-1}$, where $T_{-1} = \min_{1 \leqslant i \leqslant n}\{\inf_{t \geqslant T}\{t - \tau_i(t)\}\}$. Therefore, by Theorem 3.1.1 and Remark 3.1.2, there exists a continuous function $\alpha \in C[[T_{-1}, \infty), \mathbb{R}]$ such that

$$\alpha(t) + \sum_{i=1}^{n} p_i(t) \exp\left(-\int_{t-\tau_i(t)}^{t} \alpha(s) \, ds\right) = 0, \qquad t \geqslant T.$$

Hence

$$\alpha(t) + \left[\sum_{i=1}^{n} p_i(t)\right] \exp\left(-\int_{t-\tau(t)}^{t} \alpha(s) \, ds\right) \leqslant 0, \qquad t \geqslant T, \tag{3.4.6}$$

and so by Lemma 1.6.3,

$$m \equiv \liminf_{t \to \infty} \left[-\int_{t-\tau(t)}^{t} \alpha(s) \, ds\right] < \infty. \tag{3.4.7}$$

By rearranging the terms in (3.4.6) and then by integrating from $t - \tau(t)$ to t we find that

$$\int_{t-\tau(t)}^{t} \alpha(s) \, ds \geqslant \int_{t-\tau(t)}^{t} \left[\sum_{i=1}^{n} p_i(u)\right] \exp\left(-\int_{u-\tau(u)}^{u} \alpha(s) \, ds\right) du.$$

Therefore, in view of (3.4.7) and (3.4.5), we are led to the contradiction that

$$m \geqslant \liminf_{t \to \infty} \left(\int_{t-\tau(t)}^{t} \sum_{i=1}^{n} p_i(u) \, du\right) e^m > \frac{1}{e} \, em = m.$$

The proof is complete.

The following result extends to equations with variable coefficients and variable delays the fact that every solution of the equation

$$\dot{x}(t) + \sum_{i=1}^{n} p_i x(t - \tau_i) = 0$$

with positive coefficients and non-negative delays oscillates provided that the characteristic equation

$$\lambda + \sum_{i=1}^{n} p_i \, e^{-\lambda \tau_i} = 0$$

has no negative roots, that is, equivalently that

$$\frac{1}{\lambda} \sum_{i=1}^{n} p_i \, e^{\lambda \tau_i} > 1 \qquad \text{for all } \lambda > 0.$$

Theorem 3.4.2. *Assume that (3.4.2) holds,*

$$\lim_{t \to \infty} (t - \tau_i(t)) = \infty \qquad \text{for } i = 1, 2, \ldots, n \tag{3.4.8}$$

and that there exist indices $i_l \in \{1, 2, \ldots, n\}$ *with* $l = 1, 2, \ldots, m$ *such that*

$$\liminf_{t \to \infty} [\tau_{i_l}(t)] > 0 \qquad \text{and} \qquad \liminf_{t \to \infty} \left[\sum_{l=1}^{m} p_{i_l}(s) \right] > 0. \tag{3.4.9}$$

Then

$$\liminf_{t \to \infty} \left[\inf_{\lambda > 0} \left\{ \frac{1}{\lambda} \sum_{i=1}^{n} p_i(t) \, e^{\lambda \tau_i(t)} \right\} \right] > 1 \tag{3.4.10}$$

is a sufficient condition for the oscillation of all solutions of eqn (3.4.1).

Proof. Assume, for the sake of contradiction, that eqn (3.4.1) has an eventually positive solution $x(t)$. Then there exists $T \geqslant 0$ such that

$$x(t) > 0 \qquad \text{for } t \geqslant T_{-1}$$

where

$$T_{-1} = \min_{1 \leqslant i \leqslant n} \left\{ \inf_{t \geqslant T} \{t - \tau_i(t)\} \right\}.$$

Therefore, by Corollary 3.1.1 there exists a continuous function $\alpha \in C[[T_{-1}, \infty), \mathbb{R}]$ such that

$$\alpha(t) + \sum_{i=1}^{n} p_i(t) \exp\left(-\int_{t - \tau_i(t)}^{t} \alpha(s) \, ds \right) = 0, \qquad t \geqslant T. \tag{3.4.11}$$

Let $r(t) = \max_{1 \leqslant l \leqslant m} \tau_{i_l}(t)$. Then

$$\alpha(t) + \left[\sum_{l=1}^{m} p_{i_l}(t) \right] \exp\left(-\int_{t-r(t)}^{t} \alpha(s) \, ds \right) \leqslant 0, \qquad t \geqslant T \qquad (3.4.12)$$

and so by Lemma 1.6.3,

$$M \equiv \liminf_{t \to \infty} [-\alpha(t)] < \infty.$$

Clearly $M > 0$, for otherwise $M = 0$ and

$$0 = \liminf_{t \to \infty} \left[\sum_{i=1}^{n} p_i(t) \exp\left(-\int_{t-\tau_i(t)}^{t} \alpha(s) \, ds \right) \right] \geqslant \liminf_{t \to \infty} \left[\sum_{l=1}^{m} p_{i_l}(t) \right]$$

which contradicts (3.4.9). Now for every $\varepsilon \in (0, M)$ there exists a $T_\varepsilon \geqslant T$ such that

$$-\alpha(t) \geqslant M - \varepsilon \qquad \text{for } t \geqslant \tilde{T}_\varepsilon$$

where

$$\tilde{T}_\varepsilon = \min_{1 \leqslant i \leqslant n} \left\{ \inf_{t \geqslant T_\varepsilon} \{t - \tau_i(t)\} \right\}.$$

From (3.4.11) we find

$$\alpha(t) + \sum_{i=1}^{n} p_i(t) \, e^{(M-\varepsilon)\tau_i(t)} \leqslant 0, \qquad t \geqslant T_\varepsilon. \qquad (3.4.13)$$

On the other hand, it follows from (3.4.10) that there exists a $T_1 > 0$ and a $q > 1$ such that

$$\sum_{i=1}^{n} p_i(t) \, e^{\lambda \tau_i(t)} \geqslant q\lambda \qquad \text{for all } \lambda > 0 \qquad \text{and} \qquad t \geqslant T_1.$$

From this and (3.4.13) we see that

$$-\alpha(t) \geqslant \sum_{i=1}^{n} p_i(t) \, e^{(M-\varepsilon)\tau_i(t)} \geqslant q(M - \varepsilon)$$

for all $t \geqslant \max\{T_\varepsilon, T_1\}$.

Therefore,

$$M = \liminf_{t \to \infty} [-\alpha(t)] \geqslant q(M - \varepsilon)$$

and, as $\varepsilon \to 0$,

$$M \geqslant qM > M.$$

This is a contradiction and so the proof is complete.

The next result in this section is an extension of Theorem 2.2.1 to the case of variable coefficients and variable delays.

Corollary 3.4.1. *Assume that (3.4.2), (3.4.8) and (3.4.9) are satisfied. Then each of the following two conditions is sufficient for the oscillation of all solutions of eqn (3.4.1).*

(a)
$$\liminf_{t \to \infty} \left[\sum_{i=1}^{n} p_i(t)\tau_i(t) \right] > 1/e; \tag{3.4.14}$$

(b)
$$\liminf_{t \to \infty} \left\{ \left[\prod_{i=1}^{n} p_i(t) \right]^{1/n} \left[\sum_{i=1}^{n} \tau_i(t) \right] \right\} > 1/e. \tag{3.4.15}$$

Proof. In either case it suffices to show that (3.4.10) holds.

(a) By using the inequality $e^x \geqslant ex$ we see that for any $\lambda > 0$,

$$\frac{1}{\lambda} \sum_{i=1}^{n} p_i(t) \, e^{\lambda \tau_i(t)} \geqslant \frac{1}{\lambda} \sum_{i=1}^{n} p_i(t) \, e\lambda\tau_i(t) = e \sum_{i=1}^{n} p_i(t)\tau_i(t)$$

and (3.4.14) implies (3.4.10).

(b) By using the inequalities $e^x \geqslant ex$ and

$$\frac{1}{n} \sum_{i=1}^{n} q_i \geqslant \left(\prod_{i=1}^{n} q_i \right)^{1/n} \qquad \text{for } q_i \geqslant 0 \qquad \text{and} \qquad i = 1, 2, \ldots, n$$

we find that

$$\frac{1}{\lambda} \sum_{i=1}^{n} p_i(t) \, e^{\lambda \tau_i(t)} \geqslant \frac{n}{\lambda} \prod_{i=1}^{n} \left[p_i(t) \, e^{\lambda \tau_i(t)} \right]^{1/n}$$

$$= \frac{n}{\lambda} \left[\prod_{i=1}^{n} p_i(t) \right]^{1/n} \exp\left(\frac{\lambda}{n} \sum_{i=1}^{n} \tau_i(t) \right) \geqslant e \left[\prod_{i=1}^{n} p_i(t) \right]^{1/n} \left[\sum_{i=1}^{n} \tau_i(t) \right]$$

and (3.4.15) implies (3.4.10). The proof is complete.

We close this section with a simple oscillation theorem which, although it does not possess the sharpness of Theorem 3.4.1, nevertheless gives a new sufficient condition for the oscillation of all solutions of eqn (3.4.1).

Theorem 3.4.3. *Assume that (3.4.2) holds. Set*

$$\tau(t) = \min_{1 \leqslant i \leqslant n} \{\tau_i(t)\}, \qquad t \geqslant 0$$

and suppose that

$$\lim_{t \to \infty} (t - \tau(t)) = \infty.$$

Then

$$\limsup_{t \to \infty} \int_{t-\tau(t)}^{t} \sum_{i=1}^{n} p_i(s) \, ds > 1 \qquad (3.4.16)$$

is a sufficient condition for the oscillation of all solutions of eqn (3.4.1).

Proof. Assume, for the sake of contradiction, that eqn (3.4.1) has an eventually positive solution $x(t)$. By integrating eqn (3.4.1) from $t - \tau(t)$ to t, for t sufficiently large, we find

$$0 = x(t) - x(t - \tau(t)) + \int_{t-\tau(t)}^{t} \sum_{i=1}^{n} p_i(s) x(s - \tau_i(s)) \, ds$$

$$\geqslant x(t) - x(t - \tau(t)) + x(t - \tau(t)) \int_{t-\tau(t)}^{t} \sum_{i=1}^{n} p_i(s) \, ds$$

$$\geqslant x(t) + x(t - \tau(t)) \left[\int_{t-\tau(t)}^{t} \sum_{i=1}^{n} p_i(s) \, ds - 1 \right] > 0.$$

This is a contradiction and the proof is complete.

Clearly, the conditions (3.4.5) and (3.4.16) are independent. However, when the coefficients $p_i(t)$ are all constants, (3.4.16) implies (3.4.5).

3.5 Notes

The results in this chapter are based on the idea of the generalized characteristic equation. This equation is obtained by looking for solutions of the form

$$x(t) = \exp\left(\int_{t_0}^{t} \alpha(s) \, ds \right).$$

This transformation has been used fruitfully in oscillation theory. See for example Arino *et al.* (1984, 1987), Domshlak and Aliev (1988), Györi (1984, 1986), Ladas *et al.* (1984*b*), Ladde (1977), and Yan (1990). Section 3.4 was motivated by the work of Hunt and Yorke (1984) in which they conjectured that under conditions (3.4.10) the solutions of eqn (3.4.1) oscillate. The proof of this conjecture was given by Györi (1986). Theorems 3.3.2 and 3.3.3 are from Chuanxi and Ladas (1990*b*).

3.6 Open problems

3.6.1   Extend the concept of 'generalized characteristic equation' to delay equations of higher order and to systems of delay differential equations.

3.6.2   Is Corollary 3.2.2 true without the hypothesis that the coefficients $p_i(t)$ are non-negative functions?

3.6.3   Extend Theorem 3.2.1 by relaxing the hypothesis that the coefficients p_i, q_i, and r_i are non-negative functions.

4

LINEARIZED OSCILLATION THEORY
AND APPLICATIONS TO
MATHEMATICAL BIOLOGY

In this chapter we shall develop and apply a *linearized oscillation theory* which, in a sense, parallels the so-called linearized stability theory of differential and difference equations. As in the case of stability theory, we prove that certain non-linear delay differential equations have the same oscillatory behaviour as an associated linear equation.

The linearized oscillation theory has been motivated by applications. For example, several equations in mathematical biology can be transformed into a delay differential equation of the form

$$\dot{x}(t) + \sum_{i=1}^{n} p_i f_i(x(t - \tau_i)) = 0 \tag{4.0.1}$$

where the biologically meaningful equilibrium of the original equation is transformed into the zero equilibrium of eqn (4.0.1) and where for $i = 1, 2, \ldots, n$,

$$p_i, \tau_i \in [0, \infty),$$

$$f_i \in C[\mathbb{R}, \mathbb{R}], \qquad u f_i(u) > 0 \qquad \text{for } u \neq 0,$$

$$\lim_{u \to 0} \frac{f_i(u)}{u} = 1$$

and for some positive number δ,

$$\text{either} \qquad f_i(u) \leqslant u \qquad \text{for } u \in [0, \delta]$$

$$\text{or} \qquad f_i(u) \geqslant u \qquad \text{for } u \in [-\delta, 0].$$

With eqn (4.0.1) we associate its so-called *linearized equation*:

$$\dot{y}(t) + \sum_{i=1}^{n} p_i y(t - \tau_i) = 0. \tag{4.0.2}$$

Our goal in this chapter is to show that every solution of eqn (4.0.1) oscillates if and only if every solution of eqn (4.0.2) oscillates. We shall also apply this result to several situations in mathematical biology.

4.1 Linearized oscillation theory

In this section we develop a linearized oscillation theory which parallels the so-called linearized stability theory of differential and difference equations. Roughly speaking, we prove that certain non-linear delay differential equations have the same oscillatory character as an associated linear equation. Consider the non-linear delay differential equation

$$\dot{x}(t) + \sum_{i=1}^{n} p_i f_i(x(t - \tau_i)) = 0 \qquad (4.1.1)$$

where for $i = 1, 2, \ldots, n$,

$$p_i \in (0, \infty), \qquad \tau_i \in [0, \infty) \qquad \text{and} \qquad f_i \in C[\mathbb{R}, \mathbb{R}] \qquad (4.1.2)$$

and

$$u f_i(u) > 0 \qquad \text{for } u \neq 0. \qquad (4.1.3)$$

For easy reference in the sequel, we also list the following hypotheses on f which will be assumed only wherever this is explicitly indicated:

(H_1)
$$\liminf_{u \to 0} \frac{f_i(u)}{u} \geq 1 \qquad \text{for } i = 1, 2, \ldots, n. \qquad (4.1.4)$$

(H_2)
$$\lim_{u \to 0} \frac{f_i(u)}{u} = 1 \qquad \text{for } i = 1, 2, \ldots, n. \qquad (4.1.5)$$

(H_3) There exists a positive constant δ such that

$$\left. \begin{array}{llll} \text{either} & f_i(u) \leq u & \text{for } 0 \leq u \leq \delta & \text{and} & i = 1, 2, \ldots, n \\ \text{or} & f_i(u) \geq u & \text{for } -\delta \leq u \leq 0 & \text{and} & i = 1, 2, \ldots, n \end{array} \right\} .$$

$$(4.1.6)$$

Whenever condition (4.1.4) or (4.1.5) is satisfied, the linear equation

$$\dot{y}(t) + \sum_{i=1}^{n} p_i y(t - \tau_i) = 0 \qquad (4.1.7)$$

will be called the *linearized equation* associated with eqn (4.1.1).

The main results in this section are the following.

Theorem 4.1.1. *Assume that (4.1.2), (4.1.3), and (4.1.4) hold and suppose that every solution of the linearized equation (4.1.7) is oscillatory. Then every solution of eqn (4.1.1) also oscillates.*

Theorem 4.1.2. *Assume that* (4.1.2), (4.1.3), *and* (4.1.6) *hold and suppose that every solution of eqn* (4.1.1) *is oscillatory. Then every solution of the linearized equation* (4.1.7) *also oscillates.*

Theorem 4.1.2 is a partial converse of Theorem 4.1.1. By combining both of these theorems we obtain the following powerful linearized oscillation result.

Corollary 4.1.1. *Assume that* (4.1.2), (4.1.3), (4.1.5), *and* (4.1.6) *hold. Then every solution of eqn* (4.1.1) *oscillates if and only if every solution of its linearized equation* (4.1.7) *oscillates.*

The proof of Theorem 4.1.1 makes use of the following lemma which is interesting in its own right.

Lemma 4.1.1. *Assume that* $n \geq 1$ *and that conditions* (4.1.2) *and* (4.1.3) *hold. Then every non-oscillatory solution of eqn* (4.1.1) *tends to zero as* $t \to \infty$.

Proof. Let $x(t)$ be a non-oscillatory solution of eqn (4.1.1). We assume that $x(t)$ is eventually positive. The case where $x(t)$ is eventually negative is similar and will be omitted. Then eventually,

$$\dot{x}(t) = -\sum_{i=1}^{n} p_i f_i(x(t - \tau_i)) < 0$$

and so $L \equiv \lim_{t \to \infty} x(t)$ exists and is a non-negative number. We must show that $L = 0$. Otherwise, $L > 0$ and so $\lim_{t \to \infty} [\dot{x}(t)] = -\sum_{i=1}^{n} p_i f(L) < 0$, which implies that $\lim_{t \to \infty} x(t) = -\infty$. This is a contradiction and the proof is complete.

Proof of Theorem 4.1.1. Assume, for the sake of contradiction, that eqn (4.1.1) has a non-oscillatory solution $x(t)$. We assume that $x(t)$ is eventually positive. The case where $x(t)$ is eventually negative is similar and will be omitted. By Lemma 4.1.1 we know that $\lim_{t \to \infty} x(t) = 0$. Thus by (4.1.4),

$$\liminf_{t \to \infty} \frac{f(x(t - \tau_i))}{x(t - \tau_i)} \geq 1 \qquad \text{for } i = 1, 2, \ldots, n.$$

Let $\varepsilon \in (0, 1)$. Then there exists a T_ε such that for every $i = 1, 2, \ldots, n$ and for $t \geq T_\varepsilon$, $x(t - \tau_i) > 0$ and $f_i(x(t - \tau_i)) \geq (1 - \varepsilon)x(t - \tau_i)$. Hence from eqn (4.1.1),

$$\dot{x}(t) + \sum_{i=1}^{n} (1 - \varepsilon)p_i x(t - \tau_i) \leq 0, \qquad t \geq T_\varepsilon.$$

It follows by Theorem 3.2.2 that eqn (4.1.7) has a positive solution. This contradicts the hypothesis that every solution of eqn (4.1.7) is oscillatory and the proof is complete.

Proof of Theorem 4.1.2. Assume that (4.1.6) holds, with $f_i(u) \leqslant u$ for $0 \leqslant u \leqslant \delta$ and $i = 1, 2, \ldots, n$. The case where $f_i(u) \geqslant u$ for $-\delta \leqslant u \leqslant 0$ and $i = 1, 2, \ldots, n$ is similar and will be omitted. Now assume, for the sake of contradiction, that eqn (4.1.7) has an eventually positive solution $y(t)$. Clearly, $\lim_{t \to \infty} y(t) = 0$ and so there exists a t_0 such that $0 < y(t) < \delta$ for $t_0 - \tau \leqslant t \leqslant t_0$, where $\tau = \max_{1 \leqslant i \leqslant n} \tau_i$. With initial function equal to $y(t)$ for $t_0 - \tau \leqslant t \leqslant t_0$, eqn (4.1.1) has a solution $x(t)$ which exists at least in some small neighbourhood to the right of t_0. It suffices to show that for as long as $x(t)$ exists,

$$y(t) \leqslant x(t) < \delta. \tag{4.1.8}$$

This is because (4.1.8) would then imply that $x(t)$ exists and is positive for all $t \geqslant t_0$, which contradicts the hypothesis that every solution of eqn (4.1.1) oscillates. Observe that for as long as $x(t)$ exists and $0 \leqslant x(t) < \delta$, we have

$$\dot{x}(t) = - \sum_{i=1}^{n} p_i f_i(x(t - \tau_i)) \geqslant - \sum_{i=1}^{n} p_i x(t - \tau_i)$$

and so by Theorem 3.2.1, $y(t) \leqslant x(t)$. It also follows from eqn (4.1.1) that for as long as $x(t)$ exists and remains positive it is strictly decreasing. Therefore (4.1.8) holds for all $t \geqslant t_0$ and the proof is complete.

4.2 Time lags in a food-limited population model

The scalar autonomous ordinary differential equation

$$\frac{\mathrm{d}N(t)}{\mathrm{d}t} = rN(t)\left\{1 - \frac{N(t)}{K}\right\} \tag{4.2.1}$$

where $r, K \in (0, \infty)$ is known as the 'logistic equation' in mathematical ecology. Equation (4.2.1) is a prototype in modelling the dynamics of single-species population systems whose biomass or density is denoted by a differentiable function N of the time variable t. The constant r is called the *growth rate* and the constant K is called the *carrying capacity* of the habitat.

An implicit assumption contained in (4.2.1) is that the average growth rate $\dfrac{1}{N(t)} \dfrac{\mathrm{d}N(t)}{\mathrm{d}t}$ is a linear function of the density $N(t)$. However, it has been demonstrated by F. E. Smith (1963) that the hypothesis of linear average growth rate is not realistic (for the Daphnia populations) and he has proposed an alternative to the logistic equation for a 'food-limited' population in the form

$$\frac{dN(t)}{dt} = rN(t)\frac{K - N(t)}{K + crN(t)} \qquad (4.2.2)$$

where r, K, c are positive numbers. We refer to Pielou (1969, p. 30) for a modern derivation of (4.2.2). Equation (4.2.2) has been recently discussed by Hallam and DeLuna (1984) in their investigation of the effects of environmental toxicants on populations.

It follows by elementary techniques that the solution of (4.2.1) corresponding to $N(0) > 0$ converges monotonically to the equilibrium K as $t \to \infty$. It has been observed, however, that population densities (or sizes) usually have a tendency to fluctuate around an equilibrium, and when there is a convergence to a positive equilibrium, such a convergence is rarely monotonic (see for example Nicholson 1954a). To incorporate such oscillations in population model systems, Hutchinson (1948) suggested the following modification of (4.2.1):

$$\frac{dN(t)}{dt} = rN(t)\left\{1 - \frac{N(t - \tau)}{K}\right\} \qquad (4.2.3)$$

where $r, \tau, K \in (0, \infty)$.

Equation (4.2.3) is commonly known as the 'delay-logistic equation' and has been extensively investigated by numerous authors (see for example Wright (1955), Kakutani and Markus (1958) and Jones (1962b)). Since the 'food-limited' model characterized by (4.2.2) also leads to monotonic convergence to the positive equilibrium, we are led to modify (4.2.2) so as to incorporate oscillations. One of the ways of doing this is to assume that the average growth rate in (4.2.2) is a function of the delayed argument $t - \tau$ rather than a function of t. Thus we are led to the time-delayed 'food-limited' model

$$\dot{N}(t) = rN(t)\frac{K - N(t - \tau)}{K + crN(t - \tau)} \qquad (4.2.4)$$

where

$$r, K, \tau \in (0, \infty) \qquad \text{and} \qquad c \in [0, \infty) \qquad (4.2.5)$$

as an alternative to (4.2.2). We note that (4.2.4) is a generalization of the delay-logistic equation, since when $c = 0$, (4.2.4) reduces to (4.2.3).

In this section we consider only those solutions of (4.2.4) which correspond to initial conditions of the form

$$N(t) = \phi(t), \quad -\tau \leqslant t \leqslant 0 \quad \text{with } \phi \in C([-\tau, 0], \mathbb{R}^+) \quad \text{and} \quad \phi(0) > 0.$$
$$(4.2.6)$$

Clearly (4.2.4) with (4.2.6) has a unique solution which exists and is positive for all $t \geqslant 0$.

We wish to establish necessary and sufficient conditions for the oscillation of every positive solution of eqn (4.2.4) about the steady state K. Set $N(t) = K \, e^{x(t)}$. Then eqn (4.2.4) becomes

$$\dot{x}(t) + r \, \frac{e^{x(t-\tau)} - 1}{1 + cr \, e^{x(t-\tau)}} = 0. \qquad (4.2.7)$$

Observe that the solution $N(t)$ of (4.2.4) and (4.2.6) oscillates about K if and only if the function $x(t)$ oscillates about zero. By applying Corollary 4.1.1 to eqn (4.2.4) we obtain the following necessary and sufficient condition for the oscillation of all positive solutions of eqn (4.2.4).

Theorem 4.2.1. *Assume that* (4.2.5) *holds. Then every positive solution of eqn* (4.2.4) *oscillates about the steady state K if and only if*

$$\frac{r}{1 + cr} \, \tau > \frac{1}{e}. \qquad (4.2.8)$$

Proof. Set

$$f(u) = (1 + cr) \, \frac{e^u - 1}{1 + cr \, e^u}$$

and note that eqn (4.2.7) can be written in the form

$$\dot{x}(t) + \frac{r}{1 + cr} \, f(x(t - \tau)) = 0 \qquad (4.2.9)$$

where f satisfies the conditions $f \in C[\mathbb{R}, \mathbb{R}]$, $uf(u) > 0$ for $u \neq 0$ and

$$\lim_{u \to 0} \frac{f(u)}{u} = 1.$$

One can also see that there exists a positive number δ such that

$$cr \leqslant 1 \ \Rightarrow \ f(u) \geqslant u \qquad \text{for } u \in [-\delta, 0]$$

while

$$cr \geqslant 1 \ \Rightarrow \ f(u) \leqslant u \qquad \text{for } u \in [0, \delta].$$

Finally, we know from Theorem 2.2.3 that (4.2.8) is a necessary and sufficient condition for the oscillation of all solutions of the linearized equation

$$\dot{y}(t) + \frac{r}{1 + cr} \, y(t - \tau) = 0.$$

The result is now an immediate consequence of Corollary 4.1.1 applied to eqn (4.2.9).

4.3 Oscillations in the delay logistic equation with several delays

Consider the delay logistic equation with several delays

$$\dot{N}(t) = N(t)\left[\alpha - \sum_{i=1}^{n} \beta_i N(t - \tau_i)\right], \qquad t \geqslant 0 \tag{4.3.1}$$

where

$$\alpha, \beta_1, \ldots, \beta_n \in (0, \infty) \qquad \text{and} \qquad 0 \leqslant \tau_1 < \tau_2 < \cdots < \tau_n \equiv \tau. \tag{4.3.2}$$

This is a generalization of the delay logistic equation

$$\dot{N}(t) = rN(t)[1 - N(t - \tau)/K], \qquad t \geqslant 0 \tag{4.3.3}$$

which represents the dynamics of a single species population model. Here $N(t)$ denotes the density of the population at time t, r is the growth rate and K is the carrying capacity of the environment. The term $1 - N(t - \tau)/K$ denotes a feedback mechanism which takes τ units of time to respond to changes in the size of the population.

With eqn (4.3.1) one associates an initial function of the form

$$N(t) = \phi(t), \quad -\tau \leqslant t \leqslant 0 \text{ where } \phi \in C[[-\tau, 0], \mathbb{R}^+] \text{ and } \phi(0) > 0. \tag{4.3.4}$$

It follows by the method of steps (see also Theorem 1.1.4) that the IVP (4.3.1) and (4.3.4) has a unique solution $N(t)$ which exists and remains positive for all $t \geqslant 0$.

Equation (4.3.1) has the unique positive equilibrium (steady state)

$$N^* = \alpha \left/ \sum_{i=1}^{n} \beta_i \right. . \tag{4.3.5}$$

Our aim in this section is to obtain necessary and sufficient conditions for every positive solution of eqn (4.3.1) to oscillate about the positive equilibrium N^*. In particular we show that every positive solution of eqn (4.3.3) oscillates about its positive equilibrium K if and only if

$$r\tau > 1/e. \tag{4.3.6}$$

Let $N(t)$ be the unique positive solution of the IVP (4.3.1) and (4.3.4). Set

$$x(t) = \ln \frac{N(t)}{N^*}, \qquad t \geqslant 0. \tag{4.3.7}$$

Then $x(t)$ satisfies the DDE

$$\dot{x}(t) + \sum_{i=1}^{n} p_i f(x(t - \tau_i)) = 0, \qquad t \geqslant \tau \tag{4.3.8}$$

where

$$p_i = N^*\beta_i \qquad \text{for } i = 1, 2, \ldots, n \text{ and } f(u) = e^u - 1.$$

Clearly $x(t)$ oscillates (about zero) if and only if $N(t)$ oscillates about N^*. Also, $f \in C[\mathbb{R}, \mathbb{R}]$, $uf(u) > 0$ for $u \neq 0$, $\lim_{u \to 0} f(u)/u = 1$ and $f(u) \geqslant u$ for $u \leqslant 0$.

The linearized equation associated with eqn (4.3.1) is

$$\dot{y}(t) + \sum_{i=1}^{n} p_i y(t - \tau_i) = 0 \tag{4.3.9}$$

The following result is now a consequence of Corollary 4.1.1 and the above discussion.

Theorem 4.3.1. *Assume that* (4.3.2) *holds. Then every positive solution of eqn* (4.3.1) *oscillates about its positive equilibrium* N^* *if and only if every solution of eqn* (4.3.9) *oscillates* (*about zero*), *that is, if and only if the characteristic equation*

$$\lambda + \sum_{i=1}^{n} p_i e^{-\lambda \tau_i} = 0$$

of eqn (4.3.9) *has no real roots.*

The change of variables (4.3.7) applied to eqn (4.3.3) yields the equation

$$\dot{x}(t) + rf(x(t - \tau)) = 0 \tag{4.3.10}$$

where

$$f(u) = e^u - 1.$$

The linearized equation associated with eqn (4.3.10) is

$$\dot{y}(t) + ry(t - \tau) = 0$$

which oscillates if and only if (4.3.6) holds. Therefore the following result is true.

Corollary 4.3.1. *Assume* r, τ, *and* K *are positive constants. Then every positive solution of the logistic equation* (4.3.3) *oscillates about the steady state* K *if and only if* $r\tau e > 1$.

Remark 4.3.1. *By using the explicit conditions of Theorem 2.2.1 for the oscillation of all solutions of linear equations we find that each one of the following two conditions also implies that every solution of eqn* (4.3.1) *oscillates* *about N*:*

(a)
$$\left(\alpha \sum_{i=1}^{n} \beta_i \tau_i\right) \bigg/ \left(\sum_{i=1}^{n} \beta_i\right) > 1/e$$

(b)
$$\alpha \left(\sum_{i=1}^{n} \tau_i\right) \bigg/ \left(\prod_{i=1}^{n} \beta_i\right)^{1/n} \bigg/ \left(\sum_{i=1}^{n} \beta_i\right) > 1/e.$$

4.4 The Lasota–Wazewska model for the survival of red blood cells

The delay differential equation

$$\dot{N}(t) = -\mu N(t) + p\, e^{-\gamma N(t-\tau)}, \qquad t \geqslant 0 \tag{4.4.1}$$

has been used by Wazewska-Czyzewska and Lasota (1988) as a model for the survival of red blood cells in an animal. Here $N(t)$ denotes the number of red blood cells at time t, μ is the probability of death of a red blood cell, p and γ are positive constants related to the production of red blood cells per unit of time, and τ is the time required to produce a red blood cell.

We consider only those solutions of eqn (4.4.1) and (4.4.2) which correspond to initial conditions of the form

$$N(t) = \phi(t), \qquad -\tau \leqslant t \leqslant 0 \qquad \text{with } \phi \in C([-\tau, 0], \mathbb{R}^+) \text{ and } \phi(0) > 0. \tag{4.4.2}$$

Clearly (4.4.1) with (4.4.2) has a unique solution which exists and is positive for all $t \geqslant 0$.

The equilibrium $N*$ of eqn (4.4.1) is positive and satisfies the equation

$$N* = \frac{p}{\mu} e^{-\gamma N*}.$$

The change of variables

$$N*(t) = N* + \frac{1}{\gamma} x(t)$$

reduces eqn (4.4.1) to the delay differential equation

$$\dot{x}(t) + \mu x(t) + \mu \gamma N*[1 - e^{-x(t-\tau)}] = 0. \tag{4.4.3}$$

Equation (4.4.3) is of the form of eqn (4.1.1) with

$$n = 2, \qquad p_1 = \mu, \qquad f_1(u) = u, \qquad p_2 = \mu\gamma N* \qquad \text{and} \qquad f_2(u) = 1 - e^{-u}.$$

Clearly $N(t)$ oscillates about $N*$ if and only if $x(t)$ oscillates about zero.

One can easily see that all the hypotheses of Corollary 4.1.1 are satisfied for eqn (4.4.3) and therefore the following result holds.

Theorem 4.4.1. *The solution of* (4.4.1) *and* (4.4.2) *oscillates about* N^* *if and only if the equation*

$$\lambda + \mu + \mu\gamma N^* \, e^{-\lambda\tau} = 0$$

has no real roots, that is if and only if

$$\mu\tau\gamma N^* \, e^{\mu\tau} > 1/e.$$

4.5 Notes

The results in Sections 4.1 and 4.3 are adapted from Kulenovic *et al.* (1987*b*) and Kulenovic and Ladas (1987*b*). Theorem 4.2.1 is extracted from Gopalsamy *et al.* (1988*c*). Theorem 4.4.1 is from Kulenovic and Ladas (1987*b*).

4.6 Open problems

4.6.1   Obtain linearized oscillation results without the restriction that the p_i's in eqn (4.1.1) are all positive. Also obtain linearized oscillation results for equations with variable coefficients. (See Kulenovic *et al.* 1990.)

4.6.2   Obtain linearized oscillation results for eqn (4.1.1) with real coefficients and with the condition (4.1.5) replaced by

$$\lim_{u \to \infty} \frac{f_i(u)}{u} = 1.$$

(See Kulenovic and Ladas 1987*a*.)

4.6.3   Obtain linearized oscillation results for higher-order delay equations. (See Kulenovic and Ladas 1987*a*, 1988.)

5

OSCILLATIONS OF SYSTEMS OF DELAY EQUATIONS

This chapter deals with the oscillation of all solutions of systems of delay differential equations. Section 5.1 gives a powerful necessary and sufficient condition for the oscillation of all solutions of a linear autonomous system. Section 5.2 utilizes this result to obtain explicit conditions for the oscillation of linear autonomous equations. Section 5.3 deals with the oscillation of linear, non-autonomous equations. Section 5.4 provides sufficient conditions for the oscillation of all solutions of a non-linear system of delay logistic equations.

Concerning systems of differential equations, there are many ways in which one can define the concept of oscillation of solutions. In this book we use only the following two definitions.

Definition 5.0.1. *A solution* $x(t) = [x_1(t), \ldots, x_n(t)]^T$ *is said to oscillate if every component* $x_i(t)$ *of the solution has arbitrarily large zeros. Otherwise the solution is called non-oscillatory.*

That is, a solution oscillates, according to Definition 5.0.1, if it oscillates componentwise and is non-oscillatory if at least one component is eventually positive or eventually negative.

Before we introduce our second definition of oscillation we must define the concept of the 'signum' of a real number. Let $x \in \mathbb{R}$; then

$$
\operatorname{sgn} x = \begin{cases} 1 & \text{if } x > 0 \\ 0 & \text{if } x = 0 \\ -1 & \text{if } x < 0. \end{cases}
$$

Definition 5.0.2. *A solution* $x(t) = [x_1(t), \ldots, x_n(t)]^T$ *is said to oscillate if it is eventually trivial or if at least one component does not have eventually constant signum. Otherwise the solution is called non-oscillatory.*

According to Definition 5.0.2, a solution $[x_1(t), \ldots, x_n(t)]^T$ is non-oscillatory if it is not eventually the trivial solution and if each component $x_i(t)$ has eventually constant signum. It follows that if $x(t)$ is non-oscillatory according to Definition 5.0.2 then at least one component is eventually positive or eventually negative.

Clearly if $x(t) = [x_1(t), x_2(t), \ldots, x_n(t)]^T$ oscillates in the sense of Definition 5.0.1, then it also oscillates in the sense of Definition 5.0.2. Equivalently, if $x(t)$ does not oscillate according to Definition 5.0.2 then it does not oscillate according to Definition 5.0.1 either. But the converse is not true. For example, $x(t) = [1, \sin t]$ oscillates according to Definition 5.0.2 but is non-oscillatory according to Definition 5.0.1. However, as we shall prove in Theorem 5.1.2 (see also Remark 6.3.1) for linear, homogeneous, autonomous systems of differential equations, if all solutions of the system oscillate with respect to one of these two definitions, then they also oscillate with respect to the other.

In the sequel, oscillation for linear, homogeneous, and autonomous systems will be understood in the sense of Definition 5.0.1. On the other hand, oscillation for non-autonomous and/or non-linear systems will be understood in the sense of Definition 5.0.2. This is because with Definition 5.0.2 it is 'easier' to obtain sufficient conditions for the oscillation of all solutions.

Finally, we should observe that for scalar equations both Definitions 5.0.1 and 5.0.2 reduce to the definition of oscillation given in the introduction to Chapter 2.

5.1 Necessary and sufficient conditions for the oscillation of linear autonomous systems

Consider the linear autonomous systems of delay differential equations

$$\dot{x}(t) + \sum_{i=1}^{n} P_i x(t - \tau_i) = 0 \qquad (5.1.1)$$

where the coefficients P_i are real $m \times m$ matrices and the delays τ_i are non-negative real numbers. With eqn (5.1.1) one associates its characteristic equation

$$\det\left(\lambda I + \sum_{i=1}^{n} P_i e^{-\lambda \tau_i} \right) = 0 \qquad (5.1.2)$$

where I is the $m \times m$ identity matrix.

Our aim in this section is to establish the following extension of Theorem 2.1.1.

Theorem 5.1.1. *Assume that*

$$P_i \in \mathbb{R}^{m \times m} \quad and \quad \tau_i \in \mathbb{R}^+ \quad for \ i = 1, 2, \ldots, n. \qquad (5.1.3)$$

Then the following statements are equivalent.

(a) *Every solution of eqn (5.1.1) oscillates componentwise.*

(b) *The characteristic equation (5.1.2) has no real roots.*

Proof. The proof that (a) $\Rightarrow$ (b) is simple. In fact if λ_0 is a real root of eqn (5.1.2) then there exists a non-zero vector ξ such that

$$\left(\lambda_0 I + \sum_{i=1}^{n} P_i e^{-\lambda_0 \tau_i} \right) \xi = 0.$$

Then clearly, $x(t) = e^{\lambda_0 t} \xi$ is a solution of eqn (5.1.1) with at least one non-oscillatory component.

The proof that (b) $\Rightarrow$ (a) makes use of Laplace transform (see Section 1.3). Assume, for the sake of contradiction, that (b) holds and that eqn (5.1.1) has a non-oscillatory solution $x(t) = [x_1(t), \dots, x_m(t)]^T$. This means that one of the components of $x(t)$ is non-oscillatory. Without loss of generality we assume that the component $x_1(t)$ is eventually positive. As eqn (5.1.1) is autonomous, we may (and do) assume that $x_1(t) > 0$ for $t \geqslant -\tau$ where $\tau = \max_{1 \leqslant i \leqslant n} \tau_i$.

By Theorem 1.2.1 we know that $x(t)$ is of exponential order and so there exists $\mu \in \mathbb{R}$ such that the Laplace transform of $x(t)$, $X(s) = \int_0^\infty e^{-st} x(t) \, dt$, exists for Re $s > \mu$. By taking Laplace transforms of both sides of eqn (5.1.1) we obtain

$$F(s)X(s) = \Phi(s), \qquad \text{Re } s > \mu \tag{5.1.4}$$

where

$$F(s) = sI + \sum_{i=1}^{n} P_i e^{-s\tau_i} \tag{5.1.5}$$

and

$$\Phi(s) = x(0) - \sum_{i=1}^{n} P_i e^{-s\tau_i} \int_{-\tau_i}^{0} e^{-st} x(t) \, dt. \tag{5.1.6}$$

By hypothesis, $\det[F(s)] \neq 0$ for all $s \in \mathbb{R}$. Moreover,

$$\lim_{\substack{s \to \infty \\ s \in \mathbb{R}}} (\det[F(s)]) = \infty \tag{5.1.7}$$

and so

$$\det[F(s)] > 0 \qquad \text{for all } s \in \mathbb{R}. \tag{5.1.8}$$

Let $X_1(s)$ be the Laplace transform of the first component $x_1(t)$ of the solution $x(t)$. Then, by Cramer's rule,

$$X_1(s) = \frac{\det[D(s)]}{\det[F(s)]}, \qquad \text{Re } s > \mu \tag{5.1.9}$$

where

$$D(s) = \begin{bmatrix} \Phi_1(s) & F_{12}(s) & \cdots & F_{1m}(s) \\ \vdots & \vdots & & \vdots \\ \Phi_m(s) & F_{m2}(s) & \cdots & F_{mm}(s) \end{bmatrix},$$

$\Phi_i(s)$ is the ith component of the vector $\Phi(s)$ and $F_{ij}(s)$ is the (i, j)th component of the matrix $F(s)$. Clearly, for all $i, j = 1, 2, \ldots, m$ the functions $\Phi_i(s)$ and $F_{ij}(s)$ are entire and hence $\det[D(s)]$ and $\det[F(s)]$ are also entire functions.

Let σ_0 be the abscissa of convergence of $X_1(s)$, that is (see Section 1.3), $\sigma_0 = \inf\{\sigma \in \mathbb{R}: X_1(\sigma) \text{ exists}\}$. By using Theorem 1.3.1 and an argument similar to that in the proof of Theorem 2.1.1 we find that $\sigma_0 = -\infty$ and (5.1.9) becomes

$$X_1(s) = \frac{\det[D(s)]}{\det[F(s)]} \qquad \text{for all } s \in \mathbb{R}. \tag{5.1.10}$$

As $x_1(t) > 0$, it follows that $X_1(s) > 0$ for all $s \in \mathbb{R}$ and, by (5.1.8) and (5.1.10), $\det[D(s)] > 0$ for $s \in \mathbb{R}$. Now one can see from the definition of $D(s)$ and from (5.1.5) and (5.1.6) that there are positive constants M, α, and s_0 such that

$$\det[D(s)] \leqslant M e^{-\alpha s} \qquad \text{for } s \leqslant -s_0. \tag{5.1.11}$$

Also from (5.1.7), (5.1.8) and the fact that $\det[F(s)]$ is a polynomial in the variables $s, e^{-s\tau_1}, \ldots, e^{-s\tau_n}$, it follows that there exists a positive number m such that

$$\det[F(s)] \geqslant m \qquad \text{for } s \in \mathbb{R}. \tag{5.1.12}$$

From (5.1.10), (5.1.11), and (5.1.12) it follows that

$$X_1(s) = \int_0^\infty e^{-st} x_1(t) \, dt \geqslant \int_T^\infty e^{-st} x_1(t) \, dt \geqslant e^{-sT} \int_T^\infty x_1(t) \, dt > 0$$

and so

$$0 < \int_T^\infty x_1(t) \, dt \leqslant \frac{M}{m} e^{s(T-\alpha)} \to 0 \qquad \text{as } s \to -\infty.$$

This implies that $x_1(t) \equiv 0$ for $t \geqslant T$, which is a contradiction. The proof is complete.

In the introduction to this chapter we saw that if a solution $x(t) = [x_1(t), \ldots, x_m(t)]^T$ of eqn (5.1.1) oscillates in the sense of Definition 5.0.1, then it also oscillates in the sense of Definition 5.0.2. By using Theorem 5.1.1 we can now prove the following result.

Theorem 5.1.2. *Assume that* (5.1.3) *holds. Then the following statements are equivalent.*

(a) *Every solution of eqn* (5.1.1) *oscillates in the sense of Definition* 5.0.1 (*that is, componentwise*).

(b) *Every solution of eqn* (5.1.1) *oscillates in the sense of Definition* 5.0.2 (*that is, it is eventually trivial or at least one component does not have eventually constant signum*).

Proof. The proof that (a) $\Rightarrow$ (b) is trivial. To prove that (b) $\Rightarrow$ (a), assume for the sake of contradiction that (b) holds but $x(t)$ is a solution of eqn (5.1.1) which is non-oscillatory according to Definition 5.0.1. Then, by Theorem 5.1.1 the characteristic equation (5.1.2) has a real root λ_0. Therefore there exists a non-zero vector $\zeta \in \mathbb{R}^n$ such that $x(t) = e^{\lambda_0 t}\zeta$ is a solution of eqn (5.1.1). But this solution is non-oscillatory according to Definition 5.0.2. This is a contradiction and the proof is complete.

5.2 Explicit conditions for oscillations and non-oscillations of linear autonomous systems

Consider the delay differential system

$$\dot{x}(t) + \sum_{i=1}^{n} P_i x(t - \tau_i) = 0 \tag{5.2.1}$$

where the coefficients P_i are real $m \times m$ matrices and the delays τ_i are non-negative real numbers. In this section we utilize the logarithmic norm $\mu(P_i) = \max_{\|u\|=1} (P_i u, u)$ of the matrices P_i to obtain explicit conditions for the oscillation and for the non-oscillation of all solutions of (5.2.1). The main tool for our proofs is Theorem 5.1.1, which states that 'every solution of (5.2.1) oscillates if and only if the characteristic equation

$$\det\left(\lambda I + \sum_{i=1}^{n} P_i e^{-\lambda \tau_i} \right) = 0 \tag{5.2.2}$$

has no real roots.
    The following result is the matrix-analogue of Theorem 2.2.1.

Theorem 5.2.1. *Assume that for each* $i = 1, 2, \ldots, n$,

$$P_i \in \mathbb{R}^{m \times m}, \qquad \tau_i \geqslant 0 \qquad and \qquad \mu(-P_i) \leqslant 0. \tag{5.2.3}$$

Then each of the following two conditions is sufficient for the oscillation of all solutions of (5.2.1):

(a)
$$\sum_{i=1}^{n} -\mu(-P_i)\tau_i > 1/e \tag{5.2.4}$$

(b)
$$m\left[\prod_{i=1}^{n} (-\mu(-P_i))\right]^{1/n} \sum_{i=1}^{n} \tau_i > 1/e. \tag{5.2.5}$$

For the proof of Theorem 5.2.1, we need the following lemma which is interesting in its own right.

Lemma 5.2.1. *Assume that*

$$P_i \in \mathbb{R}^{m \times m} \quad and \quad \tau_i \geqslant 0 \quad for\ i = 1, 2, \ldots, n$$

are such that

$$\sum_{i=1}^{n} \mu(-P_i) e^{-\gamma \tau_i} < 0 \quad for\ \gamma \in \mathbb{R}^{+} \tag{5.2.6}$$

and

$$\inf_{\gamma < 0} \left[\frac{1}{\gamma} \sum_{i=1}^{n} \mu(-P_i) e^{-\gamma \tau_i}\right] > 1. \tag{5.2.7}$$

Then every solution of (5.2.1) *oscillates.*

Proof. Assume, for the sake of contradiction, that (5.2.1) has a non-oscillatory solution. Then, by Theorem 5.1.1, the characteristic equation (5.2.2) has a real root λ_0. But then there exists a vector $u \in \mathbb{R}^n$ with $\|u\| = 1$ such that

$$\left(\lambda_0 I + \sum_{i=1}^{n} P_i e^{-\lambda_0 \tau_i}\right) u = 0.$$

Hence

$$\lambda_0 = \left(-\sum_{i=1}^{n} P_i e^{-\lambda_0 \tau_i} u, u\right) = \sum_{i=1}^{n} (-P_i u, u) e^{-\lambda_0 \tau_i} \leqslant \sum_{i=1}^{n} \mu(-P_i) e^{-\lambda_0 \tau_i}$$

and so, by (5.2.6), $\lambda_0 < 0$ and

$$1 \geqslant \frac{1}{\lambda_0} \sum_{i=1}^{n} \mu(-P_i) e^{-\lambda_0 \tau_i}.$$

This contradicts (5.2.7) and completes the proof.

Proof of Theorem 5.2.1. We employ Lemma 5.2.1. As $\mu(-P_i) \leqslant 0$, (5.2.6) is satisfied and so it suffices to establish (5.2.7). First, assume that (5.2.4) holds. Then, by using the inequality $e^x \geqslant ex$, we see that for all $\gamma < 0$,

$$\frac{1}{\gamma} \sum_{i=1}^{n} \mu(-P_i) e^{-\gamma\tau_i} \geqslant \frac{1}{\gamma} \sum_{i=1}^{n} \mu(-P_i) e(-\gamma\tau_i) = e \sum_{i=1}^{n} -\mu(-P_i)\tau_i.$$

From this and (5.2.4) it follows that (5.2.7) holds. Next, assume that (5.2.5) holds. Then, by using the arithmetic mean–geometric mean inequality we see that for all $\gamma < 0$,

$$\frac{1}{\gamma} \sum_{i=1}^{n} \mu(-P_i) e^{-\gamma\tau_i} = -\frac{1}{\gamma} \sum_{i=1}^{n} [-\mu(-P_i)] e^{-\gamma\tau_i}$$

$$\geqslant -\frac{1}{\gamma} n \left[\prod_{i=1}^{n} (-\mu(-P_i)) e^{-\gamma\tau_i} \right]^{1/n}$$

$$= -\frac{1}{\gamma} n \left[\prod_{i=1}^{n} (-\mu(-P_i)) \right]^{1/n} \exp\left(-\frac{1}{n} \gamma \sum_{i=1}^{n} \tau_i \right)$$

$$\geqslant -\frac{1}{\gamma} n \left[\prod_{i=1}^{n} (-\mu(-P_i)) \right]^{1/n} e\left(-\frac{1}{n} \gamma \right) \sum_{i=1}^{n} \tau_i$$

$$= e \left[\prod_{i=1}^{n} (-\mu(-P_i)) \right]^{1/n} \sum_{i=1}^{n} \tau_i.$$

From this and (5.2.5) it follows that (5.2.7) holds. The proof is complete.

In the special case of the delay differential system with one delay,

$$\dot{x}(t) + Px(t - \tau) = 0 \tag{5.2.8}$$

where

$$P \in \mathbb{R}^{m \times m} \quad \text{and} \quad \tau \geqslant 0$$

the conditions (5.2.4) and (5.2.5) coincide and each reduces to

$$-\mu(-P)\tau > 1/e. \tag{5.2.9}$$

Note that (5.2.9) is sharp in the sense that the lower bound $1/e$ cannot be improved. Moreover, when P is a scalar, (5.2.9) is a necessary and sufficient condition for the oscillation of all solutions of eqn (5.2.8).

For eqn (5.2.8) we also have the following explicit necessary and sufficient condition for the oscillation of all solutions.

Theorem 5.2.2. *Assume that*

$$P \in \mathbb{R}^{m \times m} \qquad and \qquad \tau \geqslant 0.$$

Then the following statements are equivalent.

(a) *Every solution of eqn (5.2.8) oscillates.*

(b) *P has no real eigenvalues in the interval* $(-\infty, 1/e\tau]$.

(When $\tau = 0$, replace $1/e\tau$ by $+\infty$).

Proof. For $\tau = 0$, this result follows immediately from Theorem 5.1.1. So assume $\tau > 0$. It suffices to show that the negation of (a) is equivalent to the negation of (b). To this end, eqn (5.2.8) has a non-oscillatory solution if and only if the characteristic equation $\det(\lambda I + P e^{-\lambda \tau}) = 0$ has a real root λ_0 if and only if $\det(\lambda_0 e^{\lambda_0 \tau} I + P) = 0$ if and only if

$$\mu_0 = -\lambda_0 e^{\lambda_0 \tau} \qquad (5.2.10)$$

is a real eigenvalue of P. But (5.2.10) holds if and only if $\lambda_0 + \mu_0 e^{-\lambda_0 \tau} = 0$, that is, if and only if the equation $\lambda + \mu_0 e^{-\lambda \tau} = 0$ has a real root. This, by Theorem 2.2.3, is equivalent to $\mu_0 \leqslant 1/e\tau$, that is, the eigenvalue μ_0 of P should lie in the interval $(-\infty, 1/e\tau]$. The proof is complete.

Definition 5.2.1. *We say that eqn (5.2.1) is oscillatory, globally in the delays, if for all $\tau_i \geqslant 0$ for $i = 1, 2, \ldots, n$, every solution of eqn (5.2.1) oscillates.*

The following corollary is an immediate consequence of Theorem 5.2.2.

Corollary 5.2.1. *Equation (5.2.8) is oscillatory globally in the delay τ if and only if P has no real eigenvalues.*

On the other hand, the scalar ($m = 1$) delay differential equation

$$\dot{x}(t) + \sum_{i=1}^{n} p_i x(t - \tau_i) = 0 \qquad \text{where } p_i \in \mathbb{R} \text{ and } \tau_i \geqslant 0 \text{ for } i = 1, 2, \ldots, n$$

cannot be oscillatory globally in the delays. More generally the following result is true.

Theorem 5.2.3. *Let m be an odd natural number and assume that*

$$P_i \in \mathbb{R}^{m \times m} \qquad and \qquad \tau_i \geqslant 0 \qquad for \ i = 1, 2, \ldots, n.$$

Then (5.2.1) is not oscillatory globally in the delays.

Proof. Otherwise, every solution of the system

$$\dot{x}(t) + \left(\sum_{i=1}^{n} P_i \right) x(t) = 0 \tag{5.2.11}$$

which results from (5.2.1) by taking all the delays equal to zero must oscillate. But, since m is odd, the characteristic equation $\det(\lambda I + \sum_{i=1}^{n} P_i) = 0$ of eqn (5.2.11), which is of degree m, has a real root. This is a contradiction and the proof is complete.

Finally we present a matrix-analogue of Theorem 2.2.4.

Theorem 5.2.4. *Assume that $P, Q \in \mathbb{R}^{m \times m}$ and $\tau, \sigma \in \mathbb{R}^+$ are such that*

$$-\mu(-P) > \mu(Q) \geqslant 0, \qquad \tau \geqslant \sigma, \qquad \mu(Q)(\tau - \sigma) \leqslant 1$$

and

$$[-\mu(-P) - \mu(Q)]\tau > \frac{1}{e}[1 - \mu(Q)(\tau - \sigma)].$$

Then every solution of

$$\dot{x}(t) + Px(t - \tau) - Qx(t - \sigma) = 0$$

oscillates.

Proof. We employ Lemma 5.2.1. Condition (5.2.6) is clearly satisfied, so it remains to establish (5.2.7). That is,

$$\inf_{\gamma < 0} \left\{ \frac{1}{\gamma} [\mu(-P) e^{-\gamma\tau} + \mu(Q) e^{-\gamma\sigma}] \right\} > 1.$$

Set $p = -\mu(-P)$ and $q = \mu(Q)$. Then condition (2.2.13) of Theorem 2.2.4 is satisfied and so the equation (2.2.14), that is,

$$F(\gamma) \equiv \gamma - \mu(-P) e^{-\gamma\tau} - \mu(Q) e^{-\gamma\sigma} = 0$$

has no real root γ. As $F(-\infty) = \infty$, it follows that

$$\gamma > \mu(-P) e^{-\gamma\tau} + \mu(Q) e^{-\gamma\sigma} \qquad \text{for all } \gamma \in \mathbb{R}.$$

Thus for all $\gamma < 0$,

$$\frac{1}{\gamma} [\mu(-P) e^{-\gamma\tau} + \mu(Q) e^{-\gamma\sigma}] > 1$$

and the proof is complete.

5.3 Sufficient conditions for the oscillation of linear non-autonomous systems

Our aim in this section is to obtain sufficient conditions for the oscillation, in the sense of Definition 5.0.2, of all solutions of the linear non-autonomous system

$$\dot{x}_i(t) + \sum_{j=1}^{m} p_{ij}(t) x_j(t - \tau(t)) = 0, \qquad t \geqslant 0, \quad i = 1, 2, \ldots, m \quad (5.3.1)$$

where

$$p_{ij} \in C[\mathbb{R}^+, \mathbb{R}] \qquad \text{for } i, j = 1, 2, \ldots, m, \quad \tau \in C[\mathbb{R}^+, \mathbb{R}] \quad (5.3.2)$$

and

$$\lim_{t \to \infty} (t - \tau(t)) = \infty. \qquad (5.3.3)$$

We also assume that the coefficients $p_{ij}(t)$ of (5.3.1) satisfy the following hypothesis.

(H) There exists a (componentwise) positive vector $u = [u_1, \ldots, u_m]^{\mathrm{T}}$ and a function $p \in C[\mathbb{R}^+, \mathbb{R}^+]$ such that for all $j = 1, 2, \ldots, m$,

$$u_j p_{jj}(t) + \sum_{\substack{i=1 \\ i \neq j}}^{m} u_i |p_{ij}(t)| \leqslant -p(t) u_j, \qquad t \geqslant 0. \qquad (5.3.4)$$

The main result in this section is the following theorem which compares the oscillation of all solutions of the system (5.3.1) to the oscillation of all solutions of the scalar equation

$$\dot{y}(t) + p(t) y(t - \tau(t)) = 0. \qquad (5.3.5)$$

At the end of the section, we use this result and Theorem 5.1.1 to obtain an interesting necessary and sufficient condition for the oscillation of all solutions of the linear autonomous system

$$\dot{x}_i(t) + \sum_{j=1}^{m} p_{ij} x_j(t - \tau) = 0, \qquad t \geqslant 0, \quad i = 1, 2, \ldots, m \quad (5.3.6)$$

where

$\tau \in (0, \infty)$, $p_{ij} \in \mathbb{R}$ for $i, j = 1, 2, \ldots, m$, the matrix $P = (p_{ij})$
is irreducible, $p_{jj} > 0$ for $j = 1, 2, \ldots, m$ and $p_{ij} \leqslant 0$ for $1 \leqslant i \neq j \leqslant m$ $\Big\}$

$$(5.3.7)$$

Theorem 5.3.1. *Suppose that (5.3.2), (5.3.3) and hypothesis (H) are satisfied and assume that every solution of the scalar equation (5.3.5) oscillates. Then every solution of eqn (5.3.1) oscillates in the sense of Definition 5.0.2.*

Proof. Assume, for the sake of contradiction, that eqn (5.3.1) has a non-oscillatory solution $x(t) = [x_1(t), \ldots, x_m(t)]^T$. Then there exists a $T_0 \geq 0$ such that for all $i = 1, 2, \ldots, m$ and for $t \geq T_0$

$$\varepsilon_i \equiv \operatorname{sgn} x_i(T_0) = \operatorname{sgn} x_i(t), \qquad \varepsilon_i x_i(t) = |x_i(t)|$$

and

$$\sum_{i=1}^{m} |x_i(t)| > 0 \qquad \text{for } t \geq T_0.$$

Set

$$z(t) = \sum_{i=1}^{m} u_i \varepsilon_i x_i(t), \qquad t \geq T_0.$$

Then $z(t) > 0$ and for t sufficiently large,

$$\dot{z}(t) = \sum_{i=1}^{m} u_i \varepsilon_i \dot{x}_i(t) = -\sum_{i=1}^{m} u_i \varepsilon_i \sum_{j=1}^{m} p_{ij}(t) x_j(t - \tau(t))$$

$$\leq \sum_{j=1}^{m} \left[-u_j p_{jj}(t) x_j(t - \tau(t)) \operatorname{sgn} x_j(t - \tau(t)) + \sum_{\substack{i=1 \\ i \neq j}}^{m} u_i |p_{ij}(t)| |x_j(t - \tau(t))| \right]$$

$$\leq \sum_{j=1}^{m} -p(t) u_j |x_j(t - \tau(t))| = -p(t) z(t - \tau(t)).$$

That is, the differential inequality

$$\dot{z}(t) + p(t) z(t - \tau(t)) \leq 0$$

has an eventually positive solution. Hence, by Corollary 3.2.2, eqn (5.3.5) also has an eventually positive solution. This contradicts the hypothesis and the proof is complete.

We now apply Theorems 5.1.1 and 5.3.1 to obtain the following extensions of Theorem 5.2.2.

Theorem 5.3.2. *Assume that condition (5.3.7) is satisfied. Then the following statements are equivalent.*

(a) *Every solution of eqn (5.3.6) oscillates componentwise.*

(b) *The matrix $P = (p_{ij})$ has no real eigenvalues in the interval $(-\infty, 1/e\tau)$.*

(c) *There is a (componentwise) positive vector $u = [u_1, u_2, \ldots, u_m]^T$ such that*

$$u_j p_{jj} + \sum_{\substack{i=1 \\ i \neq j}}^{m} u_i p_{ij} > u_j/e\tau \qquad \text{for } j = 1, 2, \ldots, m. \qquad (5.3.8)$$

Proof. The fact that (a) $\Leftrightarrow$ (b) was established in Theorem 5.2.2. Now we prove that (c) $\Rightarrow$ (a). Let $p \in (1/e\tau, \infty)$. Then (5.3.8) implies that (5.3.4) is satisfied with the $p_{ij}(t)$ and $p(t)$ replaced by the constants p_{ij} and p respectively. Also every solution of (5.3.5) with $p(t)$ replaced by p oscillates because of Theorem 2.2.3 and the fact that

$$p\tau > 1/e.$$

Hence, by Theorem 5.3.1, every solution of eqn (5.3.1) oscillates in the sense of Definition 5.0.2. But, as we proved in Theorem 5.1.2, if every solution of an autonomous system oscillates in the sense of Definition 5.0.2 then it also oscillates componentwise. The proof that (c) $\Rightarrow$ (a) is complete.

Finally we prove that (b) $\Rightarrow$ (c). Here we use the facts that a matrix P and its transpose P^T have the same eigenvalues and so λ is an eigenvalue of P if and only if $-\lambda$ is an eigenvalue of $-P^T$. In view of (5.3.7), the matrix $-P^T$ is essentially positive and so $-P^T$ has a real eigenvalue λ_0 with a corresponding eigenvector $u = (u_1, u_2, \ldots, u_m)^T$ which is (componentwise) positive. Hence

$$-P^T u = \lambda_0 u$$

or equivalently,

$$p_{jj} u_j + \sum_{\substack{i=1 \\ i \neq j}}^{m} p_{ij} u_i = -\lambda_0 u_j \qquad \text{for } j = 1, 2, \ldots, m. \qquad (5.3.9)$$

But $-\lambda_0$ is an eigenvalue of P and by hypothesis (b), $-\lambda_0 > 1/e\tau$. Therefore (5.3.9) yields (5.3.8) and the proof is complete.

5.4 Oscillations of a system of delay logistic equations

Our aim in this section is to obtain sufficient conditions for the oscillation of all positive solutions of the system of delay logistic equations

$$\dot{N}_i(t) = N_i(t)\left[a_i - \sum_{j=1}^{m} b_{ij} N_j(t - \tau) \right], \qquad i = 1, 2, \ldots, m \qquad (5.4.1)$$

where

$$\tau \in (0, \infty) \qquad \text{and} \qquad a_i, b_{ij} \in \mathbb{R} \qquad \text{for } i, j = 1, 2, \ldots, m \qquad (5.4.2)$$

about its steady state

$$N^* = [N_1^*, N_2^*, \ldots, N_m^*]^{\mathrm{T}},$$

that is, about the solution of the system

$$\sum_{j=1}^{m} b_{ij} N_j^* = a_i \qquad \text{for } i = 1, 2, \ldots, m. \tag{5.4.3}$$

We also obtain sufficient conditions for the existence of a non-oscillatory solution of (5.4.1).

Together with system (5.4.1) we assume initial conditions of the form

$$N_i(t) = \phi_i(t), \qquad -\tau \leqslant t \leqslant 0 \qquad \text{where } \phi_i \in C[[-\tau, 0], \mathbb{R}^+] \text{ and } \phi_i(0) > 0 \tag{5.4.4}$$

for $i = 1, 2, \ldots, m$. Then by using an argument similar to that given in the proof of Theorem 1.1.5 we have that (5.4.1) with (5.4.4) has a unique solution $N(t) = [N_1(t), N_2(t), \ldots, N_m(t)]^{\mathrm{T}}$ valid for all $t \geqslant 0$ and such that

$$N_i(t) > 0 \qquad \text{for } t \geqslant 0 \text{ and } i = 1, 2, \ldots, m.$$

Throughout this section we assume that the system (5.4.3) has a solution N^* with positive components.

Set

$$N_i(t) = N_i^* \, e^{x_i(t)} \qquad \text{for } t \geqslant 0 \text{ and } i = 1, 2, \ldots, m.$$

Then the functions $x_i(t)$ satisfy the system of delay equations

$$\dot{x}_i(t) + \sum_{j=1}^{m} p_{ij} [e^{x_j(t-\tau)} - 1] = 0, \qquad i = 1, 2, \ldots, m \tag{5.4.5}$$

where

$$p_{ij} = b_{ij} N_j^* \qquad \text{for } i, j = 1, 2, \ldots, m.$$

We shall say that a solution $N(t) = [N_1(t), \ldots, N_m(t)]^{\mathrm{T}}$ of (5.4.1) *oscillates* about $N^* = [N_1^*, \ldots, N_m^*]^{\mathrm{T}}$ if for some $i = 1, 2, \ldots, m$ the function $N_i(t) - N_i^*$ has arbitrarily large zeros. If, on the other hand, each of the functions $N_i(t) - N_i^*$ for $i = 1, 2, \ldots, m$ is eventually different from zero, then we shall say that the solution $N(t)$ is *non-oscillatory* about N^*.

It is clear that for every ϕ satisfying (5.4.4) the solution of (5.4.1) and (5.4.4) oscillates about N^* if and only if every solution of (5.4.5) oscillates about $[0, 0, \ldots, 0]^{\mathrm{T}}$.

The first result in this section is the following.

Theorem 5.4.1. *Consider the system (5.4.1) and assume that (5.4.2) is satisfied and that (5.4.3) has a solution $N^* = [N_1^*, N_2^*, \ldots, N_m^*]^T$ with positive components. Set*

$$\mu = \min_{1 \le j \le m} \left[N_j^* \left(b_{jj} - \sum_{\substack{i=1 \\ i \ne j}}^{m} |b_{ij}| \right) \right] \tag{5.4.6}$$

and suppose that

$$\mu \tau \, e > 1. \tag{5.4.7}$$

Then for every ϕ satisfying (5.4.4) the solution of (5.4.1) and (5.4.4) oscillates about N^*.

Proof. Assume, for the sake of contradiction, that (5.4.1) with (5.4.4) has a solution $N(t)$ which is non-oscillatory about N^*. Then (5.4.5) has a solution $x(t) = [x_1(t), x_2(t), \ldots, x_m(t)]^T$ which is non-oscillatory about $[0, 0, \ldots, 0]^T$. First, we claim that

$$\lim_{t \to \infty} x_i(t) = 0, \qquad i = 1, 2, \ldots, m. \tag{5.4.8}$$

To this end, set for t sufficiently large, $\delta_i = \operatorname{sgn} x_i(t)$ for $i = 1, 2, \ldots, m$, $z_i(t) = \delta_i x_i(t)$ for $i = 1, 2, \ldots, m$, and $v(t) = \sum_{i=1}^{m} z_i(t)$. Then $z_i(t) > 0$ for $i = 1, 2, \ldots, m$ and

$$\dot{z}_i(t) + \sum_{j=1}^{m} p_{ij} \frac{\delta_j}{\delta_i} \frac{e^{\delta_j z_j(t-\tau)} - 1}{\delta_j} = 0, \qquad i = 1, 2, \ldots, m. \tag{5.4.9}$$

Hence

$$\dot{z}_i(t) + p_{ii} \frac{e^{\delta_i z_i(t-\tau)} - 1}{\delta_i} - \sum_{\substack{j=1 \\ j \ne i}}^{m} |p_{ij}| \frac{e^{\delta_j z_j(t-\tau)} - 1}{\delta_j} \le 0, \qquad i = 1, 2, \ldots, m. \tag{5.4.10}$$

By summing up (vertically) the terms in (5.4.10) for $i = 1, 2, \ldots, m$ and using (5.4.6), we find

$$\dot{v}(t) + \mu \sum_{i=1}^{m} \frac{e^{\delta_i z_i(t-\tau)} - 1}{\delta_i} \le 0. \tag{5.4.11}$$

As $\mu > 0$ and

$$\frac{e^{\delta_i z_i(t-\tau)} - 1}{\delta_i} > 0 \qquad \text{for } i = 1, 2, \ldots, m,$$

it follows that $\dot{v}(t) < 0$ and so

$$L = \lim_{t \to \infty} v(t) \tag{5.4.12}$$

exists and is non-negative. From (5.4.12) we see that $z_i(t)$ is bounded for each $i = 1, 2, \ldots, m$. Hence, from (5.4.9), $\dot{z}_i(t)$ and $\ddot{z}_i(t)$ are also bounded for each $i = 1, 2, \ldots, m$. Thus $\dot{v}(t)$ is uniformly continuous on $[T, \infty)$ ($T \in (0, \infty)$), that is, there exists an $M > 0$ such that

$$|\dot{v}(t_2) - \dot{v}(t_1)| \leq M|t_2 - t_1| \qquad \text{for all } t_1, t_2 \geq T.$$

On the other hand, because of (5.4.11) and (5.4.12) one can easily see that $\dot{v}(t)$ is integrable on $[T, \infty)$. Therefore for any $t \geq T$ and $\delta > 0$, we have

$$\dot{v}(t) = \frac{1}{\delta} \int_t^{t+\delta} \dot{v}(s) \, ds - \frac{1}{\delta} \int_t^{t+\delta} [\dot{v}(s) - \dot{v}(t)] \, ds$$

and clearly

$$\limsup_{t \to +\infty} |\dot{v}(t)| \leq \frac{1}{\delta} \limsup_{t \to +\infty} \int_t^{t+\delta} |\dot{v}(s)| \, ds + \frac{M}{\delta} \limsup_{t \to +\infty} \int_t^{t+\delta} (s - t) \, ds = \frac{M}{2} \delta.$$

Since $\delta > 0$ is an arbitrary number this yields that $\lim_{t \to +\infty} \dot{v}(t) = 0$. It follows from (5.4.11) that $\lim_{t \to \infty} z_i(t) = 0$ for each $i = 1, 2, \ldots, m$. This completes the proof of (5.4.8). Next, we rewrite (5.4.5) in the form

$$\dot{x}_i(t) + \sum_{j=1}^m \left[p_{ij} \frac{e^{x_j(t-\tau)} - 1}{x_j(t-\tau)} \right] x_j(t - \tau) = 0$$

that is,

$$\dot{x}_i(t) + \sum_{j=1}^m P_{ij}(t) x_j(t - \tau) = 0, \qquad i = 1, 2, \ldots, m \tag{5.4.13}$$

where

$$P_{ij}(t) = p_{ij} \frac{e^{x_j(t-\tau)} - 1}{x_j(t-\tau)} \qquad \text{for } i, j = 1, 2, \ldots, m$$

and

$$\lim_{t \to \infty} P_{ij}(t) = p_{ij} \qquad \text{for } i, j = 1, 2, \ldots, m. \tag{5.4.14}$$

From (5.4.13) we see that

$$\dot{z}_i(t) + P_{ii}(t) z_i(t - \tau) - \sum_{\substack{j=1 \\ j \neq i}}^m |P_{ij}(t)| z_j(t - \tau) \leq 0, \qquad i = 1, 2, \ldots, m \tag{5.4.15}$$

where, as before,

$$\delta_i = \text{sgn } x_i(t) \quad \text{and} \quad z_i(t) = \delta_i x_i(t) \quad \text{for } i = 1, 2, \ldots, m.$$

Let $\varepsilon > 0$ be chosen in such a way that

$$(\mu - \varepsilon)\tau\, e > 1 \tag{5.4.16}$$

which is possible in view of (5.4.7). Now, summing up (5.4.15) (vertically) for $i = 1, 2, \ldots, m$, we obtain

$$\dot{v}(t) + \sum_{j=1}^{m} \left[P_{jj}(t) - \sum_{\substack{i=1 \\ i \neq j}}^{m} |P_{ij}(t)| \right] z_j(t - \tau) \leqslant 0$$

which in view of (5.4.14), (5.4.6), and (5.4.7) implies that

$$\dot{v}(t) + (\mu - \varepsilon)v(t - \tau) \leqslant 0. \tag{5.4.17}$$

It follows from Corollary 2.4.1 and Theorem 2.2.3 that because of (5.4.16), inequality (5.4.17) cannot have an eventually positive solution. This is a contradiction and the proof is complete.

Theorem 5.4.2. *Consider the system* (5.4.1) *and assume that* $\tau > 0$, $b_{ij} > 0$ *for* $i, j = 1, 2, \ldots, n$ *and that* (5.4.3) *has a solution* $N^* = [N_1^*, N_2^*, \ldots, N_m^*]$ *with positive components. Suppose that*

$$\rho\tau\, e < 1 \tag{5.4.18}$$

where ρ *denotes the spectral radius of the* $m \times m$ *matrix with* (i, j) *components equal to* $b_{ij}N_j^*$. *Then* (5.4.1) *has a solution which is non-oscillatory about the steady state* N^*.

The proof of Theorem 5.4.2 will be facilitated by the next four lemmas which are interesting in their own right. In the sequel, inequalities and equalities about matrices and vectors are assumed to hold componentwise.
Consider the systems of equations

$$\dot{x}(t) + P(t)f(x(t - \tau)) = 0, \qquad t \geqslant t_0 \tag{5.4.19}$$

and

$$\dot{y}(t) + Qy(t - \tau) = 0, \qquad t \geqslant t_0 \tag{5.4.20}$$

and the inequality

$$\dot{z}(t) + P(t)f(z(t - \tau)) \leqslant 0, \qquad t \geqslant t_0 \tag{5.4.21}$$

where $\tau > 0$, $P(t)$ is an $m \times m$ matrix with positive and continuous components, Q is an $m \times m$ matrix with positive and constant components, and

$f = [f_1, f_2, \ldots, f_m]^T$ is non-decreasing and such that

$$f \in C[\mathbb{R}^m, \mathbb{R}^m] \text{ and for any } u = [u_1, \ldots, u_m]^T \left. \atop \text{with } u_i \neq 0, \ u_i f_i(u_1, \ldots, u_m) > 0 \text{ for } i = 1, \ldots, m. \right\} \qquad (5.4.22)$$

Lemma 5.4.1. *Assume that (5.4.22) holds, $\tau > 0$, $P(t) > 0$ and continuous and that inequality (5.4.21) has an eventually positive solution. Then the corresponding eqn (5.4.19) also has an eventually positive solution.*

Proof. Let $z(t)$ be an eventually positive solution of (5.4.21) and let $T \geq t_0$ be such that $z(t) > 0$ for $t \geq T - \tau$. Then $\dot{z}(t) < 0$ for $t \geq T$ and so $\lim_{t \to \infty} z(t) \equiv L$ exists and is a finite and non-negative vector. Thus, integrating (5.4.21) from T to ∞ we obtain

$$L + \int_t^\infty P(s) f(z(s - \tau)) \, \mathrm{d}s \leq z(t), \qquad t \geq T. \qquad (5.4.23)$$

Let W denote the set of non-negative and non-increasing functions w on $[T, \infty)$ such that $L \leq w(t) \leq z(t)$ for $t \geq T$. For every $w \in W$, set

$$\tilde{w}(t) = \begin{cases} w(t), & t \geq T \\ w(T) + z(t) - z(T), & T - \tau \leq t < T. \end{cases}$$

Define the mapping S on W by

$$(Sw)(t) = L + \int_t^\infty P(s) f(\tilde{w}(s - \tau)) \, \mathrm{d}s, \qquad t \geq T.$$

In view of (5.4.23), S maps W into W and all the hypotheses of the Knaster–Tarski fixed-point theorem (see Theorem 1.7.3) are satisfied. Hence there exists a point $x \in W$ such that $Sx = x$. Clearly x satisfies eqn (5.4.19) and so the proof will be complete if we show that eventually $x(t) > 0$. For $T - \tau \leq t < T$, we have

$$x(t) = x(T) + z(t) - z(T) > 0.$$

Now assume, for the sake of contradiction, that there exists a $t^* \geq T$ such that

$$x(t) > 0 \qquad \text{for } T - \tau \leq t < t^*$$

while

$$x(t^*) = 0.$$

Then by (5.4.19) and (5.4.22), $\dot{x}(t^*) = -P(t^*) f(x(t^* - \tau)) < 0$, which contradicts the fact that $x \in S$ and, consequently, $x(t) \geq 0$ for $t \geq T$. The proof is complete.

Lemma 5.4.2. *Let* $q, \tau \in (0, \infty)$ *be such that* $q\tau \, e < 1$. *Then the equation* $\mu + q \, e^{-\mu\tau} = 0$ *has a negative root.*

Proof. Set $F(\mu) = \mu + q \, e^{-\mu\tau}$ and observe that

$$F(0)F\left(-\frac{1}{\tau}\right) = q\left(-\frac{1}{\tau} + q \, e\right) = q \, \frac{q\tau \, e - 1}{\tau} < 0,$$

from which the result follows.

Lemma 5.4.3. *Let* A *be an* $m \times m$ *matrix and let* τ, λ_0, $\mu_0 \in \mathbb{R}$ *and* $\xi \in \mathbb{R}^n$ *be such that*

$$A\xi = \lambda_0 \xi$$

and

$$\mu_0 + \lambda_0 \, e^{-\mu_0\tau} = 0.$$

Then

$$z(t) = e^{\mu_0 t}\xi$$

is a solution of

$$\dot{z}(t) + Az(t - \tau) = 0.$$

Proof. $\dot{z}(t) + Az(t - \tau) = \mu_0 \, e^{\mu_0 t}\xi + A \, e^{\mu_0(t - \tau)}\xi = e^{\mu_0(t - \tau)}(A + \mu_0 \, e^{\mu_0\tau}I)\xi$

$$= e^{\mu_0(t - \tau)}(A - \lambda_0 I)\xi = 0.$$

The next lemma gives sufficient conditions for the system (5.4.19) to have a non-oscillatory solution.

Lemma 5.4.4. *Consider the system* (5.4.19) *and assume that* $P(t)$ *is a continuous* $m \times m$ *matrix such that*

$$\lim_{t \to \infty} P(t) = Q$$

where the constant matrix Q *has positive components and is such that*

$$\rho(Q)\tau \, e < 1 \qquad (5.4.24)$$

where $\tau > 0$ *and* $\rho(Q)$ *is the spectral radius of* Q. *Suppose that* f *satisfies* (5.4.22) *and that there exists a* $\delta > 0$ *such that either*

$$f(u) \leqslant u \qquad for \ 0 < u \leqslant \delta \qquad (5.4.25)$$

or

$$f(u) \geqslant u \qquad for \quad -\delta \leqslant u < 0. \qquad (5.4.26)$$

Then the system (5.4.19) has a non-oscillatory solution.

Proof. We assume that (5.4.25) holds and we prove that (5.4.19) has an eventually positive solution. When (5.4.26) holds, one can set $v(t) = -x(t)$ and show, by a similar argument, that (5.4.19) has an eventually negative solution. To this end, choose $\varepsilon > 0$ and $T \geqslant t_0$ such that

$$\rho(Q(\varepsilon))\tau \, e < 1$$

and

$$0 \leqslant P(t) \leqslant Q(\varepsilon) \qquad for \ t \geqslant T$$

where $Q(\varepsilon)$ is the $m \times m$ matrix with ij-components equal to $q_{ij} + \varepsilon$ and $\rho(Q(\varepsilon))$ denotes the spectral radius of $Q(\varepsilon)$. Let ξ be a positive eigenvector of $Q(\varepsilon)$ associated with the eigenvalue $\rho(Q(\varepsilon))$. Such an eigenvector exists because $Q(\varepsilon)$ is a positive matrix and $\rho(Q(\varepsilon))$ is the largest eigenvalue of $Q(\varepsilon)$.

Let $\mu_0(\varepsilon)$ be a negative root of the equation

$$\mu + \rho(Q(\varepsilon)) \, e^{-\mu\tau} = 0$$

as is guaranteed by Lemma 5.4.2 and let $z(t) = e^{\mu_0(\varepsilon)t}\xi$ be the positive solution of the equation

$$\dot{z}(t) + Q(\varepsilon)z(t - \tau) = 0$$

which is guaranteed by Lemma 5.4.3. Then

$$0 = \dot{z}(t) + Q(\varepsilon)z(t - \tau) \geqslant \dot{z}(t) + P(t)z(t - \tau) \geqslant \dot{z}(t) + P(t)f(z(t - \tau)).$$

Hence, by Lemma 5.4.1, the system (5.4.19) has also an eventually positive solution. The proof is complete.

Proof of Theorem 5.4.2. As we saw in the introduction to this section, the transformation

$$N_i(t) = N_i^* \, e^{x_i(t)} \qquad for \ i = 1, 2, \ldots, m \ and \ t \geqslant 0$$

reduces system (5.4.1) to system (5.4.5). On the other hand, $N(t)$ is non-oscillatory about N^* if and only if $x(t)$ is non-oscillatory about $[0, 0, \ldots, 0]^T$. Thus it suffices to show that system (5.4.5) has a non-oscillatory solution. For each $u = [u_1, u_2, \ldots, u_m]^T$, set

$$f(u) = [e^{u_1} - 1, e^{u_2} - 1, \ldots, e^{u_m} - 1]^T$$

and observe that f is non-decreasing and

$$(e^{u_i} - 1)u_i > 0 \qquad \text{for } u_i \neq 0$$

and

$$e^u - 1 \geqslant u \qquad \text{for } u < 0$$

which implies that f satisfies conditions (5.4.22) and (5.4.26). By applying Lemma 5.4.4 to eqn (5.4.5) we see that (5.4.5) has a non-oscillatory solution. The proof is complete.

5.5 Notes

The results in Section 5.1 are adapted from Arino and Györi (1990) and the results of Sections 5.2 and 5.3 are based on the work of Ferreira and Györi (1987). The concept of oscillation 'globally in the delays' was introduced by Ferreira (1990) and Theorem 5.2.3 is from Arino and Ferreira (1989). The results in Section 5.4 are from Gopalsamy *et al.* (1990*c*).

For some related work on the oscillation of systems see Gopalsamy (1984, 1987). See also Marusiak and Norkin (1981).

5.6 Open problems

5.6.1 Obtain oscillation results for linear systems with asymptotically constant coefficients. (See Section 2.4 for the scalar case.)

5.6.2 Obtain linearized oscillation results for non-linear systems. (See Section 4.1 for the scalar case.)

5.6.3 Study the oscillatory character of solutions of various systems of delay equations which appear in mathematical biology. (See Gopalsamy *et al.* 1990*c*.)

6

OSCILLATIONS OF NEUTRAL
DIFFERENTIAL EQUATIONS

A neutral delay differential equation is a differential equation in which the highest-order derivative of the unknown function appears in the equation both with and without delays. For example,

$$\frac{d}{dt} [y(t) + P(t)y(t - \tau)] + Q(t)y(t - \sigma) = 0 \qquad (6.0.1)$$

where

$$P, Q \in C[[t_0, \infty), \mathbb{R}] \qquad \text{and} \qquad \tau, \sigma \in [0, \infty) \qquad (6.0.2)$$

is a first-order neutral delay differential equation.

Let $\gamma = \max\{\tau, \sigma\}$ and let $t_1 \geqslant t_0$. By a solution of eqn (6.0.1) on $[t_1, \infty)$ we mean a function $y \in C[[t_1 - \gamma, \infty), \mathbb{R}]$ such that $y(t) + P(t)y(t - \tau)$ is continuously differentiable for $t \geqslant t_1$ and such that eqn (6.0.1) is satisfied for $t \geqslant t_1$.

Let $t_1 \geqslant t_0$ be a given *initial point* and let $\phi \in C[[t_1 - \gamma, t_1], \mathbb{R}]$ be a given *initial function*. Then, as we proved by the method of steps in Section 1.1, eqn (6.0.1) has a unique solution on $[t_1, \infty)$ satisfying the initial condition

$$y(t) = \phi(t) \qquad \text{for } t_1 - \gamma \leqslant t \leqslant t_1. \qquad (6.0.3)$$

As usual, when we say that every solution of eqn (6.0.1) oscillates, we mean that for every initial point $t_1 \geqslant t_0$ and for every initial function $\phi \in C[[t_1 - \gamma, t_1], \mathbb{R}]$ the unique solution of (6.0.1) and (6.0.3) on $[t_1, \infty)$ has arbitrarily large zeros. If it is false that every solution of eqn (6.0.1) oscillates, then there exists a $t_1 \geqslant t_0$, an initial function $\phi \in C[[t_1 - \gamma, t_1], \mathbb{R}]$ and a $T \geqslant t$ such that the solution of (6.0.1) and (6.0.2) is either positive or negative for $t \geqslant T$.

In general, the theory of neutral delay differential equations presents complications, and results which are true for non-neutral equations may not be true for neutral equations. Snow (1965) has shown, for example, that even though the characteristic roots of a neutral differential equation may all have negative real parts, it is still possible for some solutions to be unbounded (see also Slemrod and Infante 1972). Besides its theoretical interest, the study of the asymptotic and oscillatory behaviour of solutions of neutral differential equations has some importance in applications. Neutral delay differential equations appear in networks containing lossless transmission lines (as in high-speed computers where the lossless transmission lines are used to interconnect switching circuits), in the study of vibrating masses attached to an elastic bar and also as the Euler equation in some variational problems (see Driver 1984, Hale 1977, Brayton and Willoughby 1967, and the references cited therein).

Our aim in this chapter is to present some of the oscillation results that have recently been obtained for neutral equations.

6.1 Oscillations and asymptotic behaviour of scalar neutral delay differential equations

In this section we study the oscillation of all solutions of the neutral delay differential equation

$$\frac{\mathrm{d}}{\mathrm{d}t}[y(t) + py(t - \tau)] + qy(t - \sigma) = 0 \tag{6.1.1}$$

where

$$p \in \mathbb{R}, \quad \tau, q \in (0, \infty) \quad \text{and} \quad \sigma \in [0, \infty). \tag{6.1.2}$$

We also investigate the asymptotic behaviour of the non-oscillatory solutions of eqn (6.1.1).

One can easily see that if $y(t)$ is a solution of eqn (6.1.1) then the functions

$$z(t) = y(t) + py(t - \tau) \tag{6.1.3}$$

$$w(t) = z(t) + pz(t - \tau) \tag{6.1.4}$$

are also solutions of the same equation. Furthermore, $z(t)$ is a differentiable solution, while $w(t)$ is twice differentiable.

The following result is useful and interesting in its own right.

Lemma 6.1.1. *Suppose that* (6.1.2) *holds and assume that eqn* (6.1.1) *has an eventually positive solution* $y(t)$. *Let* $z(t)$ *and* $w(t)$ *be defined by* (6.1.3) *and* (6.1.4) *respectively. Then the following statements hold.*

(a) $z(t)$ *is a strictly decreasing and differentiable solution of eqn* (6.1.1) *and either*

$$\lim_{t \to \infty} z(t) = -\infty \tag{6.1.5}$$

or

$$\lim_{t \to \infty} z(t) = 0. \tag{6.1.6}$$

(b) *The following statements are equivalent:*

  (i) (6.1.5) *holds*;
  (ii) $p < -1$;
  (iii) $\lim_{t\to\infty} y(t) = \infty$;
  (iv) $w(t)$ *is a twice differentiable solution of eqn* (6.1.1) *such that*

$$w(t) > 0, \quad \dot{w}(t) > 0, \quad \ddot{w}(t) > 0 \quad and \quad \lim_{t\to\infty} w(t) = \infty. \quad (6.1.7)$$

(c) *The following statements are equivalent:*

  (i) (6.1.6) *holds*;
  (ii) $p > -1$;
  (iii) $\lim_{t\to\infty} y(t) = 0$;
  (iv) $w(t)$ *is a twice differentiable solution of eqn* (6.1.1) *such that*

$$w(t) > 0, \quad \dot{w}(t) < 0, \quad \ddot{w}(t) > 0 \quad and \quad \lim_{t\to\infty} w(t) = 0. \quad (6.1.8)$$

Proof.
  (a) We have

$$\dot{z}(t) = -qy(t - \sigma) < 0 \qquad (6.1.9)$$

and so $z(t)$ is strictly decreasing. If (6.1.5) is not true, then there exists $l \in \mathbb{R}$ such that $\lim_{t\to\infty} z(t) = l$. By integrating (6.1.9) from t to ∞, we see that $l - z(t) = -\int_t^\infty y(s - \sigma)\, ds$. This implies that $y \in L^1[t, \infty)$ and so $z \in L^1[t, \infty)$. But then l must be zero and the proof of (a) is complete.

The proofs of (b) and (c) follow immediately from the following discussion. First assume that (6.1.5) holds. Then p must be negative and $y(t)$ unbounded. Therefore, there exists $t_0 \geq 0$ such that $z(t_0) < 0$ and $y(t_0) = \max_{t \leq t_0} y(t) > 0$. Then $0 > z(t_0) = y(t_0) + py(t_0 - \tau) \geq y(t_0)(1 + p)$, which implies that $p < -1$. But $z(t) = y(t) + py(t - \tau) > py(t - \tau)$ and (6.1.5) implies that $\lim_{t\to\infty} y(t) = \infty$. Observe that under the hypothesis (6.1.5) we have $\dot{w}(t) = -qz(t - \sigma) > 0$ and $\ddot{w}(t) = -q\dot{z}(t - \sigma) > 0$, which imply that (6.1.7) holds. Now assume that (6.1.6) holds. If $p \geq 0$, then from (6.1.3) we see that $\lim y(t) = 0$. On the other hand, if $p \in (-1, 0)$, it follows from Lemma 1.5.1 that $\lim_{t\to\infty} y(t) = (\lim_{t\to\infty} z(t))/(1 + p) = 0$. Finally, if $p \leq -1$ then $y(t) \geq -py(t - \tau) \geq y(t - \tau)$ and so $y(t)$ is bounded from below by a positive constant, say m. Then from (6.1.9), $\dot{z}(t) \leq -qm$, which yields the contradiction that $z(t) \to -\infty$ as $t \to \infty$. Thus (6.1.6) implies that $p > -1$ and that $\lim_{t\to\infty} y(t) = 0$. Furthermore in this case, $\dot{w}(t) = -qz(t - \sigma) < 0$, $\ddot{w}(t) = -q\dot{z}(t - \sigma) > 0$ and $w(t)$ decreases to zero. In particular, $w(t) > 0$. On the basis of the above discussion, the proofs of (b) and (c) are now obvious.

The following results are immediate consequences of Lemma 6.1.1.

Theorem 6.1.1. *Assume that* (6.1.2) *holds and let* $y(t)$ *be a non-oscillatory solution of eqn* (6.1.1). *Then the following are true:*

(a) $p < -1$ *if and only if* $\lim_{t \to \infty} |y(t)| = \infty$.

(b) $p > -1$ *if and only if* $\lim_{t \to \infty} y(t) = 0$.

Theorem 6.1.2. *Consider the NDDE*

$$\frac{d}{dt}[y(t) - y(t - \tau)] + qy(t - \sigma) = 0 \qquad (6.1.10)$$

where

$$\tau, \sigma \in [0, \infty) \qquad and \qquad q \in (0, \infty).$$

Then for every solution of eqn (6.1.10) *oscillates.*

The following result gives sufficient conditions for the oscillation of all solutions of eqn (6.1.1) when (6.1.2) holds and $p \neq -1$.

Theorem 6.1.3. *Assume that* (6.1.2) *holds*

$$p \neq -1 \qquad and \qquad \frac{q}{1+p}(\sigma - \tau) < 1/e. \qquad (6.1.11)$$

Then every solution of eqn (6.1.1) *is oscillatory.*

Proof. Assume, for the sake of contradiction, that eqn ((6.1.1) has an eventually positive solution $y(t)$. Then, by Lemma 6.1.1, $w(t) > 0$, $\dot{w}(t)$ is eventually increasing and $\dot{w}(t) + p\dot{w}(t - \tau) + qw(t-\sigma) = 0$. Hence

$$(1 + p)\dot{w}(t - \tau) + qw(t - \sigma) \leq 0$$

and so

$$\dot{w}(t) + \frac{q}{1 + p} w(t - (\sigma - \tau)) \leq 0 \qquad \text{if } 1 + p > 0 \qquad (6.1.12)$$

or

$$\dot{w}(t) - \frac{q}{-(1+p)} w(t + (\tau - \sigma)) \geq 0 \qquad \text{if } 1 + p < 0. \qquad (6.1.13)$$

But by Theorems 2.3.3 and 2.3.4, when (6.1.11) holds, the inequalities (6.1.12) and (6.1.13) cannot have eventually positive solutions. This contradicts the fact that $w(t) > 0$. The proof is complete.

6.2 Oscillations of scalar neutral equations with mixed arguments

Consider the neutral equation with mixed (delayed and advanced) arguments

$$\frac{d}{dt}[y(t) + py(t - \tau)] + qy(t - \sigma) = 0 \qquad (6.2.1)$$

where p, q, τ and σ are real numbers. The main results in this section are the following two theorems, which give necessary and sufficient conditions for the oscillation of all unbounded and all bounded solutions of eqn (6.2.1) by means of the characteristic equation

$$F(\lambda) \equiv \lambda + p\lambda\, e^{-\lambda\tau} + q\, e^{-\lambda\sigma} = 0 \qquad (6.2.2)$$

of eqn (6.2.1)

Theorem 6.2.1. *Assume that p, q, τ and σ are real numbers. Then the following statements are equivalent.*

(a) *Every unbounded solution of eqn (6.2.1) oscillates.*

(b) *The characteristic equation (6.2.2) has no positive roots and zero is not a multiple root of eqn (6.2.2).*

Theorem 6.2.2. *Assume that p, q, τ and σ are real numbers. Then the following statements are equivalent.*

(a) *Every bounded solution of eqn (6.2.1) oscillates.*

(b) *The characteristic equation (6.2.2) has no roots in $(-\infty, 0]$.*

The following necessary and sufficient condition for the oscillation of all solutions of eqn (6.2.1) ensues as a corollary.

Corollary 6.2.1. *Assume that p, q, τ and σ are real numbers. Then the following statements are equivalent.*

(a) *Every solution of eqn (6.2.1) oscillates.*

(b) *The characteristic equation (6.2.2) has no real roots.*

When the arguments τ and σ are not both delays, it is not known whether the unbounded solution of eqn (6.2.1) are exponentially bounded (see Hale 1977). Therefore one cannot apply Laplace transforms to prove Theorem 6.2.1. On the other hand, Theorem 6.2.2 can be proved, by Laplace transforms, by slightly modifying the proof of Theorem 2.1.1 (see also Section 6.3).

Our proof of Theorem 6.2.1 is elementary in nature but lengthy. One can give a similar proof for Theorem 6.2.2, but for economy we refer the reader to Ladas and Schultz (1990). See also Section 6.3 and the notes at the end of this chapter.

The following 'duality lemma', whose proof is obvious, will enable us to reduce the required number of cases we have to consider in our proof of Theorem 6.2.1.

(Duality) Lemma 6.2.1. *Suppose that p is a non-zero real number. Then y(t) is a solution of eqn (6.2.1) if and only if y(t) is a solution of*

$$\frac{d}{dt}\left[y(t) + \frac{1}{p}y(t - (-\tau))\right] + \frac{q}{p}y(t - (\sigma - \tau)) = 0.$$

The following result, which we shall use several times in the proof of Theorem 6.2.1, can be proved by direct substitution into eqn (6.2.1).

Lemma 6.2.2. *Let y(t) be a solution of eqn (6.2.1) for $t \geqslant t_o$ and let α and β be any constants. Then*

$$x(t) = \int_{t-\alpha}^{t-\beta} y(s)\,ds$$

is also a solution for $t \geqslant t_0 + \max\{\alpha, \beta\}$.

Proof of Theorem 6.2.1.

(a) $\Rightarrow$ (b). Clearly, if there is a root in $(0, \infty)$ then an unbounded non-oscillatory solution exists. Also, $\lambda = 0$ cannot be a double root, for if 0 were a double root then $q = 0$ and $p = -1$ would reduce eqn (6.2.1) to

$$\frac{d}{dt}\left[y(t) - y(t - \tau)\right] = 0$$

which has the unbounded non-oscillatory solution $y(t) = t$.

(b) $\Rightarrow$ (a). Assume, for the sake of contradiction, that eqn (6.2.1) has an unbounded eventually positive solution $y(t)$. First assume $p = 0$. Then clearly q must be negative, for otherwise $y(t)$ would be bounded. Also $\sigma \neq 0$, for otherwise the characteristic equation would have a positive root. Hence there remain the following cases to consider:

    (i) $q < 0$ and $\sigma > 0$ (ii) $q < 0$ and $\sigma < 0$.

Case (i). We have $F(0)F(\infty) < 0$, which contradicts the hypothesis that the characteristic equation has no real roots.

Case (ii). The proof when $p = 0$ is identical to the proof of Theorem 2.1.2 (case 1) and will be omitted. However, the reader is encouraged to review it because the same type of argument will appear several times as we are searching to obtain a contradiction in the various regions of the parameters p, q, τ and σ.

The case when $\tau = 0$ and $p \neq -1$ follows in a manner analogous to the case when $p = 0$. On the other hand, the case $\tau = 0$ and $p = -1$ is trivial. So we will assume $p\tau \neq 0$.

When $q = 0$, eqn (6.2.1) reduces to $d/dt[y(t) + py(t - \tau)] = 0$, which implies $y(t) + py(t - \tau) = c$. Clearly $y(t)$ cannot be positive and unbounded if $p > 0$. Also (6.2.2) reduces to $F(\lambda) = \lambda(1 + p\,e^{-\lambda\tau}) = 0$ and it follows that $p > -1$ when $\tau > 0$, for otherwise (6.2.2) has a positive root or 0 is a double root. Furthermore, because of the Duality Lemma, we need only consider $-1 < p < 0$ and $\tau > 0$ to complete the proof when $q = 0$. To this end, let $\{t_n\}$ be a sequence of points such that $\lim_{n \to \infty} t_n = \infty$, $y(t_n) = \max_{s \leqslant t_n} y(s)$ and $\lim_{n \to \infty} y(t_n) = \infty$. Observe that

$$c = y(t_n) + py(t_n - \tau) \geqslant (1 + p)y(t_n) \to \infty \qquad \text{as } n \to \infty,$$

which is impossible.

Finally, by utilizing the Duality Lemma, one can see that the following eight cases remain to complete the proof of the theorem.

	p	q	σ	τ
1	$+$	$+$	$+$	$+, 0$
2	$+$	$+$	$+$	$-$
3	$+$	$-$	$+$	$+, 0$
4	$+$	$-$	$+$	$-$
5	$-$	$+$	$+$	$+, 0$
6	$-$	$+$	$+$	$-$
7	$-$	$-$	$+$	$+, 0$
8	$-$	$-$	$+$	$-$

Our goal is to obtain a contradiction in each case.

Cases 1 and 2: $p > 0$, $q > 0$, $\tau > 0$, $\sigma \geqslant 0$ or $p > 0$, $q > 0$, $\tau > 0$, $\sigma < 0$. As $y(t)$ is unbounded, Lemma 6.2.1 implies that $p < -1$, which is a contradiction.

Cases 3 and 7: $p > 0$, $q < 0$, $\tau > 0$, $\sigma \geqslant 0$ or $p < 0$, $q < 0$, $\tau > 0$, $\sigma \geqslant 0$. We have $F(0)F(\infty) < 0$, which implies that the characteristic equation has positive root, which is a contradiction.

In each of the remaining cases the proof is in the spirit of the proof of Theorem 2.2.1 (case 1). More precisely, in each of these cases we construct a sequence of solutions of eqn (6.2.1), $\{w_n(t)\}_{n=0}^{\infty}$ such that

$$w_n(t) > 0, \qquad \dot{w}_n(t) > 0 \quad \text{and} \quad \ddot{w}_n(t) > 0 \qquad \text{for } n = 0, 1, 2, \ldots$$

The proof continues by defining the sets

$$\Lambda_n = \{\lambda \geqslant 0 : -\dot{w}_n(t) + \lambda w_n(t) \leqslant 0\} \qquad \text{for } n = 0, 1, 2, \ldots.$$

Note that $0 \in \Lambda_n$ for all n. Also, if $0 \leqslant a \leqslant b$ and $b \in \Lambda_n$ then $a \in \Lambda_n$. That is, Λ_n is a non-empty subinterval of non-negative real numbers. The proof is completed by showing that the following contradictory properties hold:

(P$_1$) There exist non-negative numbers λ_1, λ_2 such that $\lambda_1 \in \bigcap_{n=1}^{\infty} \Lambda_n$, and $\lambda_2 \notin \bigcup_{n=1}^{\infty} \Lambda_n$

(P$_2$) There exists a positive number μ such that if $\lambda \in \Lambda_n$, then $\lambda + \mu \in \Lambda_{n+1}$.

Since this is a contradiction there cannot exist an eventually positive solution to eqn (6.2.1). Hence every solution oscillates and the proof of the theorem will be complete.

For the proof in the following cases let $y(t)$ be an eventually positive solution of eqn (6.2.1) and set $z(t) = y(t) + py(t - \tau)$ and $w(t) = z(t) + pz(t - \tau)$. Then an analysis similar to that in the proof of Lemma 6.1.1 shows that $w(t)$ is a twice differentiable solution of eqn (6.2.1) and

$$w(t) > 0, \qquad \dot{w}(t) > 0 \quad \text{and} \quad \ddot{w}(t) > 0.$$

As the characteristic equation (6.2.2) has no positive roots, it follows that there exists an $m \in (0, \infty)$ such that for all $\lambda \geqslant 0$,

$$-\lambda - \lambda p\, e^{-\lambda\tau} - q\, e^{-\lambda\sigma} \leqslant -m \qquad \text{if } q > 0$$

while

$$\lambda + p\lambda\, e^{-\lambda\tau} + q\, e^{-\lambda\sigma} \leqslant -m \qquad \text{if } q < 0.$$

We now prove the remaining cases.

Case 4: $p > 0$, $q < 0$, $\tau > 0$ and $\sigma < 0$. The dual of Case 4 is $p > 0$, $q < 0$, $\tau < 0$ and $\sigma < 0$ ($\sigma < \tau$), which we now consider. Set

$$w_n(t) = \begin{cases} w(t), & n = 0 \\ w_{n-1} + pw_{n-1}(t - \tau), & n = 1, 2, \ldots \end{cases}$$

and

$$\Lambda_n = \{\lambda \geqslant 0 : -\dot{w}_n(t) + \lambda w_n(t) \leqslant 0\}.$$

It follows as in Lemma 6.1.1 that for $n = 1, 2, \ldots$,

$$\dot{w}_n(t) + p\dot{w}_n(t - \tau) + qw_n(t - \sigma) = 0$$

$$\ddot{w}_n(t) = -q\dot{w}_{n-1}(t - \sigma)$$

$$w_n(t) > 0, \qquad \dot{w}_n(t) > 0 \qquad \text{and} \qquad \ddot{w}_n(t) > 0.$$

From the above we have $\dot{w}_n(t - \tau) + [q/(1 + p)]w_n(t - \sigma) \geq 0$, which implies

$$\dot{w}_n(t) + \frac{q}{1 + p} w_n(t + (\tau - \sigma)) \geq 0 \tag{6.2.3}$$

and so

$$-\dot{w}_n(t) + \frac{-q}{1 + p} w_n(t) \leq 0. \tag{6.2.4}$$

Applying Lemma 1.6.6 to (6.2.3) yields $\lambda_2 = \ln\left(\dfrac{4(1 + p)^2}{q^2(\tau - \sigma)^2}\right)\Big/(\tau - \sigma) \notin \bigcap_{n=1}^{\infty} \Lambda_n$, while (6.2.4) implies $\lambda_1 = -q/(1 + p) \in \bigcup_{n=1}^{\infty} \Lambda_n$. Let $\lambda \in \Lambda_n$ and set $\phi_n(t) = e^{-\lambda t} w_n(t)$ and $\mu = m/(1 + p e^{-\lambda_2 \tau})$. Now

$$\dot{w}_{n+1}(t) + (\lambda + \mu)w_{n+1}(t) = qw_n(t - \sigma) + (\lambda + \mu)[w_n(t) + pw_n(t - \tau)]$$

$$= q\phi_n(t - \sigma) e^{\lambda(t - \sigma)} + (\lambda + \mu)[e^{\lambda t}\phi_n(t) + p e^{\lambda(t - \tau)}\phi_n(t - \tau)]$$

$$\leq \phi_n(t - \sigma) e^{\lambda t}[q e^{-\lambda \sigma} + \lambda + \lambda p e^{-\lambda \tau} + \mu + \mu p e^{-\lambda_2 \tau}]$$

$$\leq \phi_n(t - \sigma) e^{\lambda t}[-m + m] = 0,$$

which completes the proof in this case.

Case 5: $p < 0, q > 0, \tau > 0$ and $\sigma \geq 0$. First assume that $\sigma < \tau$. Set

$$w_n(t) = \begin{cases} w(t), & n = 0 \\ -[w_{n-1}(t) + pw_{n-1}(t - \tau)], & n = 1, 2, \ldots \end{cases}$$

and

$$\Lambda_n = \{\lambda \geq 0 : -\dot{w}_n(t) + \lambda w_n(t) \leq 0\}.$$

It follows that for $n = 1, 2, \ldots$,

$$\dot{w}_n(t) + p\dot{w}_n(t - \tau) + qw_n(t - \sigma) = 0 \tag{6.2.5}$$

$$\ddot{w}_n(t) = qw_{n-1}(t - \sigma)$$

$$w_n(t) > 0, \qquad \dot{w}_n(t) > 0 \qquad \text{and} \qquad \ddot{w}_n(t) > 0.$$

From (6.2.5) we have $\dot{w}_n(t - \tau) + p\dot{w}_n(t - \tau) + qw_n(t - \sigma) \leq 0$ and from

Lemma 6.1.1 it follows that $p < -1$. By combining these results we obtain

$$\dot{w}_n(t) - \frac{q}{-(1+p)} w_n(t + (\tau - \sigma)) \geq 0$$

and

$$-\dot{w}_n(t) + \frac{q}{-(1+p)} w_n(t) \leq 0.$$

It now follows that

$$\lambda_1 = \frac{q}{-(1+p)} \in \bigcap_{n=1}^{\infty} \Lambda_n \quad \text{and} \quad \lambda_2 = \ln \left(\frac{4(1+p)^2}{q^2(\tau - \sigma)^2} \right) \bigg/ (\tau - \sigma) \notin \bigcup_{n=1}^{\infty} \Lambda_n.$$

Let $\lambda \in \Lambda_n$ and set $\phi_n(t) = e^{-\lambda t} w_n(t)$ and $\mu = m/(-p)$. Now

$$-\dot{w}_{n+1}(t) + (\lambda + \mu) w_{n+1}(t)$$
$$= -q w_n(t - \sigma) + (\lambda + \mu)[-w_n(t) - p w_n(t - \tau)]$$
$$\leq \phi_n(t - \sigma) e^{\lambda t}[-q e^{-\lambda\sigma} - \lambda - \lambda p e^{-\lambda\tau} - \mu - \mu p e^{-\lambda\tau}]$$
$$\leq \phi_n(t - \sigma) e^{\lambda t}[-m + m] = 0$$

This completes the proof in Case 5 for $\sigma < \tau$.

For $\sigma \geq \tau$, set

$$w_n(t) = \begin{cases} w(t), & n = 0 \\ -[w_{n-1}(t) + p w_{n-1}(t - \tau)] + q \int_{t-\sigma}^{t-\tau} w_{n-1}(s)\, ds, & n = 1, 2, \ldots \end{cases}$$

$$(6.2.6)$$

and observe that

$$\dot{w}_{n+1}(t) = q w_n(t - \tau) > 0. \tag{6.2.7}$$

From (6.2.6) we obtain $w_{n+1}(t) < [-p + q(\sigma - \tau)] w_n(t - \tau)$. This together with (6.2.7) gives

$$-\dot{w}_{n+1}(t) + \frac{q}{[-p + q(\sigma - \tau)]} w_{n+1}(t) \leq 0$$

which implies $\lambda_1 = q/(-p + q(\sigma - \tau)) \in \bigcap_{n=1}^{\infty} \Lambda_n$. Now

$$\dot{w}_n(t) + p \dot{w}_n(t - \tau) + q w_n(t - \sigma) = 0$$

implies

$$\dot{w}_n(t) + p \dot{w}_n(t - \tau) \leq 0. \tag{6.2.8}$$

By combining (6.2.7) and (6.2.8) we find

$$qw_{n-1}(t-\tau) + p\dot{w}_n(t-\tau) \leqslant 0.$$

(6.2.9)

Integrating (6.2.9) from t to $t + \tau$ we obtain

$$q\tau w_{n-1}(t-\tau) + pw_n(t) - pw_n(t-\tau) \leqslant 0.$$

The above implies

$$qw_{n-1}(t-\tau) < -(p/\tau)w_n(t).$$

(6.2.10)

From (6.2.7) and (6.2.10) we now obtain $-\dot{w}_n(t) + (-p/\tau)w_n(t) > 0$, which implies $\lambda_2 = -p/\tau \notin \bigcup_{n=1}^{\infty} \Lambda_n$. Let $\lambda \geqslant \lambda_1$ and set $\phi_n(t) = e^{-\lambda t}w_n(t)$ and $\mu = m/-p + (q/\lambda_1)$. Observe that

$$-\dot{w}_{n+1}(t) + (\lambda + \mu)w_{n+1}(t)$$

$$= -qw_n(t-\tau) + (\lambda + \mu)[-w_n(t) - pw_n(t-\tau) + q \int_{t-\sigma}^{t-\tau} w_n(s)\,ds]$$

$$\leqslant -q\phi_n(t-\tau)e^{\lambda(t-\tau)} + (\lambda + \mu)$$

$$\times [-\phi_n(t)e^{\lambda t} - p\phi_n(t-\tau)e^{\lambda(t-\tau)} + q\int_{t-\sigma}^{t-\tau} e^{\lambda s}\phi_n(s)\,ds]$$

$$\leqslant \phi_n(t-\tau)e^{\lambda t}[-q\,e^{-\lambda\tau} - \lambda - \lambda p\,e^{-\lambda t} - \mu - \mu p\,e^{-\lambda t}$$

$$+ q\,e^{-\lambda\tau} - q\,e^{-\lambda\sigma} + \frac{\mu q}{\lambda}(e^{-\lambda\tau} - e^{-\lambda\sigma})]$$

$$\leqslant \phi_n(t-\tau)e^{\lambda t}[-\lambda - \lambda p\,e^{-\lambda\tau} - q\,e^{-\lambda\sigma} - \mu$$

$$- \mu p\,e^{-\lambda\tau} + \frac{\mu q}{\lambda}(e^{-\lambda\tau} - e^{-\lambda\sigma})]$$

$$\leqslant \phi_n(t-\tau)e^{\lambda t}[-m + m] = 0.$$

The proof is complete.

Case 6: $p < 0,\ q > 0,\ \tau > 0$ and $\sigma < 0$. Set

$$w_n(t) = \begin{cases} w(t), & n = 0 \\ -[w_{n-1}(t) + pw_{n-1}(t-\tau)] - q\displaystyle\int_{t-\tau}^{t-\sigma} w_{n-1}(s)\,ds, & n = 1, 2, \ldots \end{cases}$$

(6.2.11)

and define Λ_n as in Case 5. Now for $n = 1, 2, \ldots$,

$$\dot{w}_n(t) + p\dot{w}_n(t - \tau) + qw_n(t - \sigma) = 0 \tag{6.2.12}$$

$$\dot{w}_{n+1}(t) = qw_n(t - \tau) > 0. \tag{6.2.13}$$

Again it follows as in Lemma 6.1.1 that $p < -1$. This together with (6.2.12) yields

$$\dot{w}_n(t) - \frac{q}{-(1 + p)} w_n(t + (-\sigma)) \geqslant 0 \quad \text{and} \quad -\dot{w}_n(t) + \frac{q}{-(1 + p)} w_n(t) \leqslant 0.$$

We now have

$$\lambda_1 = \frac{q}{-(1 + p)} \in \bigcap_{n=1}^{\infty} \Lambda_n \quad \text{and} \quad \lambda_2 = \left[\ln\left(\frac{4(1 + p)^2}{(q\sigma)^2}\right) \middle/ (-\sigma) \right] \notin \bigcup_{n=1}^{\infty} \Lambda_n.$$

To complete the proof in this case repeat exactly the same argument as in Case 5 when $\sigma \geqslant \tau$.

Case 8: $p < 0$, $q < 0$, $\tau > 0$ and $\sigma < 0$. Let V be the set of all C^2 solutions of eqn (6.2.1) which satisfy

$$v(t) > 0, \qquad \dot{v}(t) > 0, \qquad \ddot{v}(t) > 0 \quad \text{and} \quad \lim_{t \to \infty} v(t) = \infty.$$

Set $\Lambda(v) = \{\lambda \geqslant 0 : -\dot{v}(t) + \lambda v(t) \leqslant 0\}$. Observe that

$$\dot{v}(t) + p\dot{v}(t - \tau) + qv(t - \sigma) = 0. \tag{6.2.14}$$

From (6.2.14) we obtain

$$\dot{v}(t) - (-q)v(t + (-\sigma)) \geqslant 0 \tag{6.2.15}$$

and so

$$-\dot{v}(t) + (-q)v(t) \leqslant 0. \tag{6.2.16}$$

Hence $-q \in \Lambda(v)$ and $\lambda^* = \ln(4/(q\sigma)^2)/-\sigma \notin \Lambda(v)$. Now let $\lambda_0 = -q$ and set $\mu = m/(1 - p\,e^{-\lambda^* \sigma})$. We shall prove by induction that if

$$\lambda_n = \lambda_{n-1} + \mu, \qquad n = 1, 2, \ldots,$$

and if

$$w_n(t) = \begin{cases} w(t), & n = 0 \\ w_{n-1}(t) + pw_{n-1}(t) - \lambda_{n-1}p \displaystyle\int_{t-\tau}^{t-\sigma} w_{n-1}(s)\,ds, & n = 1, 2, \ldots \end{cases}$$

then $w_n \in V$ and $\lambda_n \in \Lambda(w_n)$. As $\Lambda(w_n)$ is bounded from above, this will be a

contradiction and will complete the proof in this case. To this end, set
$\phi_n(t) = e^{-\lambda_n t} w_n(t)$ and observe that

$$-\dot{w}_{n+1}(t) + (\lambda_n + \mu) w_{n+1}(t)$$
$$= q w_n(t - \sigma) + \lambda_n p[w_n(t - \sigma) - w_n(t - \tau)] +$$
$$+ (\lambda_n + \mu)\left[w_n(t) + p w_n(t - \tau) - \lambda_n p \int_{t-\tau}^{t-\sigma} w_n(s)\, ds\right]$$
$$\leqslant q w_n(t - \sigma) + \lambda_n p w_n(t - \sigma) + \lambda_n w_n(t)$$
$$- \lambda_n^2 p \int_{t-\tau}^{t-\sigma} w_n(s)\, ds + \mu\left[w_n(t) - \lambda_n p \int_{t-\tau}^{t-\sigma} w_n(s)\, ds\right]$$
$$\leqslant \phi_n(t - \sigma)\, e^{\lambda_n t}[q\, e^{-\lambda_n \sigma} + \lambda_n + \lambda_n p\, e^{-\lambda_n \tau} + \mu(1 - p\, e^{-\lambda^* \sigma})]$$
$$\leqslant \phi_n(t - \sigma)\, e^{\lambda_n t}[-m + m] = 0.$$

The proof of Theorem 6.2.1 is complete.

Remark 6.2.1. *In several instances, in the proofs of Theorem 6.2.1, we found points*

$$\lambda_1 \in \bigcap_{n=1}^{\infty} \Lambda_n \qquad \text{and} \qquad \lambda_2 \notin \bigcup_{n=1}^{\infty} \Lambda_n.$$

The values of λ_1 and λ_2 were expressed in terms of the coefficients, delays and advances of eqn (6.2.1). Clearly, when they are such that $\lambda_1 \geqslant \lambda_2$, this is a contradiction. Utilizing this idea we can obtain 'easily verifiable' sufficient conditions for the oscillation of all bounded and all unbounded solutions of eqn (6.2.1).

6.3 Necessary and sufficient conditions for the oscillation of systems of neutral equations

Consider the linear autonomous system of neutral delay differential equations

$$\frac{d}{dt}\left[x(t) + \sum_{j=1}^{l} P_j x(t - \tau_j)\right] + \sum_{i=1}^{n} Q_i x(t - \sigma_i) = 0 \qquad (6.3.1)$$

where the coefficients P_j and Q_i are real $m \times m$ matrices and the delays τ_j and σ_i are non-negative real numbers. With eqn (6.3.1) one associates its characteristic equation

$$\det\left(\lambda I + \lambda \sum_{j=1}^{l} P_j e^{-\lambda \tau_j} + \sum_{i=1}^{n} Q_i e^{-\lambda \sigma_i}\right) = 0 \qquad (6.3.2)$$

where I is the $m \times m$ identity matrix.

A slight modification in the proof of Theorem 5.1.1 shows that the following result is also true.

Theorem 6.3.1. *Assume that for $j = 1, 2, \ldots, l$ and $i = 1, 2, \ldots, n$,*

$$P_j, Q_i \in \mathbb{R}^{m \times m}, \qquad \tau_j \in (0, \infty) \qquad and \qquad \sigma_i \in [0, \infty).$$

Then the following statements are equivalent.

(a) *Every solution of (6.3.1) oscillates componentwise.*

(b) *The characteristic equation (6.3.2) has no real roots.*

If we modify the functions $F(s)$ and $\Phi(s)$ which were defined in (5.1.5) and (5.1.6) by setting

$$F(s) = sI + s \sum_{j=1}^{l} P_j e^{-s\tau_j} + \sum_{i=1}^{n} Q_i e^{-s\sigma_i} \tag{6.3.3}$$

$$\Phi(s) = x(0) + \sum_{j=1}^{l} P_j x(-\tau_j) - s \sum_{j=1}^{l} P_j e^{-s\sigma_j} \int_{-\tau_j}^{0} e^{-st} x(t) \, dt$$

$$- \sum_{i=1}^{n} Q_i e^{-s\sigma_i} \int_{-\sigma_i}^{0} e^{-st} x(t) \, dt \tag{6.3.4}$$

then the proof of Theorem 6.3.1 is the same as that of Theorem 5.1.1, so we omit it.

One should notice that the proofs, by Laplace transform, of Theorems 2.1.1, 5.2.1 and 6.1.1 make use of the fact that the non-oscillatory solutions are exponentially bounded. When the τ_j's and σ_i's are not all delays, it is not known whether the (non-oscillatory) solutions have this property. If we could prove that the non-oscillatory solutions of a neutral differential equation with mixed arguments are exponentially bounded, then the conclusion of Theorem 6.3.1. would also be true for such equations. We strongly believe that this result is true and so we state it as an open problem in Section 6.12. Although we cannot prove this open problem in its generality, we can establish it in the following special case.

Theorem 6.3.2. *Assume that for $j = 1, 2, \ldots, l$ and for $i = 1, 2, \ldots, n$,*

$$P_j, Q_i \in \mathbb{R}^{m \times m} \qquad and \qquad \tau_j, \sigma_i \in \mathbb{R}. \tag{6.3.5}$$

Let μ_0 be a real number. Then the following statements are equivalent.

(a) *Every solution $x(t)$ of eqn (6.3.1) with Lyapunov exponent*

$$\mu[x] \equiv \limsup_{t \to \infty} \frac{\ln \|x(t)\|}{t} \leq \mu_0 \tag{6.3.6}$$

oscillates.

(b) *The characteristic equation (6.3.2) has no roots in $(-\infty, \mu_0]$.*

Proof. The proof that (a) implies (b) is simple. In fact, if $\lambda_0 \in (-\infty, \mu_0]$ is a root of (6.3.2) then there exists a non-zero vector ξ such that $x(t) = e^{\lambda_0 t}\xi$ is a solution of (6.3.1) with at least one non-oscillatory component. Clearly,

$$\mu[x] = \lim_{t \to \infty} \sup \frac{\ln\|e^{\lambda_0 t}\xi\|}{t} \leqslant \lambda_0 \leqslant \mu_0,$$

which is a contradiction because $x(t)$ is non-oscillatory. So if (a) holds, (6.3.2) cannot have a root in $(-\infty, \mu_0]$.

The proof that (b) implies (a) is a simple modification of the proof of Theorem 5.1.1 with the functions F and Φ as given by (6.3.3) and (6.3.4) respectively. We should also observe that $\det F(s) \neq 0$ for all $s \leqslant \mu_0$, so (5.1.10) is valid for all $s \leqslant \mu_0$. The details are omitted here.

The following corollary is a consequence of Theorem 6.3.2 and the observation that bounded solutions have Lyapunov exponent equal to zero.

Corollary 6.3.1. *Assume that (6.3.5) holds. Then every bounded solution of eqn (6.3.1) oscillates if and only if the characteristic equation (6.3.2) has no roots in $(-\infty, 0]$.*

Remark 6.3.1. *It follows as in Theorem 5.1.2 that if all solutions of eqn (6.3.1) oscillate with respect to definition 5.0.1. or 5.0.2 then they also oscillate with respect to the other.*

6.4 Oscillations of scalar neutral equations with variable coefficients

In this section we obtain sufficient conditions for the oscillation of all solutions of the neutral delay differential equation with variable coefficients

$$\frac{d}{dt}[y(t) + P(t)y(t - \tau)] + Q(t)y(t - \sigma) = 0, \qquad t \geqslant t_0 \qquad (6.4.1)$$

where

$$P \in C[[t_0, \infty), \mathbb{R}], \qquad Q \in C[[t_0, \infty), \mathbb{R}^+] \qquad \text{and} \qquad \tau, \sigma \in \mathbb{R}^+. \qquad (6.4.2)$$

The main results in this section are the following theorems.

Theorem 6.4.1. *Assume that (6.4.2) holds with $P(t) \equiv -1$ and suppose that*

$$\int_{t_0}^{\infty} Q(s) \, ds = \infty.$$

$$(6.4.3)$$

Then every solution of the neutral equation

$$\frac{d}{dt} [y(t) - y(t - \tau)] + Q(t) y(t - \sigma) = 0, \qquad t \geqslant t_0 \qquad (6.4.4)$$

is oscillatory.

Theorem 6.4.2. *Assume that (6.4.2) and (6.4.3) hold, $P(t) \leqslant -1$, and*

$$\liminf_{t \to \infty} \int_{t+\sigma}^{t+\tau} \left[\frac{Q(s - \tau)}{-P(s - \sigma)} \right] ds > 1/e. \qquad (6.4.5)$$

Then every solution of eqn (6.4.1) is oscillatory.

Theorem 6.4.3. *Assume that (6.4.2) and (6.4.3) hold, $-1 \leqslant P(t) \leqslant 0$, and*

$$\liminf_{t \to \infty} \int_{t-\sigma}^{t} Q(s) \, ds > 1/e. \qquad (6.4.6)$$

Then every solution of eqn (6.4.1) is oscillatory.

Theorem 6.4.4. *Assume that (6.4.2) holds with $P(t) \equiv p$, $p \neq \pm 1$, $Q(t)$ is τ-periodic and*

$$\frac{1}{1 + p} \left[\liminf_{t \to \infty} \int_{t-\sigma}^{t-\tau} Q(s) \, ds \right] > 1/e. \qquad (6.4.7)$$

Then every solution of the neutral equation

$$\frac{d}{dt} [y(t) + p y(t - \tau)] + Q(t) y(t - \sigma) = 0 \qquad (6.4.8)$$

is oscillatory.

Before we present the proofs of the theorems, we establish the following two lemmas which are interesting in their own right.

Lemma 6.4.1. *Assume that (6.4.2) and (6.4.3) hold. Let $y(t)$ be an eventually positive solutions of eqn (6.4.1) and set*

$$z(t) = y(t) + P(t) y(t - \tau).$$

Then the following statements are true:

(a) $z(t)$ is an eventually decreasing function;

(b) *if* $P(t) \leqslant -1$ *then* $z(t) < 0$;   (6.4.9)

(c) *if* $-1 \leqslant P(t) \leqslant 0$ *then* $z(t) > 0$ *and* $\lim_{t \to \infty} z(t) = 0$.   (6.4.10)

Proof.

(a) We have

$$\dot{z}(t) = -Q(t)y(t - \sigma) \leqslant 0 \qquad (6.4.11)$$

and so $z(t)$ is an eventually decreasing function.

(b) Otherwise, $z(t) > 0$ and so

$$y(t) \geqslant -P(t)y(t - \tau) \geqslant y(\tau - \tau)$$

which implies that $y(t)$ is bounded from below by a positive constant, say m. From (6.4.11) we see that

$$\dot{z}(t) \leqslant -mQ(t)$$

which in view of (6.4.3) implies that $\lim_{t \to \infty} z(t) = -\infty$. This is a contradiction and so the proof of (6.4.9) is complete.

(c) First we claim that $z(t) > 0$. Otherwise, $z(t) < 0$, which implies that $y(t) \leqslant -P(t)y(t - \tau) \leqslant y(t - \tau)$ and so $y(t)$ is a bounded function. Hence $z(t)$ is also bounded, and since $z(t)$ is decreasing, $z(t)$ tends to a finite limit l, that is,

$$\lim_{t \to \infty} z(t) = l \in \mathbb{R}. \qquad (6.4.12)$$

But then because of (6.4.3),

$$\liminf_{t \to \infty} y(t) = 0. \qquad (6.4.13)$$

Otherwise, by integrating (6.4.11) from t_1 to ∞, with t_1 sufficiently large, we are led to the contradiction that $\lim_{t \to \infty} z(t) = -\infty$.

In view of (6.4.12) and (6.4.13), by applying Lemma 1.5.2, we find that $l = 0$. This implies that $z(t) > 0$, which contradicts our hypothesis that $z(t) < 0$. Therefore we have established that $z(t) > 0$. But then (6.4.12) and (6.4.13) are also true and by Lemma 1.5.1, $l = 0$. The proof is complete.

Lemma 6.4.2. *Assume that (6.4.2) holds with $P(t) \equiv p \neq 1$ and suppose that (6.4.3) is satisfied. Let $y(t)$ be an eventually positive solution of eqn (6.4.8) and set*

$$z(t) = y(t) + py(t - \tau). \qquad (6.4.14)$$

Then the following statements hold.

(a) $z(t)$ is a decreasing function and either

$$\lim_{t \to \infty} z(t) = -\infty \qquad (6.4.15)$$

or

$$\lim_{t \to \infty} z(t) = 0. \qquad (6.4.16)$$

(b) *The following statements are equivalent:*

    (i) (6.4.15) *holds*;
    (ii) $p < -1$;
    (iii) $\lim_{t \to \infty} y(t) = \infty.$ (6.4.17)

(c) *The following statements are equivalent:*

    (i) (6.4.16) *holds*;
    (ii) $p > -1$;
    (iii) $\lim_{t \to \infty} y(t) = 0.$ (6.4.18)

Proof. From eqn (6.4.1) we obtain

$$\dot{z}(t) = -Q(t)y(t - \sigma) \qquad (6.4.19)$$

and so eventually $\dot{z}(t) \leqslant 0$. Hence, either (6.4.15) holds or

$$\lim_{t \to \infty} z(t) \equiv l \in \mathbb{R}. \qquad (6.4.20)$$

If (6.4.20) holds, then by integrating (6.4.19) from t_1 to ∞, with t_1 sufficiently large, we find

$$l - z(t_1) = -\int_{t_1}^{\infty} Q(s)y(s - \sigma)\, ds. \qquad (6.4.21)$$

In view of (6.4.3) this implies that $\lim\inf_{t \to \infty} y(t) = 0$ and so by Lemma 1.5.1, $l = (1 + p) \cdot 0 = 0$. The proof of (a) is complete.

    Now we turn to the proofs of (b) and (c). First assume that (6.4.15) holds. Then p must be negative and $y(t)$ unbounded. Therefore there exists a $t^* \geqslant t_0$ such that $z(t^*) < 0$ and $y(t^*) = \max_{s \leqslant t^*} y(s) > 0$. Then

$$0 > z(t^*) = y(t^*) + p y(t^* - \tau) \geqslant y(t^*)(1 + p)$$

which implies that $p < -1$. Also

$$z(t) = y(t) + p y(t - \tau) > p y(t - \tau)$$

and (6.4.15) implies that (6.4.17) holds. Now assume that (6.4.16) holds. If $p \geqslant 0$, then from (6.4.14) it follows that (6.4.18) holds. Next, assume that $p \in (-1, 0)$. Then by Lemma 1.5.3, (6.4.18) holds. Finally if $p \leqslant -1$, then

$$y(t) > -p y(t - \tau) \geqslant y(t - \tau)$$

which shows that $y(t)$ is bounded from below by a positive constant, say m. Then (6.4.21) yields

$$l - z(t_1) + \frac{m}{2} \int_{t_1}^{\infty} Q(s)\, ds \leqslant 0,$$

which is a contradiction. Therefore if (6.4.16) holds, $p > -1$. On the basis of the above discussion, the proofs of (b) and (c) follow immediately.

Proof of Theorem 6.4.1. This is an immediate consequence of either Lemma 6.4.1 or Lemma 6.4.2.

Proof of Theorem 6.4.2. Assume, for the sake of contradiction, that eqn (6.4.1) has an eventually positive solution $y(t)$. Set $z(t) = y(t) + P(t)y(t - \tau)$. Then by Lemma 6.2.1,

$$z(t) < 0. \tag{6.4.22}$$

Observe that

$$z(t) > P(t)y(t - \tau)$$

and so

$$\frac{1}{P(t + \tau - \sigma)} Q(t)z(t + \tau - \sigma) < Q(t)y(t - \sigma) = -\dot{z}(t)$$

or

$$\dot{z}(t) - \left[\frac{Q(t)}{-P(t + \tau - \sigma)} \right] z(t + (\tau - \sigma)) < 0. \tag{6.4.23}$$

In view of (6.4.5) and Theorem 2.3.4(2), the advanced differential inequality (6.4.23) cannot have an eventually negative solution. This contradicts (6.4.22) and the proof is complete.

Proof of Theorem 6.4.3. Otherwise, eqn (6.4.1) has an eventually positive solution $y(t)$. Set $z(t) = y(t) + P(t)y(t - \tau)$. Then by Lemma 6.4.1,

$$z(t) > 0. \tag{6.4.24}$$

As $z(t) > y(t)$, it follows from (6.4.19) that

$$\dot{z}(t) + Q(t)z(t - \sigma) \leqslant 0. \tag{6.4.25}$$

In view of (6.4.6) and Theorem 2.3.3, the delay inequality (6.4.25) cannot have an eventually positive solution. This contradicts (6.4.24) and the proof is complete.

Proof of Theorem 6.4.4. Otherwise, eqn (6.4.1) has an eventually positive solution $y(t)$. Set $z(t) = y(t) + py(t - \tau)$ and $w(t) = z(t) + pz(t - \tau)$. One can show by direct substitution that $z(t)$ and $w(t)$ are differentiable solutions of eqn (6.4.8). That is,

$$\dot{z}(t) + p\dot{z}(t - \tau) + Q(t)z(t - \sigma) = 0 \qquad (6.4.26)$$

$$\dot{w}(t) + p\dot{w}(t - \tau) + Q(t)w(t - \sigma) = 0. \qquad (6.4.27)$$

By Lemma 6.4.2, $z(t)$ is decreasing and either (6.4.15) or (6.4.16) holds. In either case we claim that

$$\dot{w}(t - \tau) \geqslant \dot{w}(t). \qquad (6.4.28)$$

Indeed,

$$\dot{w}(t) = -Q(t)z(t - \sigma) \leqslant -Q(t)z(t - \sigma - \tau)$$
$$= -Q(t - \tau)z(t - \sigma - \tau) = \dot{w}(t - \tau)$$

Furthermore, it follows from Lemma 6.4.2 that as long as $p \neq \pm 1$,

$$w(t) > 0. \qquad (6.4.29)$$

By using (6.4.28) in (6.4.27) we obtain

$$(1 + p)\dot{w}(t - \tau) + Q(t)w(t - \sigma) \leqslant 0$$

in view of the τ-periodicity of Q we find

$$\dot{w}(t) + \frac{1}{1 + p} Q(t)w(t - (\sigma - \tau)) \leqslant 0 \qquad \text{if } 1 + p > 0 \qquad (6.4.30)$$

$$\dot{w}(t) - \frac{1}{-(1 + p)} Q(t)w(t + (\tau - \sigma)) \geqslant 0 \qquad \text{if } 1 + p < 0. \qquad (6.4.31)$$

In view of (6.4.7) and Theorems 2.3.3 and 2.3.4, the delay differential inequalities (6.4.30) and (6.4.31) cannot have eventually positive solutions. This contradicts (6.4.29) and completes the proof of the theorem.

6.5 Differential inequalities and comparison theorems

Our aim in this section is to establish a comparison result for the positive solutions of the scalar neutral delay differential inequalities

$$\dot{x}(t) - \sum_{j=1}^{l} p_j(t)\dot{x}(t - \tau_j(t)) + \sum_{i=1}^{n} q_i(t)x(t - \sigma_i(t)) \leqslant 0, \qquad a.e. \text{ on } [t_0, T)$$

$$(6.5.1)$$

$$\dot{y}(t) - \sum_{j=1}^{l} \bar{p}_j(t)\dot{y}(t - \tau_j(t)) + \sum_{i=1}^{n} \bar{q}_i(t)y(t - \sigma_i(t)) \geqslant 0, \qquad a.e. \text{ on } [t_0, T)$$

$$(6.5.2)$$

where for $j = 1, 2, \ldots, l$ and $i = 1, 2, \ldots, n$,

$$p_j, \bar{p}_j, q_i, \bar{q}_i \in C[[t_0, T), \mathbb{R}^+], \qquad \tau_j, \sigma_i \in C[[t_0, T), (0, \infty)], \qquad (6.5.3)$$

$$\bar{p}_j(t) \leqslant p_j(t) \qquad \text{and} \qquad \bar{q}_i(t) \leqslant q_i(t) \qquad \text{for } t \in [t_0, T). \qquad (6.5.4)$$

Let $t_{-1} = \min\{T_{-1}, \tilde{T}_{-1}\}$, where

$$T_{-1} = \min_{1 \leqslant i \leqslant n} \left\{ \inf_{t_0 \leqslant t \leqslant T} \{t - \sigma_i(t)\} \right\} \quad \text{and} \quad \tilde{T}_{-1} = \min_{1 \leqslant j \leqslant l} \left\{ \inf_{t_0 \leqslant t \leqslant T} \{t - \tau_j(t)\} \right\}.$$

Then the initial interval associated with the above inequalities is $t_{-1} \leqslant t \leqslant t_0$.

The following comparison theorem, which extends Theorem 3.2.1 to neutral equations, is the main result in this section.

Theorem 6.5.1. *Assume that* (6.5.3) *and* (6.5.4) *hold. Let* x *and* $y \in C[[t_{-1}, T), \mathbb{R}]$ *be solutions of* (6.5.1) *and* (6.5.2) *respectively, which are absolutely continuous with locally bounded derivatives on* $[\tilde{T}_{-1}, T)$. *Furthermore, assume that the following hypotheses are satisfied:*

(i) $x(t) > 0$ *on* $[t_{-1}, T)$ *and* $\dot{x}(t) \leqslant 0$ *a.e. on* $[\tilde{T}_{-1}, T)$; $\qquad (6.5.5)$

(ii) $y(t) > 0$ *on* $[t_{-1}, t_0]$; $\qquad (6.5.6)$

(iii) $\dfrac{\dot{x}(t)}{x(t)} \leqslant \dfrac{\dot{y}(t)}{y(t)}$, *a.e. on* $[\tilde{T}_{-1}, t_0]$; $\qquad (6.5.7)$

(iv) $\dfrac{x(t)}{x(t_0)} \geqslant \dfrac{y(t)}{y(t_0)}$ *on* $[t_{-1}, t_0]$. $\qquad (6.5.8)$

Then

$$\frac{x(t)}{x(t_0)} \leqslant \frac{y(t)}{y(t_0)} \qquad \text{on } [t_0, T). \qquad (6.5.9)$$

Proof. The proof is accomplished in two steps. In the first step we assume that $y(t)$ is positive in the interval $[t_0, T_1)$, for some $T_1 \in (t_0, T)$, and we prove that

$$\frac{x(t)}{x(t_0)} \leqslant \frac{y(t)}{y(t_0)} \qquad \text{on } [t_0, T_1). \qquad (6.5.10)$$

In the second step we prove that $y(t) > 0$ on $[t_0, T)$, that is, we prove that $T_1 = T$.

Step 1. Assume that for some $T_1 \in (t_0, T)$, $y(t) > 0$ on $[t_0, T_1)$. Clearly such a T_1 exists because of (6.5.6) and the continuity of y. Set

$$\alpha(t) = -\frac{\dot{x}(t)}{x(t)} \qquad \text{and} \qquad \beta(t) = -\frac{\dot{y}(t)}{y(t)} \qquad \text{for } \tilde{T}_{-1} \leqslant t \leqslant T_1.$$

Then one can see that α and β are locally integrable functions. Let $\beta^+(t) = \max\{0, \beta(t)\}$ and define the function $\mu(t)$ by

$$\mu(t) = m(\{t_0 \leqslant s \leqslant t : \alpha(s) \leqslant \beta^+(s)\})$$

that is, $\mu(t)$ is the Lebesgue measure of the set $\{t_0 \leqslant s \leqslant t : \alpha(s) \leqslant \beta^+(s)\}$. We claim that $\mu(t) = 0$ for all $t \in [t_0, T_1)$, that is, $\alpha(t) \geqslant \beta^+(t)$ a.e. on $[t_0, T_1)$. Otherwise, the set

$$\bigcup = \{t \in [t_0, T_1) : \mu(t) > 0\}$$

is non-empty. Let $t_1 = \inf \bigcup$. Since $\mu(t)$ is a monotone non-decreasing function, it is clear that

$$\mu(t) = 0 \quad \text{for } t_0 \leqslant t \leqslant t_1, \quad \text{and} \quad \mu(t) > 0 \quad \text{for } t_1 < t < T.$$

As in Section 3.1, for $t_1 \leqslant t \leqslant T_1$,

$$\alpha(t) \geqslant \sum_{j=1}^{l} p_j(t)\alpha(t - \tau_j(t)) \exp\left(\int_{t-\tau_j(t)}^{t} \alpha(s)\, ds\right)$$

$$+ \sum_{i=1}^{n} q_i(t) \frac{x(h_i(t))}{x(t_0)} \exp\left(\int_{H_i(t)}^{t} \alpha(s)\, ds\right) \quad (6.5.11)$$

$$\beta(t) \leqslant \sum_{j=1}^{l} \bar{p}_j(t)\beta(t - \tau_j(t)) \exp\left(\int_{t-\tau_j(t)}^{t} \beta(s)\, ds\right)$$

$$+ \sum_{i=1}^{n} \bar{q}_i(t) \frac{y(h_i(t))}{y(t_1)} \exp\left(\int_{H_i(t)}^{t} \beta(s)\, ds\right) \quad (6.5.12)$$

where for $i = 1, 2, \ldots, n$,

$$h_i(t) = \min\{t_1, t - \sigma_i(t)\} \quad \text{and} \quad H_i(t) = \max\{t_1, t - \sigma_i(t)\}.$$

Then clearly, for almost every $t \in [t_1, T_1)$, $\beta^+(t)$ satisfied the inequality

$$\beta^+(t) \leqslant \sum_{j=1}^{l} \bar{p}_j(t)\beta^+(t - \tau_j(t)) \exp\left(\int_{t-\tau_j(t)}^{t} \beta^+(s)\, ds\right)$$

$$+ \sum_{i=1}^{n} \bar{q}_i(t) \frac{y(h_i(t)}{y(t_1)} \exp\left(\int_{H_i(t)}^{t} \beta^+(s)\, ds\right) \quad (6.5.12')$$

Since $\tau_j(t)$ and $\sigma_i(t)$ are positive on $[t_0, T)$ for $j = 1, \ldots, l$ and $i = 1, \ldots, n$, it follows that there exists a $\delta > 0$ such that for all $t \in [t_1, t_1 + \delta)$, $t - \tau_j(t) < t_1$ for $j = 1, \ldots, l$ and $H_i(t) = t_1$ for $i = 1, \ldots, n$. On the other hand, $\alpha(s) \geqslant \beta(s)$ a.e. on $[\tilde{T}_{-1}, t_1)$, and by (6.5.8) and the definition of $h_i(t)$ we have

$$\frac{x(h_i(t))}{x(t_1)} \geqslant \frac{y(h_i(t))}{y(t_1)} > 0 \quad \text{for } t_0 \leqslant t < T_1 \quad \text{and} \quad i = 1, \ldots, n.$$

Define the non-negative a.e. function $a(t)$ by

$$a(t) = \sum_{j=1}^{l} \bar{p}_j(t)\beta^+(t - \tau_j(t)) \exp\left(\int_{t-\tau_j(t)}^{t_1} \beta^+(s)\, ds\right) + \sum_{i=1}^{n} \bar{q}_i(t) \frac{y(h_i(t))}{y(t_1)}$$

for almost every $t \in [t_1, t_1 + \delta)$. Then from (6.5.11) it follows that for almost every $t \in [t_1, t_1 + \delta)$,

$$\alpha(t) \geq \sum_{j=1}^{l} p_j(t)\alpha(t - \tau_j(t)) \exp\left(\int_{t-\tau_j(t)}^{t_1} \alpha(s)\, ds\right) \exp\left(\int_{t_1}^{t} \alpha(s)\, ds\right)$$

$$+ \sum_{i=1}^{n} q_i(t) \frac{x(h_i(t))}{x(t_1)} \exp\left(\int_{t_1}^{t} \alpha(s)\, ds\right)$$

$$\geq \left[\sum_{j=1}^{l} \bar{p}_j(t)\beta^+(t - \tau_j(t)) \exp\left(\int_{t-\tau_j(t)}^{t_1} \beta^+(s)\, ds\right) + \sum_{i=1}^{n} \bar{q}_i(t) \frac{y(h_i(t))}{y(t_1)}\right]$$

$$\times \exp\left(\int_{t_1}^{t} \alpha(s)\, ds\right).$$

That is,

$$\alpha(t) \geq a(t) \exp\left(\int_{t_1}^{t} \alpha(s)\, ds\right) \qquad \text{a.e. on } [t_1, t_1 + \delta). \qquad (6.5.13)$$

In a similar manner, it follows from (6.5.12′) that

$$\beta^+(t) \leq a(t) \exp\left(\int_{t_1}^{t} \beta^+(s)\, ds\right) \qquad \text{a.e. on } [t_1, t_1 + \delta). \qquad (6.5.14)$$

By integrating (6.5.13) and (6.5.14) we find that for all $t \in [t_1, t_1 + \delta)$,

$$\int_{t_1}^{t} \alpha(u)\, du \geq \int_{t_1}^{t} a(u) \exp\left(\int_{t_1}^{u} \alpha(s)\, ds\right) du$$

and
$$\qquad\qquad (6.5.15)$$

$$\int_{t_1}^{t} \beta^+(u)\, du \leq \int_{t_1}^{t} a(u) \exp\left(\int_{t_1}^{u} \beta^+(s)\, ds\right) du.$$

Hence,

$$\int_{t_1}^{t} \alpha(u)\, du \geq \int_{t_1}^{t} \beta^+(u)\, du, \qquad t_1 \leq t < t_1 + \delta.$$

From this, (6.5.12) and (6.5.13), it follows that

$$\alpha(t) \geq \beta^+(t) \qquad \text{a.e. on } [t_1, t_1 + \delta).$$

Hence, for all $t \in (t_1, t_1 + \delta)$,

$$\mu(t) = m(\{t_0 \leq s \leq t : \alpha(s) \leq \beta^+(s)\})$$

$$= \mu(t_1) + m(\{t_1 \leq s \leq t : \alpha(s) \leq \beta^+(s)\}) = 0$$

which contradicts the definition of t_1. Therefore $\mu(t) = 0$ for all $t \in [t_1, T_1)$, so

$$\alpha(t) \geq \beta^+(t) \geq \beta(t) \qquad \text{a.e. on } [t_1, T_1).$$

This implies that

$$\ln \frac{x(t_0)}{x(t)} = \int_{t_0}^{t} \alpha(s)\, ds \geq \int_{t_0}^{t} \beta(s)\, ds = \ln \frac{y(t_0)}{y(t)},$$

which completes the first step toward the proof of (6.5.10).

Step 2. We claim that $y(t) > 0$ on $[t_0, T)$. Otherwise, there exists a $T_1 \in (t_0, T)$ such that

$$y(t) > 0 \qquad \text{for } t_0 \leqslant t < T_1 \text{ and } y(T_1) = 0.$$

But in this case, it follows from the first step that (6.5.10) holds, so

$$y(t) \geqslant \frac{y(t_0)}{x(t_0)} x(t), \qquad t_0 \leqslant t < T_1.$$

Hence, by the continuity of $x(t)$ and $y(t)$ we have

$$0 = y(T_1) = \lim_{t \to T_1^-} y(t) \geqslant \frac{y(t_0)}{x(t_0)} \left[\lim_{t \to T_1^-} x(t) \right] = \frac{y(t_0)}{x(t_0)} x(T_1) > 0,$$

which is a contradiction and the proof is complete.

The following corollary of Theorem 6.5.1 will be useful in the next section, where we establish a linearized oscillation theorem for neutral equations.

Corollary 6.5.1. *Assume that*

$$p_j, q_i, \tau_j, \sigma_i \in [0, \infty) \qquad \text{for } j = 1, 2, \ldots, l \text{ and } i = 1, 2, \ldots, n.$$

Set

$$\tau = \max\{\tau_1, \ldots, \tau_l\}, \qquad \sigma = \max\{\sigma_1, \ldots, \sigma_n\} \qquad \text{and} \qquad t_{-1} = -\max\{\tau, \sigma\}.$$

Let $x, y \in C[[t_{-1}, T), \mathbb{R}]$ *satisfy the following conditions:*

(i) $\dot{x}(t)$ *and* $\dot{y}(t)$ *are locally integrable and bounded functions on* $[-\sigma, T)$.

(ii) *For almost every* $t \in [t_0, T)$,

$$\dot{x}(t) - \sum_{j=1}^{l} p_j \dot{x}(t - \tau_j) + \sum_{i=1}^{n} q_i x(t - \sigma_i) \leqslant 0$$

$$\dot{y}(t) - \sum_{j=1}^{l} p_j \dot{y}(t - \tau_j) + \sum_{i=1}^{n} q_i y(t - \sigma_i) \geqslant 0.$$

(iii) $x(t) > 0$ *on* $[t_{-1}, T)$ *and* $\dot{x}(t) < 0$ *a.e. on* $[-\sigma, T)$.

(iv) $y(t) \equiv \delta > 0$ *on* $[t_{-1}, 0]$.

Then

$$y(t) \geqslant 0 \qquad \text{for } 0 \leqslant t < T.$$

By utilizing Theorem 6.5.1 and 6.3.1 we obtain the following extension of Theorem 3.2.2.

Theorem 6.5.2. *Let $p_j, \tau_j, q_i \in (0, \infty)$ and $\sigma_i \in [0, \infty)$ for $j = 1, 2, \ldots, l$ and $i = 1, 2, \ldots, n$, and assume that*

$$\sum_{i=1}^{l} p_i \leqslant 1. \tag{6.5.11}$$

Then the following statements are equivalent.

(a) *The neutral differential equation*

$$\frac{d}{dt}\left[x(t) - \sum_{j=1}^{l} p_j x(t - \tau_j) \right] + \sum_{i=1}^{n} q_i x(t - \sigma_i) = 0 \tag{6.5.17}$$

has an eventually positive solution.

(b) *The characteristic equation*

$$\lambda - \lambda \sum_{j=1}^{l} p_j e^{-\lambda\tau_i} + \sum_{i=1}^{n} q_i e^{-\lambda\sigma_i} = 0 \tag{6.5.18}$$

has a real root in $(-\infty, 0]$.

(c) *The inequality*

$$\dot{v}(t) - \sum_{j=1}^{l} p_j \dot{v}(t - \tau_j) + \sum_{i=1}^{n} q_i v(t - \sigma_i) \leqslant 0 \tag{6.5.19}$$

has a continuously differentiable solution which is eventually positive and decreasing.

(d) *There exists an $\varepsilon_0 \in (0, 1)$ such that for every $\varepsilon \in [0, \varepsilon_0]$ the inequality*

$$\dot{w}(t) - \sum_{j=1}^{l} (1 - \varepsilon)p_j \dot{w}(t - \tau_j) + \sum_{i=1}^{n} (1 - \varepsilon)q_i w(t - \sigma_i) \leqslant 0 \tag{6.5.20}$$

has a continuously differentiable solution which is eventually positive and decreasing.

Proof. The proof that (a) $\Leftrightarrow$ (b) is a simple consequence of Theorem 6.3.1. Now we prove that (a) $\Leftrightarrow$ (c). Let $x(t)$ be an eventually positive solution of eqn (6.5.17) and set

$$v(t) = x(t) - \sum_{j=1}^{l} p_j x(t - \tau_j).$$

Then it follows in a manner similar to that in the proof of Lemma 6.1.1(c) that $v(t)$ is a continuously differentiable solution of the equation

$$\dot{v}(t) - \sum_{j=1}^{l} p_j \dot{v}(t - \tau_j) + \sum_{i=1}^{n} v(t - \sigma_i) = 0$$

and that, because of (6.5.16), $\lim_{t \to \infty} v(t) = 0$. It follows that $v(t)$ is eventually positive and the proof that (a) $\Rightarrow$ (c) is complete. The proof that (c) $\Rightarrow$ (a) follows in a manner similar to that in Theorem 6.5.1 and Corollary 6.5.1. The proof that (d) $\Rightarrow$ (c) is obvious by taking $\varepsilon = 0$. Finally, one can show that (c) $\Rightarrow$ (d) by using an argument similar to that in the proof of Theorem 3.2.2. The proof is complete.

6.6 Linearized oscillations for neutral equations

Consider the non-linear neutral delay differential equation

$$\frac{d}{dt}[x(t) - p(t)g(x(t-\tau))] + q(t)h(x(t-\sigma)) = 0, \qquad t \geq t_0 \quad (6.6.1)$$

under the following hypotheses:

$$p, q \in C[[t_0, \infty), \mathbb{R}^+], \qquad g, h \in C[\mathbb{R}, \mathbb{R}], \qquad \tau \in (0, \infty), \qquad \sigma \in [0, \infty), \tag{6.6.2}$$

$$\lim_{t \to \infty} p(t) \equiv p_0 \in [0, 1), \qquad \lim_{t \to \infty} q(t) \equiv q_0 \in (0, \infty), \tag{6.6.3}$$

$$ug(u) > 0 \text{ for } u \neq 0, \qquad g(u) \leq u \text{ for } u \geq 0 \qquad \text{and} \qquad g(u) \geq u \text{ for } u \leq 0,$$

$$\lim_{u \to 0} \frac{g(u)}{u} = 1, \tag{6.6.4}$$

$$uh(u) > 0 \qquad \text{for } u \neq 0 \qquad \text{and} \qquad \lim_{u \to 0} \frac{h(u)}{u} = 1. \tag{6.6.5}$$

With eqn (6.6.1) we associate the linear equation

$$\frac{d}{dt}[y(t) - p_0 y(t-\tau)] + q_0 y(t-\sigma) = 0. \tag{6.6.6}$$

Our aim in this section is to establish conditions for the oscillation of all solutions of eqn (6.6.1) in terms of the oscillation of all solutions of eqn (6.6.6) and vice versa.

We recall that every solution of eqn (6.6.6) oscillates if and only if its characteristic equation

$$\lambda - \lambda p_0 e^{-\lambda \tau} + q_0 e^{-\lambda \sigma} = 0 \tag{6.6.7}$$

has no real roots. As $p_0 \in [0, 1)$, eqn (6.6.7) has no roots in $[0, \infty)$ and so every solution of eqn (6.6.6) oscillates if and only if eqn (6.6.7) has no negative roots.

Theorem 6.6.1. *Assume that* (6.6.2)–(6.6.5) *are satisfied and that eqn* (6.6.7) *has no real roots. Then every solution of eqn* (6.6.1) *oscillates.*

Proof. Assume, for the sake of contradiction, that eqn (6.6.1) has a non-oscillatory solution $x(t)$. We assume that $x(t)$ is eventually positive. The case where $x(t)$ is eventually negative is similar and is omitted. Set

$$z(t) = x(t) - p(t)g(x(t-\tau)).$$

Then eventually,

$$\dot{z}(t) = -q(t)h(x(t - \sigma)) \leqslant 0 \tag{6.6.8}$$

and so $z(t)$ is a decreasing function. We claim that $x(t)$ is bounded. Otherwise there exists a sequence of points $\{t_n\}$ such that

$$\lim_{n \to \infty} t_n = \infty, \qquad \lim_{t_n \to \infty} x(t_n) = \infty \qquad \text{and} \qquad x(t_n) = \max_{s \leqslant t_n} x(s).$$

Then in view of (6.6.3) and (6.6.4),

$$z(t_n) = x(t_n) - p(t_n)g(x(t_n - \tau)) \geqslant x(t_n) - p(t_n)x(t_n - \tau)$$

$$\geqslant x(t_n)[1 - p(t_n)] \to \infty \qquad \text{as } n \to \infty,$$

which contradicts the fact that $z(t)$ is decreasing. Thus $x(t)$ is bounded and so

$$l \equiv \lim_{t \to \infty} z(t) \in \mathbb{R}. \tag{6.6.9}$$

By integrating both sides of (6.6.8) from t_1 to ∞, with t_1 sufficiently large, we find

$$\int_{t_1}^{\infty} q(s)h(x(s - \sigma))\, \mathrm{d}s = z(t_1) - l. \tag{6.6.10}$$

Now in view of the conditions on q and h, (6.6.10) implies that

$$\liminf_{t \to \infty} x(t) = 0. \tag{6.6.11}$$

We write

$$z(t) = x(t) - P(t)x(t - \tau)$$

where

$$0 \leqslant P(t) = p(t)\frac{g(x(t - \tau))}{x(t - \tau)} \leqslant p(t).$$

Then, from (6.6.9), (6.6.11), and (6.6.3), and by applying Lemma 1.5.2 to $z(t)$, we obtain,

$$l = \lim_{t \to \infty} [x(t) - p(t)g(x(t - \tau))] = 0. \tag{6.6.12}$$

Let $\varepsilon \in (0, 1 - p_0)$ be given. Then for t sufficiently large, it follows from (6.6.12), (6.6.3), and (6.6.4) that

$$0 < x(t) < p(t)g(x(t - \tau)) + \varepsilon \leqslant (p_0 + \varepsilon)x(t - \tau) + \varepsilon.$$

From this and Lemma 1.5.3 we conclude that

$$\lim_{t \to \infty} x(t) = 0. \tag{6.6.13}$$

Set

$$P(t) = p(t) \frac{g(x(t - \tau))}{x(t - \tau)} \quad \text{and} \quad Q(t) = q(t) \frac{h(x(t - \sigma))}{x(t - \sigma)}.$$

Then in view of the hypotheses and (6.6.13),

$$\lim_{t \to \infty} P(t) = p_0 \in [0, 1), \qquad \lim_{t \to \infty} Q(t) = q_0 \in (0, \infty) \qquad (6.6.14)$$

and

$$\frac{d}{dt} [x(t) - P(t)x(t - \tau)] + Q(t)x(t - \sigma) = 0.$$

By integrating both sides from t to ∞, for t sufficiently large, and by using (6.6.12) we find,

$$x(t) - P(t)x(t - \tau) + \int_t^\infty Q(s)x(s - \sigma) \, ds = 0. \qquad (6.6.15)$$

Set

$$w(t) = \int_t^\infty Q(s)x(s - \sigma) \, ds. \qquad (6.6.16)$$

Then eventually $w(t) > 0$ and $\dot{w}(t) = -Q(t)x(t - \sigma) < 0$. Hence

$$x(t) = -\frac{\dot{w}(t + \sigma)}{Q(t + \sigma)} \qquad (6.6.17)$$

and by substituting (6.6.16) and (6.6.17) into (6.6.15) we obtain, for t sufficiently large,

$$\dot{w}(t) - P(t - \sigma) \frac{Q(t)}{Q(t - \tau)} \dot{w}(t - \tau) + Q(t)w(t - \sigma) = 0. \qquad (6.6.18)$$

Clearly

$$\lim_{t \to \infty} \left[P(t - \sigma) \frac{Q(t)}{Q(t - \tau)} \right] = p_0.$$

First, assume that $p_0 > 0$. Then for any $\varepsilon \in (0, 1)$, (6.6.18) yields

$$\dot{w}(t) - (1 - \varepsilon)p_0\dot{w}(t - \tau) + (1 - \varepsilon)q_0w(t - \sigma) \leq 0. \qquad (6.6.19)$$

By applying Theorem 6.5.2, it follows that eqn (6.6.7) has a real root. This is a contradiction and the proof is complete when $p_0 > 0$. Next, assume that $p_0 = 0$. Then the hypothesis that (6.6.7) has no real roots is, by Theorem

2.2.3, equivalent to

$$q_0 \sigma \, e > 1.$$

Choose $\varepsilon \in (0, 1)$ such that

$$(1 - \varepsilon) q_0 \sigma \, e > 1. \tag{6.6.20}$$

As $\dot{w}(t) < 0$, (6.6.19) yields

$$\dot{w}(t) + (1 - \varepsilon) q_0 w(t - \sigma) \leqslant 0. \tag{6.6.21}$$

But in view of (6.6.20) and Theorem 2.3.3 it is impossible for (6.6.21) to have an eventually positive solution. This contradicts the fact that eventually $w(t) > 0$ and the proof is complete.

The following result is a partial converse to Theorem 6.6.1.

Theorem 6.6.2. *Consider the neutral differential equation*

$$\frac{d}{dt} [x(t) - p_0 x(t - \tau)] + q(t) h(x(t - \sigma)) = 0, \qquad t \geqslant t_0 \tag{6.6.22}$$

where

and
$$p_0 \in (0, 1), \quad \tau \in (0, \infty), \quad \sigma \in [0, \infty), \quad q \in C[[t_0, \infty), \mathbb{R}^+] \tag{6.6.23}$$
$$h \in C[\mathbb{R}, \mathbb{R}].$$

Assume that there exist positive numbers q_0 and δ such that

$$0 \leqslant q(t) \leqslant q_0 \qquad \text{for } t \geqslant t_0 \tag{6.6.24}$$

and

$$\left. \begin{array}{llll} either & 0 \leqslant h(u) \leqslant u & for \ 0 \leqslant u \leqslant \delta \\ or & 0 \geqslant h(u) \geqslant u & for \ -\delta \leqslant u \leqslant 0 \end{array} \right\}. \tag{6.6.25}$$

Suppose that eqn (6.6.7) has a real root. Then eqn (6.6.22) has a positive solution on $[t_0 - r, \infty)$ where $r = \max\{\tau, \sigma\}$.

Proof. Assume that (6.6.25) holds with $0 \leqslant h(u) \leqslant u$ for $0 \leqslant u \leqslant \delta$. The case that $0 \geqslant h(u) \geqslant u$ for $-\delta \leqslant u \leqslant 0$ is similar and is omitted. Choose $d \in (0, \delta)$ and consider the solution $x(t)$ of eqn (6.6.22) with initial condition

$$x(t) = d \qquad \text{for } t_0 - r \leqslant t \leqslant t_0.$$

Then by Theorem 1.1.4, $x(t)$ is absolutely continuous on $[t_0 - r, \infty)$ and satisfies the equation

$$\dot{x}(t) - p_0 \dot{x}(t - \tau) + q(t) h(x(t - \sigma)) = 0 \qquad \text{a.e. on } [t_0, \infty). \tag{6.6.26}$$

We prove that $x(t) > 0$ for $t \geqslant t_0 - r$. First, we claim that for as long as $x(t) > 0$, it remains strictly less than δ. Otherwise, there exists a T_1 such that

$$0 < x(t) < \delta \qquad \text{for } t_0 - r \leqslant t < T_1 \text{ and } x(T_1) = \delta.$$

Set

$$z(t) = x(t) - p_0 x(t - \tau).$$

Then

$$\dot{z}(t) = -q(t) h(x(t - \sigma)) \leqslant 0, \qquad t_0 \leqslant t \leqslant T_1$$

and so $z(T_1) \leqslant z(t_0)$. Hence

$$\delta = x(T_1) \leqslant p_0 x(T_1 - \tau) + x(t_0) - p_0 x(t_0 - \tau) < p_0 \delta + \delta - p_0 \delta = \delta$$

which is a contradiction. Now assume, for the sake of contradiction, that there exists a $T > t_0$ such that

$$0 < x(t) < \delta \qquad \text{for } t_0 - r \leqslant t < T \text{ and } x(T) = 0. \qquad (6.6.27)$$

From (6.6.26), (6.6.24), (6.6.25) and (6.6.27) we have,

$$\dot{x}(t) - p_0 \dot{x}(t - \tau) + q_0 x(t - \sigma) \geqslant 0 \qquad \text{a.e. on } [t_0, \infty).$$

By our assumption, eqn (6.6.7) has a real root, say λ_0. As $p_0 \in (0, 1)$ and $q_0 > 0$, it is easily seen that $\lambda_0 < 0$. Therefore $y(t) = e^{\lambda_0 t}$ is a positive, continuously differentiable and decreasing solution of eqn (6.6.6). By Corollary 6.5.1 it follows that $x(t) > 0$ for all $t \geqslant t_0$ and the proof is complete.

By combining Theorems 6.6.1 and 6.6.2 we obtain the following linearized oscillation result for neutral differential equations.

Corollary 6.6.1. *Assume that the conditions* (6.6.5), (6.6.23), (6.6.24), *and* (6.6.25) *are satisfied and suppose that*

$$\lim_{t \to \infty} q(t) \equiv q_0 \in (0, \infty).$$

Then every solution of eqn (6.6.22) *oscillates if and only if every solution of eqn* (6.6.6) *oscillates if and only if eqn* (6.6.7) *has no negative real roots.*

6.7 Existence of positive solutions

Consider the neutral delay differential equation

$$\frac{d}{dt} [y(t) + P(t) y(t - \tau)] + Q(t) y(t - \sigma) = 0 \qquad (6.7.1)$$

where

$$P \in C^1[[t_0, \infty), \mathbb{R}], \quad Q \in C[[t_0, \infty), \mathbb{R}], \quad \tau \in (0, \infty) \quad \text{and} \quad \sigma \in [0, \infty).$$
$$\text{(6.7.2)}$$

In this section we establish several sufficient conditions for eqn (6.7.1) to have a positive solution on $[t_1, \infty)$ for every $t_1 \geq t_0$. Conditions for the existence of a positive solution are relatively scarce in the literature. The main tool in our proofs is the Banach contraction principle.

Our motivation for the first theorem stems from the observation that if we look for a positive solution of eqn (6.7.1) of the form

$$y(t) = \exp\left(-\int_{t_1}^t \lambda(s) \, ds \right), \qquad t \geq t_1$$

then $\lambda(t)$ satisfies the equation

$$\lambda(t) = [-P(t)\lambda(t - \tau) + \dot{P}(t)] \exp\left(\int_{t-\tau}^t \lambda(s) \, ds \right) + Q(t) \exp\left(\int_{t-\sigma}^t \lambda(s) \, ds \right).$$
$$\text{(6.7.3)}$$

Theorem 6.7.1. *Assume that (6.7.2) holds and that there exists a positive number μ such that*

$$|P(t)| \mu \, e^{\mu\tau} + |\dot{P}(t)| \, e^{\mu\tau} + |Q(t)| \, e^{\mu\sigma} \leq \mu \qquad \text{for } t \geq t_0. \qquad \text{(6.7.4)}$$

Then for every $t_1 \geq t_0$, eqn (6.7.1) has a positive solution on $[t_1, \infty)$.

Proof. Let $m = \max\{\tau, \sigma\}$. Define the set of functions

$$\Lambda = \{\lambda \colon \lambda(t) \equiv 0 \text{ for } t_1 - m \leq t < t_1, \ \lambda(t) \text{ is continuous on } (t_1 + n\tau,$$
$$t_1 + (n + 1)\tau) \text{ for } n = 0, 1, 2, \ldots \text{ and } |\lambda(t)| \leq \mu \text{ for } t > t_1\}.$$

For $\lambda_1, \lambda_2 \in \Lambda$ define

$$d(\lambda_1, \lambda_2) = \sup_{t \geq t_1 - m} |\lambda_1(t) - \lambda_2(t)| \, e^{-\eta t}$$

where η is chosen so large that

$$\frac{\mu}{\eta} + |P(t)| \, e^{(\mu - \eta)\tau} < \tfrac{1}{2} \qquad \text{for } t \geq t_1. \qquad \text{(6.7.5)}$$

Then (Λ, d) is a complete metric space. Define

$$(T\lambda)(t) = \begin{cases} [-P(t)\lambda(t - \tau) + \dot{P}(t)] \exp\left(\int_{t-\tau}^t \lambda(s) \, ds \right) \\ \qquad\qquad + Q(t) \exp\left(\int_{t-\sigma}^t \lambda(s) \, ds \right), \qquad t > t_1 \\ (T\lambda)(t_1), \qquad\qquad\qquad\qquad\qquad\qquad t_1 - m \leq t < t_1. \end{cases}$$

Then $T\lambda$ is continuous on $(t_1 + n\tau, t_1 + (n+1)\tau)$ for $n = 0, 1, 2, \ldots$ and for $t > t_1$,

$$|(T\lambda)(t)| \leqslant |P(t)|\mu\, e^{\mu\tau} + |\dot{P}(t)|\, e^{\mu\tau} + |Q(t)|\, e^{\mu\sigma} \leqslant \mu,$$

which shows that T maps Λ into Λ.

Next, we claim that T is a contraction mapping. Before we establish this property of T we need to observe that, for $\lambda_1, \lambda_2 \in \Lambda$ and for $t \geqslant t_1$, by applying the mean value theorem we find

$$\left|\exp\left(\int_{t-\tau}^{t} \lambda_1(s)\, ds\right) - \exp\left(\int_{t-\tau}^{t} \lambda_2(s)\, ds\right)\right| \leqslant e^{\mu\tau} \int_{t-\tau}^{t} |\lambda_1(s) - \lambda_2(s)|\, ds.$$

Also observe that

$$\left|\lambda_1(t-\tau)\exp\left(\int_{t-\tau}^{t} \lambda_1(s)\, ds\right) - \lambda_2(t-\tau)\exp\left(\int_{t-\tau}^{t} \lambda_2(s)\, ds\right)\right|$$

$$= \left|\lambda_1(t-\tau)\left[\exp\left(\int_{t-\tau}^{t} \lambda_1(s)\, ds\right) - \exp\left(\int_{t-\tau}^{t} \lambda_2(s)\, ds\right)\right.\right.$$

$$\left.\left. + \exp\left(\int_{t-\tau}^{t} \lambda_2(s)\, ds\right)[\lambda_1(t-\tau) - \lambda_2(t-\tau)]\right|\right.$$

$$\leqslant \mu\, e^{\mu\tau} \int_{t-\tau}^{t} |\lambda_1(s) - \lambda_2(s)|\, ds + e^{\mu\tau}|\lambda_1(t-\tau) - \lambda_2(t-\tau)|.$$

Hence if $\lambda_1, \lambda_2 \in \Lambda$ and $t \geqslant t_1$,

$|(T\lambda_1)(t) - (T\lambda_2)(t)|$

$$\leqslant |P(t)|\left|\lambda_1(t-\tau)\exp\left(\int_{t-\tau}^{t} \lambda_1(s)\, ds\right) - \lambda_2(t-\tau)\exp\left(\int_{t-\tau}^{t} \lambda_2(s)\, ds\right)\right|$$

$$+ |\dot{P}(t)|\left|\exp\left(\int_{t-\tau}^{t} \lambda_1(s)\, ds\right) - \exp\left(\int_{t-\tau}^{t} \lambda_2(s)\, ds\right)\right|$$

$$+ |Q(t)|\left|\exp\left(\int_{t-\sigma}^{t} \lambda_1(s)\, ds\right) - \exp\left(\int_{t-\sigma}^{t} \lambda_2(s)\, ds\right)\right|$$

$$\leqslant |P(t)|\mu\, e^{\mu\tau} \int_{t-\tau}^{t} |\lambda_1(s) - \lambda_2(s)|\, ds + |P(t)|\, e^{\mu\tau}|\lambda_1(t-\tau) - \lambda_2(t-\tau)|$$

$$+ |\dot{P}(t)|\, e^{\mu\tau} \int_{t-\tau}^{t} |\lambda_1(s) - \lambda_2(s)|\, ds + |Q(t)|\, e^{\mu\sigma} \int_{t-\sigma}^{t} |\lambda_1(s) - \lambda_2(s)|\, ds$$

$$= |P(t)|\mu\, e^{\mu\tau} \int_{t-\tau}^{t} (|\lambda_1(s) - \lambda_2(s)|\, e^{-\eta s})\, e^{\eta s}\, ds$$

$$+ |P(t)| \, e^{\mu\tau}(|\lambda_1(t-\tau) - \lambda_2(t-\tau)| \, e^{-\eta(t-\tau)}) \, e^{\eta(t-\tau)}$$

$$+ |\dot{P}(t)| \, e^{\mu\tau} \int_{t-\tau}^{t} (|\lambda_1(s) - \lambda_2(s)| \, e^{-\eta s}) \, e^{\eta s} \, ds$$

$$+ |Q(t)| \, e^{\mu\tau} \int_{t-\sigma}^{t} (|\lambda_1(s) - \lambda_2(s)| \, e^{-\eta s}) \, e^{\eta s} \, ds$$

$$\leqslant |P(t)| \mu \, e^{\mu\tau} d(\lambda_1, \lambda_2) \frac{1}{\eta} (e^{\eta t} - e^{\eta(t-\tau)}) + |P(t)| \, e^{\mu\tau} d(\lambda_1, \lambda_2) \, e^{\eta(t-\tau)}$$

$$+ |\dot{P}(t)| \, e^{\mu\tau} d(\lambda_1, \lambda_2) \frac{1}{\eta} (e^{\eta t} - e^{\eta(t-\tau)})$$

$$+ |Q(t)| \, e^{\mu\sigma} d(\lambda_1, \lambda_2) \frac{1}{\eta} (e^{\eta t} - e^{\eta(t-\sigma)})$$

$$\leqslant \frac{1}{\eta} d(\lambda_1, \lambda_2) \, e^{\eta t} [|P(t)| \mu \, e^{\mu\tau} + \eta |P(t)| \, e^{(\mu-\eta)\tau} + |\dot{P}(t)| \, e^{\mu\tau} + |Q(t)| \, e^{\eta\sigma}].$$

From this inequality and by using (6.7.4) and (6.7.5) we see that for $t \geqslant t_1$,

$$|(T\lambda_1)(t) - (T\lambda_2)(t)| \, e^{-\eta t} \leqslant \frac{1}{\eta} d(\lambda_1, \lambda_2)[\mu + \eta |P(t)| \, e^{\mu(\tau-\eta)}]$$

$$= d(\lambda_1, \lambda_2)\left[\frac{\mu}{\eta} + |P(t)| \, e^{(\mu-\eta)\tau}\right] < \tfrac{1}{2} d(\lambda_1, \lambda_2).$$

It follows that

$$d(T\lambda_1, T\lambda_2) \leqslant \tfrac{1}{2} d(\lambda_1, \lambda_2).$$

Therefore there exists a (unique) solution $\lambda \in \Lambda$ of $T\lambda = \lambda$. To complete the proof it suffices to show that the positive function

$$y(t) = \exp\left(-\int_{t_1}^{t} \lambda(s) \, ds\right), \qquad t \geqslant t_1 - \tau$$

is a solution of eqn (6.7.1) on $[t_1, \infty)$. To this end, observe that for $t \geqslant t_1$,

$$\dot{y}(t) = -\lambda(t) y(t).$$

On the other hand, for $t \geqslant t_1$,

$$y(t - \tau) = \exp\left(-\int_{t_1}^{t-\tau} \lambda(s) \, ds\right) = \exp\left(-\int_{t_1}^{t} \lambda(s) \, ds\right) \exp\left(\int_{t-\tau}^{t} \lambda(s) \, ds\right)$$

$$= y(t) \exp\left(\int_{t-\tau}^{t} \lambda(s) \, ds\right).$$

Similarly,

$$y(t - \sigma) = y(t) \exp\left(\int_{t-\sigma}^{t} \lambda(s)\, ds\right).$$

Hence for $t \geq t_1$,

$$\frac{d}{dt}[y(t) + P(t)y(t - \tau)] + Q(t)y(t - \sigma)$$

$$= -\lambda(t)y(t) - P(t)\lambda(t - \tau)y(t - \tau) + \dot{P}(t)y(t) \exp\left(\int_{t-\tau}^{t} \lambda(s)\, ds\right)$$

$$+ Q(t)y(t) \exp\left(\int_{t-\sigma}^{t} \lambda(s)\, ds\right) = y(t)[-\lambda(t) + (T\lambda)(t)] = 0$$

and the proof is complete.

The next lemma will be used to obtain several sufficient conditions for the existence of a positive solution of eqn (6.7.1). As in the case of Theorem 6.7.1, the motivation behind this result stems from the observation that if we look for a positive solution of eqn (6.7.1) in the form

$$y(t) = \exp\left(\int_{t_1}^{t} \lambda(s)\, ds\right), \qquad t \geq t_1,$$

then $\lambda(t)$ satisfies the equation

$$\lambda(t) = [-P(t)\lambda(t - \tau) - \dot{P}(t)] \exp\left(-\int_{t-\tau}^{t} \lambda(s)\, ds\right)$$

$$- Q(t) \exp\left(-\int_{t-\sigma}^{t} \lambda(s)\, ds\right).$$

Lemma 6.7.1. *Assume that (6.7.2) holds and that the functions P, $\dot{P}$, and Q are all bounded on $[t_0, \infty)$. Let $m = \max\{\tau, \sigma\}$ and suppose that for $t_1 \geq t_0$ there exists a positive number μ such that for every function λ in the set*

$$\Lambda = \{\lambda: \lambda(t) \equiv 0 \text{ for } t_1 - m \leq t < t_1, \lambda(t) \text{ is continuous on } (t_1 + n\tau,$$
$$t_1 + (n + 1)\tau \text{ for } n = 0, 1, 2, \dots \text{ and } 0 \leq \lambda(t) \leq \mu \text{ for } t \geq t_1\}$$

the following inequality holds:

$$0 \leq [-P(t)\lambda(t - \tau) - \dot{P}(t)] \exp\left(-\int_{t-\tau}^{t} \lambda(s)\, ds\right) - Q(t) \exp\left(-\int_{t-\sigma}^{t} \lambda(s)\, ds\right)$$

$$\leq \mu, \qquad t \geq t_1. \tag{6.7.6}$$

Then eqn (6.7.1) has a (positive) increasing solution on $[t_1, \infty)$.

Proof. Let B and η be positive constants chosen in such a way that

$$|P(t)|\mu + |\dot{P}(t)| + |Q(t)| \leqslant B$$

and

$$\frac{B}{\eta} + |P(t)|\,e^{-\eta\tau} \leqslant \tfrac{1}{2} \qquad \text{for } t \geqslant t_1. \tag{6.7.7}$$

For $\lambda_1, \lambda_2 \in \Lambda$ define

$$d(\lambda_1, \lambda_2) = \sup_{t \geqslant t_1} |\lambda_1(t) - \lambda_2(t)|\,e^{-\eta t}.$$

Then (Λ, d) is a complete metric space. Define

$$(T\lambda)(t) = \begin{cases} [-P(t)\lambda(t-\tau) - \dot{P}(t)]\exp\left(-\int_{t-\tau}^{t}\lambda(s)\,ds\right) \\[2mm] \qquad - Q(t)\exp\left(-\int_{t-\sigma}^{t}\lambda(s)\,ds\right), & t > t_1 \\[4mm] 0, & t_1 - m \leqslant t < t_1 \end{cases}$$

where $m = \max\{\tau, \sigma\}$. Then $T\lambda$ is continuous on $(t_1 + n\tau, t_1 + (n+1)\tau)$ for $n = 0, 1, 2, \ldots$ and by (6.7.6),

$$0 \leqslant (T\lambda)(t) \leqslant \mu.$$

Hence $T\lambda$ maps Λ into Λ. Moreover, T is a contraction mapping. For if $\lambda_1, \lambda_2 \in \Lambda$ and $t \geqslant t_1$

$$|(T\lambda_1)(t) - (T\lambda_2)(t)|$$

$$\leqslant |P(t)|\left|\lambda_1(t-\tau)\exp\left(-\int_{t-\tau}^{t}\lambda_1(s)\,ds\right) - \lambda_2(t-\tau)\exp\left(-\int_{t-\tau}^{t}\lambda_2(s)\,ds\right)\right|$$

$$+ |\dot{P}(t)|\left|\exp\left(-\int_{t-\tau}^{t}\lambda_1(s)\,ds\right) - \exp\left(-\int_{t-\tau}^{t}\lambda_2(s)\,ds\right)\right|$$

$$+ |Q(t)|\left|\exp\left(-\int_{t-\sigma}^{t}\lambda_1(s)\,ds\right) - \exp\left(-\int_{t-\sigma}^{t}\lambda_2(s)\,ds\right)\right|$$

$$= |P(t)|\left|\lambda_1(t-\tau)\left[\exp\left(-\int_{t-\tau}^{t}\lambda_1(s)\,ds\right) - \exp\left(-\int_{t-\tau}^{t}\lambda_2(s)\,ds\right)\right]\right.$$

$$\left. + \exp\left(-\int_{t-\tau}^{t}\lambda_2(s)\,ds\right)[\lambda_1(t-\tau) - \lambda_2(t-\tau)]\right|$$

$$+ |\dot{P}(t)|\left|\exp\left(-\int_{t-\tau}^{t}\lambda_1(s)\,ds\right) - \exp\left(-\int_{t-\tau}^{t}\lambda_2(s)\,ds\right)\right|$$

$$+ |Q(t)| \left| \exp\left(-\int_{t-\sigma}^{t} \lambda_1(s)\, ds \right) - \exp\left(-\int_{t-\sigma}^{t} \lambda_2(s)\, ds \right) \right|$$

$$\leqslant |P(t)| \mu \int_{t-\tau}^{t} |\lambda_1(s) - \lambda_2(s)|\, ds + |P(t)||\lambda_1(t - \tau) - \lambda_2(t - \tau)|$$

$$+ |\dot{P}(t)| \int_{t-\tau}^{t} |\lambda_1(s) - \lambda_2(s)|\, ds + |Q(t)| \int_{t-\sigma}^{t} |\lambda_1(s) - \lambda_2(s)|\, ds$$

$$\leqslant d(\lambda_1, \lambda_2) \left\{ [|P(t)|\mu + |\dot{P}(t)|] \int_{t-\tau}^{t} e^{\eta s}\, ds + |Q(t)| \int_{t-\sigma}^{t} e^{\eta s}\, ds + |P(t)|\, e^{\eta(t-\tau)} \right\}$$

$$\leqslant d(\lambda_1, \lambda_2)\, e^{\eta t} \frac{1}{\eta} [|P(t)|\mu + |\dot{P}(t)| + |Q(t)| + \eta|P(t)|\, e^{-\eta\tau}]$$

$$\leqslant d(\lambda_1, \lambda_2)\, e^{\eta t} \left[\frac{B}{\eta} + |P(t)|\, e^{-\eta\tau} \right].$$

From this inequality and by using (6.7.7) we see that,

$$d(T\lambda_1, T\lambda_2) \leqslant \tfrac{1}{2} d(\lambda_1, \lambda_2).$$

Hence there exists a (unique) solution $\lambda \in \Lambda$ of $T\lambda = \lambda$. To complete the proof it suffices to observe by direct substitution into eqn (6.7.1) that

$$y(t) = \exp\left(\int_{t_1}^{t} \lambda(s)\, ds \right), \qquad t \geqslant t_1$$

is a positive (increasing) solution of eqn (6.7.1) on $[t_1, \infty)$. The proof is complete.

Theorem 6.7.2. *Assume that* (6.7.2) *holds and suppose that the functions* $P(t)$, $\dot{P}(t)$ *and* $Q(t)$ *are all bounded on* $[t_0, \infty)$. *Then in each of the following three cases and for every* $t_1 \geqslant t_0$, *eqn* (6.7.1) *has a positive (increasing) solution on* $[t_1, \infty)$.

(a) *For* $t \geqslant t_0$ *and for some positive number* α *the following inequalities hold:*

$$P(t) \leqslant 0, \qquad \dot{P}(t) + Q(t) \leqslant 0, \qquad (\sigma - \tau)Q(t) \geqslant 0$$

$$[-1 + |P(t)|]\alpha + |\dot{P}(t)| + |Q(t)| \leqslant 0.$$

(b) *For* $t \geqslant t_0$ *and for some positive number* β *the following inequalities hold:*

$$|P(t)|\beta + \dot{P}(t) + Q(t) \leqslant 0, \qquad (\sigma - \tau)Q(t) \geqslant 0$$

$$[-1 + |P(t)|]\beta + |\dot{P}(t)| + |Q(t)| \leqslant 0.$$

(c) *For* $t \geqslant t_0$ *and for some positive number* γ *the following inequalities hold:*

$$P(t) \leqslant 0, \qquad \dot{P}(t) \leqslant 0, \qquad Q(t) \leqslant 0$$

$$[1 + P(t)]\gamma + \dot{P}(t) + Q(t) \geqslant 0.$$

Proof. In view of Lemma 6.7.1, it suffices to show that in each of the three cases (a), (b), and (c), condition (6.7.6) is satisfied. First, assume that (a) holds. We prove that (6.7.6) holds with the μ in the definition of the set Λ of Lemma 6.7.1 taken equal to α. To simplify the notation, set

$$I = [-P(t)\lambda(t - \tau) - \dot{P}(t)] \exp\left(-\int_{t-\tau}^{t} \lambda(s)\,ds\right) - Q(t)\exp\left(-\int_{t-\sigma}^{t} \lambda(s)\,ds\right).$$

We must prove that $0 \leqslant I \leqslant \alpha$. Indeed,

$$I \leqslant |P(t)|\alpha + |\dot{P}(t)| + |Q(t)| \leqslant \alpha.$$

Furthermore,

$$I \geqslant -\dot{P}(t)\exp\left(-\int_{t-\tau}^{t} \lambda(s)\,ds\right) - Q(t)\exp\left(-\int_{t-\sigma}^{t} \lambda(s)\,ds\right)$$

$$\geqslant Q(t)\left[\exp\left(-\int_{t-\tau}^{t} \lambda(s)\,ds\right) - \exp\left(-\int_{t-\sigma}^{t} \lambda(s)\,ds\right)\right] \geqslant 0$$

and the proof in this case is complete.

Next, assume that (b) holds. Here the μ in the definition of Λ is taken equal to β. Clearly, $I \leqslant \beta$. Furthermore,

$$I \geqslant [-|P(t)|\beta - \dot{P}(t)]\exp\left(-\int_{t-\tau}^{t} \lambda(s)\,ds\right) - Q(t)\exp\left(-\int_{t-\sigma}^{t} \lambda(s)\,ds\right)$$

$$\geqslant Q(t)\left[\exp\left(-\int_{t-\tau}^{t} \lambda(s)\,ds\right) - \exp\left(-\int_{t-\sigma}^{t} \lambda(s)\,ds\right)\right] \geqslant 0$$

and the proof in case (b) is complete.

Finally, assume that (c) holds. Here the μ in the definition of Λ is taken equal to γ. Clearly in this case $I \geqslant 0$ and

$$I \leqslant -\gamma P(t) - \dot{P}(t) - Q(t) \leqslant \gamma.$$

The proof of the Theorem is complete.

6.8 Oscillations in neutral delay logistic differential equations

In the classical delay logistic equation (4.3.3) considered in Section 4.3 the per capita growth rate

$$r(t) = \dot{N}(t)/N(t)$$

is given by $r[1 - N(t - \tau)/K]$, where $N(t)$ denotes the density of the population at time t, r is the growth rate and K is the carrying capacity of the environment. The term $r[1 - N(t - \tau)/K]$ denotes the feedback mechanism which takes τ units of time to respond to changes in the size of the population.

In this section we assume that the per capita growth rate $r(t)$ is given by $r(t) = r_\sigma(t) + r_\tau(t)$, where

$$r_\sigma(t) = r[1 - N(t - \tau)/K]$$

is the per capita growth rate associated with density dependence, and

$$r_\tau(t) = c \frac{\dot{N}(t - \tau)}{N(t - \tau)}$$

is the per capita growth rate associated with per capita growth rate at time $t - \tau$. This leads to the neutral delay differential equation

$$\dot{N}(t) = N(t)\left\{r\left[1 - \frac{N(t - \sigma)}{K}\right] + c\frac{\dot{N}(t - \tau)}{N(t - \tau)}\right\}, \qquad t \geq 0 \qquad (6.8.1)$$

where $r, K \in (0, \infty)$, $\sigma, \tau \in [0, \infty)$, and $c \in (-\infty, \infty)$.

An alternative model may be obtained by considering the growth rates rather than the per capita growth rates. This leads to the neutral delay differential equation

$$\dot{N}(t) = N(t)\{r[1 - N(t - \sigma)/K] + c\dot{N}(t - \tau)\}, \qquad t \geq 0. \qquad (6.8.2)$$

With eqns (6.8.1) and (6.8.2) one associates an initial condition of the form

$$N(t) = \phi(t), \qquad -\gamma \leq t \leq 0 \qquad \text{with } \gamma = \max\{\tau, \sigma\}$$

where the function ϕ satisfies the following condition:

$$\left.\begin{array}{l} \phi \in C[[-\gamma, 0], \mathbb{R}^+], \qquad \phi(t) > 0 \qquad \text{for } -\gamma \leq t \leq 0 \qquad \text{and} \\[1em] \phi(t) \text{ is absolutely continuous with locally bounded} \\ \text{derivative on } -\tau \leq t \leq 0. \end{array}\right\} \qquad (6.8.3)$$

By Theorem 1.1.5 we know that the initial value problem (6.8.1) and (6.8.3) has a unique solution which exists and remains positive on $[0, \infty)$. The same is true for the initial value problem (6.8.2) and (6.8.3). By introducing the change of variable

$$x(t) = \ln \frac{N(t)}{K},$$

eqns (6.8.1) and (6.8.2) are transformed to

$$\frac{d}{dt}[x(t) - cx(t - \tau)] + r[e^{x(t-\sigma)} - 1] = 0, \qquad t \geqslant 0 \qquad (6.8.4)$$

and

$$\frac{d}{dt}\{x(t) - c[e^{x(t-\tau)} - 1]\} + r[e^{x(t-\sigma)} - 1] = 0, \qquad t \geqslant 0 \qquad (6.8.5)$$

respectively.

In this section we consider the oscillatory properties of the positive solutions of eqns (6.8.1) and (6.8.2) about the unique positive steady state K. Clearly, every positive solution of (6.8.1) and (6.8.2) oscillates about K if and only if every solution of (6.8.4) and (6.8.5) oscillates.

First we establish a necessary and sufficient condition for the oscillation of every positive solution of eqn (6.8.1) about K.

Theorem 6.8.1. *Assume that*

$$r, K, \tau \in (0, \infty), \qquad \sigma \in [0, \infty), \qquad and \qquad c \in (0, 1). \qquad (6.8.6)$$

Then every positive solution of eqn (6.8.1) oscillates about the positive steady state K if and only if every solution of the equation

$$\frac{d}{dt}[x(t) - cx(t - \tau)] + rx(t - \sigma) = 0 \qquad (6.8.7)$$

oscillates or, equivalently, if and only if the characteristic equation

$$\lambda - \lambda c\,e^{-\lambda\tau} + r\,e^{-\lambda\sigma} = 0 \qquad (6.8.8)$$

has no negative real roots.

Proof. Set $p_0 = c$, $q(t) \equiv r$, and $h(u) = e^u - 1$. Then eqn (6.8.4) can be written in the form (6.6.22) and clearly the conditions of Corollary 6.6.1 are satisfied. The result is now an immediate consequence of Corollary 6.6.1 applied to eqn (6.8.4).

Next we establish a sufficient condition for the oscillation about K of every positive and bounded solution of eqn (6.8.2).

Theorem 6.8.2. *Assume that (6.8.6) is satisfied and that*

$$e \, r\sigma > 1. \tag{6.8.9}$$

Then every bounded positive solution of eqn (6.8.2) oscillates about the positive steady state K.

We need the following lemma.

Lemma 6.8.1. *Assume that (6.8.6) is satisfied. Then every positive bounded solution of eqn (6.8.2) which does not oscillate about K tends to K as $t \to +\infty$.*

Proof. Suppose $N(t)$ is a bounded positive solution of eqn (6.8.2) which is non-oscillatory about K. First we show that $\liminf_{t \to +\infty} N(t) > 0$. Clearly this is true if eventually $N(t) > K$. Therefore we consider the case where eventually $N(t) < K$. From eqn (6.8.2) we see that

$$N(t) = N(0) \exp\left\{ \int_0^t [r(1 - N(s - \sigma)/K) + c\dot{N}(s - \tau)] \, ds \right\}, \quad t \geq 0,$$

that is,

$$N(t) = N(0) \, e^{-cN(-\tau)} \exp\left\{ r \int_0^t [1 - N(s - \sigma)/K] \, ds \right\} e^{cN(t-\tau)}, \quad t \geq 0.$$

Since eventually $N(t) < K$, $\liminf_{t \to +\infty} N(t) = 0$ if and only if $\int_0^\infty [1 - N(s - \sigma)/K] \, ds = -\infty$, or, equivalently, if and only if $\lim_{t \to +\infty} N(t) = 0$. This is a contradiction, so $\liminf_{t \to +\infty} N(t) > 0$.

Set $x(t) = \ln N(t)/K$. Then $x(t)$ is a bounded non-oscillatory solution of (6.8.5). Suppose that $x(t)$ is eventually positive. The case where $x(t)$ is eventually negative is similar and is omitted. Now we claim that $\lim_{t \to +\infty} x(t) = 0$. Since $x(t)$ is a bounded and eventually positive solution of (6.8.5), it follows that for t sufficiently large, either

$$x(t) - c(e^{x(t-\tau)} - 1) > 0 \quad \text{and} \quad \frac{d}{dt}[x(t) - c(e^{x(t-\tau)} - 1)] < 0 \tag{6.8.10}$$

or

$$x(t) - c(e^{x(t-\tau)} - 1) < 0 \quad \text{and} \quad \frac{d}{dt}[x(t) - c(e^{x(t-\tau)} - 1)] < 0. \tag{6.8.11}$$

If (6.8.10) holds, then it can be easily seen that

$$\lim_{t \to +\infty} \frac{d}{dt}[x(t) - c(e^{x(t-\tau)} - 1)] = 0$$

and hence (6.8.5) implies $\lim_{t \to +\infty} (e^{x(t-\tau)} - 1) = 0$, that is, $\lim_{t \to +\infty} x(t) = 0$. On the other hand, if (6.8.11) holds, then we note that $x(t)$ is a bounded function and thus $\lim_{t \to \infty} [x(t) - c(e^{x(t-\tau)} - 1)]$ is a finite number. In this case, it follows from (6.8.5) that $\lim_{t \to +\infty} x(t) = 0$. The proof is complete.

We are now ready to prove Theorem 6.8.2.

Proof of Theorem 6.8.2. For the sake of contradiction we assume that eqn (6.8.2) has a bounded positive solution $N(t)$ which is not oscillatory about K. Then $x(t) = \ln N(t)/K$ is a bounded and non-oscillatory solution of (6.8.5) and $\lim_{t \to +\infty} x(t) = 0$. Therefore there exists a $t_0 \geq 0$ such that for all $t \geq t - \max\{\tau, \sigma\}$, $x(t) > 0$ and

$$-1 \leq P(t) \equiv -c \frac{e^{x(t-\tau)} - 1}{x(t - \tau)} \leq 0.$$

Set

$$Q(t) = r \frac{e^{x(t-\sigma)} - 1}{x(t - \sigma)} \geq 0, \qquad t \geq t_0.$$

Then

$$\liminf_{t \to +\infty} \int_{t-\sigma}^{t} Q(s) \, \mathrm{d}s > 1/e.$$

On the other hand, from eqn (6.8.5) we see that $x(t)$ satisfies the linear neutral equation

$$\frac{\mathrm{d}}{\mathrm{d}t} [x(t) + P(t)x(t - \tau)] + Q(t)x(t - \sigma) = 0, \qquad t \geq t_0, \qquad (6.8.12)$$

where $P(t)$ and $Q(t)$ satisfy all the conditions of Theorem 6.4.3. Hence $x(t)$ oscillates, which is a contradiction. The proof is complete.

6.9 Oscillations in non-autonomous equations with several delays

Consider the non-autonomous equation with several delays

$$\frac{\mathrm{d}}{\mathrm{d}t} \left[x(t) - \sum_{j=1}^{l} p_j(t)x(t - \tau_j(t)) \right] + \sum_{i=1}^{n} q_i(t)x(t - \sigma_i(t)) = 0, \qquad t \geq t_0$$

$$(6.9.1)$$

where

$$p_j, \tau_j, q_i, \sigma_i \in C[[t_0, \infty), \mathbb{R}^+] \qquad \text{for } j = 1, 2, \ldots, l \text{ and } i = 1, 2, \ldots, n,$$
(6.9.2)

and for all $j = 1, 2, \ldots, l$ and $i = 1, 2, \ldots, n$,

$$\sup_{t \geqslant t_0} \tau_j(t) < \infty \qquad \text{and} \qquad \sup_{t \geqslant t_0} \sigma_i(t) < \infty. \tag{6.9.3}$$

In this section we obtain sufficient conditions for the oscillation of all solutions of eqn (6.9.1) under the assumption that for t sufficiently large,

$$\sum_{j=1}^{l} p_j(t) \leqslant 1 \qquad \text{and} \qquad \sup\left\{\sum_{i=1}^{n} q_i(s) : s \geqslant t\right\} > 0. \tag{6.9.4}$$

The following lemma will be very useful.

Lemma 6.9.1. *Assume that (6.9.2), (6.9.3), and (6.9.4) are satisfied. Let $x(t)$ be an eventually positive solution of eqn (6.9.1) and set*

$$v(t) = x(t) - \sum_{j=1}^{l} p_j(t) x(t - \tau_j(t)). \tag{6.9.5}$$

Then $v(t)$ is eventually positive and decreasing and

$$\dot{v}(t) = -\sum_{i=1}^{n} q_i(t) x(t - \sigma_i(t)). \tag{6.9.6}$$

Proof. Clearly (6.9.6) holds, and for t sufficiently large, $\dot{v}(t) \leqslant 0$. Hence $v(t)$ is eventually decreasing. Because of the condition (6.9.4) on the q_i's, $v(t)$ cannot be eventually identically zero. It remains to show that $v(t)$ is eventually positive. Otherwise, $v(t)$ is eventually negative and there exists a $t_1 \geqslant t_0$ such that (6.9.4) holds and for $t \geqslant t_1$,

$$x(t) > 0 \qquad \text{and} \qquad v(t) \leqslant v(t_1) < 0.$$

Then (6.9.5) yields

$$x(t) \leqslant v(t_1) + \sum_{j=1}^{l} p_j(t) x(t - \tau_j(t)). \tag{6.9.7}$$

But by Lemma 1.5.4, inequality (6.9.7) cannot have an eventually positive solution. This is a contradiction and the proof is complete.

The following result shows that, under the conditions (6.9.2), (6.9.3), and (6.9.4), every solution of eqn (6.9.1) oscillates, provided that the same is true for the non-neutral equation

$$\dot{y}(t) + \sum_{i=1}^{n} q_i(t) y(t - \sigma_i(t)) = 0. \tag{6.9.8}$$

Theorem 6.9.1. *Assume that (6.9.2), (6.9.3), and (6.9.4) hold and suppose that every solution of eqn (6.9.8) oscillates. Then every solution of eqn (6.9.1) also oscillates.*

Proof. Assume, for the sake of contradiction, that eqn (6.9.1) has an eventually positive solution $x(t)$. Then by Lemma 6.9.1, $v(t)$ is eventually positive. Also $x(t) > v(t)$, so (6.9.6) yields

$$\dot{v}(t) + \sum_{i=1}^{n} q_i(t) v(t - \sigma_i(t)) \leq 0. \tag{6.9.9}$$

By Corollary 3.2.2, it follows that eqn (6.9.8) also has an eventually positive solution. This is a contradiction and the proof is complete.

Remark 6.9.1. *One can use Theorem 6.9.1 together with any explicit sufficient conditions for the oscillation of all solutions of eqn (6.9.8) to obtain explicit sufficient conditions for the oscillation of all solutions of eqn (6.9.1).*

We now describe a technique which one can use to obtain successively improved oscillation results for eqn (6.9.1). In this process, Theorem 6.9.1 may be thought of as being the first theorem. The second theorem in this succession is obtained as follows. By substituting (6.9.5) into (6.9.6), we find

$$\dot{v}(t) + \sum_{i=1}^{n} q_i(t) v(t - \sigma_i(t))$$

$$+ \sum_{i=1}^{n} q_i(t) \sum_{j=1}^{l} p_j(t - \sigma_i(t)) x(t - \sigma_i(t) - \tau_j(t - \sigma_i(t))) = 0. \tag{6.9.10}$$

Under the hypotheses of Lemma 6.9.1 we have $0 < v(t) < x(t)$, so (6.9.10) yields the inequality

$$\dot{v}(t) + \sum_{i=1}^{n} q_i(t) v(t - \sigma_i(t))$$

$$+ \sum_{i=1}^{n} q_i(t) \sum_{j=1}^{l} p_j(t - \sigma_i(t)) v(t - \sigma_i(t) - \tau_j(t - \sigma_i(t))) \leq 0.$$

The following result, which improves Theorem 6.9.1, is now obvious.

Theorem 6.9.2. *Assume that* (6.9.2), (6.9.3), *and* (6.9.4) *hold and suppose that every solution of the equation*

$$\dot{y}(t) + \sum_{i=1}^{n} q_i(t) y(t - \sigma_i(t))$$

$$+ \sum_{i=1}^{n} q_i(t) \sum_{j=1}^{l} p_j(t - \sigma_i(t)) y(t - \sigma_i(t) - \tau_j(t - \sigma_i(t))) = 0$$

oscillates. Then every solution of eqn (6.9.1) *also oscillates.*

The following result is an immediate consequence of Theorem 6.9.2 and Theorem 2.2.1(a) applied to the neutral equation with constant coefficients and constant delays

$$\frac{d}{dt}\left[x(t) - \sum_{j=1}^{l} p_j x(t - \tau_j) \right] + \sum_{i=1}^{n} q_i x(t - \sigma_i) = 0. \qquad (6.9.11)$$

Corollary 6.9.1. *Assume that the coefficients and the delays of eqn* (6.9.11) *are non-negative real numbers such that*

$$\sum_{j=1}^{l} p_j \leqslant 1$$

and

$$\left(\sum_{i=1}^{n} q_i \sigma_i \right)\left(1 + \sum_{j=1}^{l} p_j \right) + \left(\sum_{i=1}^{n} q_i \right)\left(\sum_{j=1}^{l} p_j \tau_j \right) > 1/e.$$

Then every solution of eqn (6.9.11) *oscillates.*

If we continue in the direction which led to Theorem 6.9.2 with the simpler equation

$$\frac{d}{dt}[x(t) - px(t - \tau)] + q(t)x(t - \sigma(t)) = 0, \qquad t \geqslant t_0 \qquad (6.9.12)$$

we obtain the following result.

Theorem 6.9.3. *Assume that* $p \in [0, 1]$, $\tau \in (0, \infty)$, $q, \sigma \in C[[t_0, \infty), \mathbb{R}^+]$,

$$\sigma_0 \equiv \sup_{t \geqslant t_0} \sigma(t) < \infty, \qquad q_0 \equiv \sup_{t \geqslant t_0} q(t) < \infty$$

and that

$$\liminf_{t \to \infty} \left\{ q(t)\left[\frac{\sigma(t)}{1 - p} + \frac{p\tau}{(1 - p)^2} \right] \right\} > \frac{1}{e}. \qquad (6.9.13)$$

Then every solution of eqn (6.9.12) *oscillates.*

Proof. Assume, for the sake of contradiction, that eqn (6.9.12) has an eventually positive solution $x(t)$. Set

$$v(t) = x(t) - px(t - \tau).$$

Then by Lemma 6.9.1 there is a $t_1 \geqslant t_0$ such that $0 < v(t) < x(t)$ for $t \geqslant t_1$ and

$$\dot{v}(t) + q(t)x(t - \sigma(t)) = 0, \qquad t \geqslant t_1 - r, \qquad (6.9.14)$$

where $r = \max\{\tau, \sigma_0\}$. Observe that $x(t) = v(t) + px(t - \tau)$ for $t \geqslant t_1 - r$, and by induction, for $N \geqslant 1$, we find

$$x(t) = \sum_{l=0}^{N-1} p^l v(t - l\tau) + p^N x(t - N\tau), \qquad t \geqslant t_1 + (N - 1)\tau - r.$$

From this and (6.9.14) we see that the inequality

$$\dot{v}(t) + q(t) \sum_{l=0}^{N-1} p^l v(t - \sigma(t) - l\tau) \leqslant 0 \qquad \text{for } N \geqslant 1$$

has an eventually positive solution $v(t)$. Hence, by Corollary 3.2.2, the equation

$$\dot{u}(t) + q(t) \sum_{l=0}^{N-1} p^l u(t - \sigma(t) - l\tau) = 0 \qquad \text{for } N \geqslant 1$$

also has an eventually positive solution. In view of Corollary 3.4.1(a) it follows that for every $N \geqslant 1$,

$$\liminf_{t \to \infty} \left\{ q(t) \sum_{l=0}^{N-1} p^l [\sigma(t) + l\tau] \right\} \leqslant 1/e. \qquad (6.9.15)$$

By using the identities

$$\sum_{l=0}^{\infty} p^l = \frac{1}{1 - p} \qquad \text{and} \qquad \sum_{l=0}^{\infty} lp^l = \frac{p}{(1 - p)^2},$$

(6.9.15) implies that

$$\liminf_{t \to \infty} \left\{ q(t) \left[\frac{\sigma(t)}{1 - p} + \frac{p\tau}{(1 - p)^2} \right] \right\} \leqslant \frac{1}{e},$$

which contradicts (6.9.13) and completes the proof.

6.10 Oscillations in systems of neutral equations

In this section we obtain sufficient conditions for the oscillation (component-wise) of every solution of the system of neutral equations

$$\frac{d}{dt}[x(t) - Px(t - \tau)] + \sum_{k=1}^{n} Q_k x(t - \sigma_k) = 0. \qquad (6.10.1)$$

Here P is an $m \times m$ diagonal matrix with diagonal entries $p_1, p_2, \ldots, p_m$ and Q_k is an $m \times m$ matrix for each $k = 1, 2, \ldots, n$ such that

$$\left.\begin{array}{ll} 0 \leqslant p_i \leqslant 1 & \text{for } i = 1, 2, \ldots, m \\ \tau, \sigma_k \in [0, \infty) \quad \text{and} \quad q_{ij}^{(k)} \in \mathbb{R} & \text{for } k = 1, 2, \ldots, n \text{ and } i, j = 1, 2, \ldots, m \end{array}\right\}.$$

$$(6.10.2)$$

Our main result is the following theorem which gives explicit sufficient conditions for the oscillation (componentwise) of every solution of eqn (6.10.1). As we shall see, a similar result holds for non-autonomous systems where oscillations are understood in the sense of Definition 5.0.2.

Theorem 6.10.1. *Assume that (6.10.2) holds. Set*

$$q_k = \min_{1 \leqslant i \leqslant n}\left[q_{ii}^{(k)} - \sum_{\substack{j=1 \\ j \neq i}}^{m} |q_{ji}^{(k)}| \right] \qquad \text{for } k = 1, 2, \ldots, n. \qquad (6.10.3)$$

Suppose that

$$q_k \geqslant 0 \qquad \text{for } k = 1, 2, \ldots, n \qquad (6.10.4)$$

and that every solution of the scalar delay equation

$$\dot{u}(t) + \sum_{k=1}^{n} q_k u(t - \sigma_k) = 0 \qquad (6.10.5)$$

oscillates. Then every solution of (6.10.1) oscillates componentwise.

Proof. Assume, for the sake of contradiction, that (6.10.1) has a non-oscillatory solution in the sense of Definition 5.0.1. Then, by Theorem 5.1.2, eqn (6.10.1) has a non-oscillatory solution $x(t) = [x_1(t), \ldots, x_m(t)]^T$ in the sense of Definition 5.0.2. That is, $x(t)$ is not eventually zero, and for t sufficiently large, each component $x_i(t)$ has eventually constant signum. For t sufficiently large, set

$$\delta_i = \text{sgn } x_i(t) \qquad \text{and} \qquad y_i(t) = \delta_i x_i(t) \qquad \text{for } i = 1, 2, \ldots, m.$$

Then for $i = 1, 2, \ldots, m$ and for t sufficiently large, it follows from (6.10.1) that

$$\frac{d}{dt}[y_i(t) - p_i y_i(t - \tau)] + \sum_{k=1}^{n} \sum_{j=1}^{m} q_{ij}^{(k)} x_j(t - \sigma_k)\delta_i = 0$$

or, equivalently,

$$\frac{d}{dt}[y_i(t) - p_i y_i(t - \tau)] + \sum_{k=1}^{n}\left[q_{ii}^{(k)} y_i(t - \sigma_k) + \sum_{j \neq i} q_{ij}^{(k)}\delta_i x_j(t - \sigma_k)\right] = 0.$$

Hence for $i = 1, 2, \ldots, m$ and for t sufficiently large,

$$\frac{d}{dt}[y_i(t) - p_i y_i(t - \tau)] + \sum_{k=1}^{n}\left[q_{ii}^{(k)} y_i(t - \sigma_k) - \sum_{j \neq i} |q_{ij}^{(k)}| y_j(t - \sigma_k)\right] \leqslant 0.$$

$$(6.10.6)$$

Set

$$v(t) = \sum_{i=1}^{m} y_i(t) - \sum_{i=1}^{m} p_i y_i(t - \tau)$$

and

$$w(t) = \sum_{i=1}^{m} y_i(t).$$

Summing-up (vertically) both sides of (6.10.6) for $i = 1, 2, \ldots, m$ and using the definition of q_k we find that for t sufficiently large,

$$\dot{v}(t) + \sum_{i=1}^{n} q_k w(t - \sigma_k) \leqslant 0. \tag{6.10.7}$$

As $w(t) > 0$ and $q_k \geqslant 0$, it follows that $v(t)$ is a decreasing function. Hence, either

$$\lim_{t \to +\infty} v(t) = -\infty \tag{6.10.8}$$

or

$$\lim_{t \to +\infty} v(t) = L \in \mathbb{R}. \tag{6.10.9}$$

First, we claim that (6.10.8) is impossible. Otherwise, $v(t) < 0$ and at least one of the components $y_i(t)$ would be unbounded. But then eventually,

$$w(t) = \sum_{i=1}^{m} y_i(t) \leqslant \sum_{i=1}^{m} p_i y_i(t - \tau) \leqslant \sum_{i=1}^{n} y_i(t - \tau) = w(t - \tau).$$

This implies that $w(t)$ is bounded, which is a contradiction. Thus (6.10.9) holds. We now claim that $L = 0$. Indeed, by integrating (6.10.7) from t_0 to t and by letting $t \to +\infty$, we find

$$L - v(t_0) + \sum_{i=1}^{m} \int_{t}^{\infty} q_k w(s - \sigma_k)\, ds \leqslant 0,$$

which implies that $w \in L^1(t_0, \infty)$. Then $y_i \in L^1(t_0, \infty)$ for $i = 1, 2, \ldots, m$, so $v \in L^1(t_0, \infty)$. But then $L = 0$, which proves our claim. As $v(t)$ decreases to zero, it follows that

$$v(t) > 0 \qquad \text{and} \qquad v(t) \leqslant w(t). \qquad (6.10.10)$$

Then (6.10.7) implies that the eventually positive function $v(t)$ satisfies the inequality

$$\dot{v}(t) + \sum_{k=1}^{n} q_k v(t - \tau_k) \leqslant 0. \qquad (6.10.11)$$

By Corollary 3.2.2, it follows that eqn (6.10.5) has an eventually positive solution. This contradicts the hypothesis and the proof is complete.

Remark 6.10.1. *It is not difficult to see that Theorem 6.10.1 holds verbatim for systems of the form* (6.10.1) *with the Q_k's continuous $m \times m$ matrices. In this case the coefficients q_k of eqn* (6.10.5) *are the functions*

$$q_k(t) = \min_{1 \leqslant i \leqslant m} \left[q_{ii}^{(k)}(t) - \sum_{\substack{j=1 \\ j \neq i}}^{m} |q_{ji}^{(k)}(t)| \right] \qquad \text{for } k = 1, 2, \ldots, n$$

and oscillation is in the sense of Definition 5.0.2.

Remark 6.10.2. *In the special case where the diagonal matrix P in eqn* (6.10.1) *is a multiple of the identity matrix, that is, when*

$$p_1 = p_2 = \cdots = p_n \equiv p \in [0, 1], \qquad (6.10.12)$$

we have

$$v(t) = w(t) - pw(t - \tau). \qquad (6.10.13)$$

By substituting (6.10.13) repeatedly into (6.10.7) we find, after N steps, that eventually

$$\dot{v}(t) + \sum_{k=1}^{n} q_k \left[\sum_{i=0}^{N} p^i v(t - \sigma_k - i\tau) \right] \leqslant 0.$$

Hence by using Corollary 2.4.1 we obtain the following extension of Theorem 6.10.1.

Theorem 6.10.2. *Assume that* (6.10.2) *and* (6.10.4) *are satisfied and that the following hypothesis holds:*

(H) *There exists a non-negative integer N such that every solution of the delay equation*

$$\dot{u}(t) + \sum_{k=1}^{n} q_k \left(\sum_{i=0}^{N} p^i u(t - \sigma_k - i\tau) \right) = 0 \qquad (6.10.14)$$

oscillates.

Then every solution of (6.10.1) *oscillates componentwise.*

By virtue of Theorem 2.2.1, hypothesis (H) is, for example, satisfied when for some $N \geqslant 0$,

$$\sum_{k=1}^{n} q_k \left[\sum_{i=0}^{N} p^i (\sigma_k + i\tau) \right] > 1/e$$

or, equivalently,

$$\sum_{k=1}^{n} \left(q_k \sigma_k \sum_{i=0}^{N} p^i + q_k \tau \sum_{i=0}^{N} p^i i \right) > 1/e. \qquad (6.10.15)$$

But it can be easily seen that (6.10.15) is satisfied if and only if

$$\sum_{i=0}^{\infty} p^i \left(\sum_{k=1}^{n} q_k \sigma_k \right) + \left(\sum_{i=1}^{\infty} p^i i \right) \left(\sum_{k=1}^{n} q_k \right) \tau > 1/e,$$

or, equivalently,

$$\frac{1}{1-p} \sum_{k=1}^{n} q_k \sigma_k + \frac{\tau}{(1-p)^2} \sum_{k=1}^{n} q_k > \frac{1}{e}. \qquad (6.10.16)$$

The following result is now an immediate consequence of the above discussion.

Corollary 6.10.1. *Assume that* (6.10.2), (6.10.4), *and* (6.10.16) *are satisfied. Then every solution of* (6.10.1) *oscillates componentwise.*

6.11 Notes

Zahariev and Bainov (1980) seems to be the first paper dealing with oscillations of neutral equations. See also Karakostas and Staikos (1984). A systematic development of the oscillation theory of neutral equations was initiated by Ladas and Sficas (1986a, b).

The results in Section 6.1 are from Ladas and Sficas (1986a). Section 6.2 is based on the work of Ladas and Schultz (1989). Theorem 6.3.1 is adapted

from Arino and Györi (1990). The results in Section 6.4 are extracted from
Ladas and Sficas (1986*a*), Grammatikopoulos *et al.* (1986*a*), and Chuanxi
and Ladas (1989*b*). The results in Section 6.5 are from Györi (1987). Section
6.6 is adapted from Györi (1989*b*). See also Ladas (1989). The results in
Section 6.7 are from Chuanxi and Ladas (1989*a*). Equations (6.8.1) and
(6.8.2) were introduced by Györi and Witten (1990) and by Gopalsamy and
Zhang (1990) respectively. Theorem 6.8.1 is a special case of a result in Györi
(1989*b*). See also Györi (1987, 1989*c*). Theorem 6.8.2 is adapted from
Gopalsamy and Zhang (1990) but its proof is new. Section 6.9 is based on
Györi (1989*d*) and Section 6.10 is extracted from Györi and Ladas (1988).

In addition to the references cited above, there is a large number of
interesting publications on this subject. See for example Arino and Ferreira
(1989), Bainov *et al.* (1988), Graef *et al.* (1989), Grammatikopoulos *et al.*
(1988*b*), Kuang (1990*a*), Kuang and Feldstein (1990), Sficas and Stavroulakis
(1987), Wang (1988*a*, *b*), Yan (1990), Grace and Lalli (1990), Georgiou *et al.*
(1990), Györi and Wu (1990), Farrell *et al.* (1990), Chuanxi *et al* (1990*c*),
Grove *et al.* (1988*b*), and Schultz (1990).

6.12 Open problems

6.12.1   Extend Theorems 6.2.1 and 6.2.2 to equations with several delays.

6.12.2   Utilize Theorem 6.3.1 to obtain explicit conditions for the oscillation
of all solutions of eqn (6.3.1).

6.12.3 (Conjecture)   Assume that for $j = 1, 2, \ldots, l$ and for $i = 1, 2, \ldots, n$.

$$P_j, Q_i \in \mathbb{R}^{m \times m} \qquad \text{and} \qquad \tau_j, \sigma_i \in \mathbb{R}.$$

The following statements are equivalent.

(a) Every solution of eqn (6.3.1) oscillates componentwise.

(b) The characteristic equation (6.3.2) has no real roots.

6.12.4 (Conjecture)   Assume that (6.3.5) holds. Then every unbounded
solution of eqn (6.3.1) oscillates if and only if the characteristic equation
(6.3.2) has no positive roots and zero is not a multiple root of (6.3.2).

6.12.5   Study the oscillatory nature of solutions of eqn (6.4.4) when the
coefficient $Q(t)$ oscillates or when condition (6.4.3) is not satisfied.

6.12.6   Extend some of the results in Section 6.4 to equations where the
coefficient $P(t)$ is in ranges different from those described there.

6.12.7   Obtain linearized oscillation results for eqn (6.6.1) when the coefficient $p(t) < 0$ for $t \geq t_0$ or when $p(t) \geq 1$ for $t \geq t_0$.

6.12.8   Extend the results in Section 6.9 to equations with positive and negative p's and/or equations with positive and negative q's.

6.12.9   Consider the neutral differential equation

$$\frac{d}{dt}[y(t) + y(t - \tau)] + Q(t)y(t - \sigma) = 0, \qquad t \geqslant t_0$$

where

$$\tau \in (0, \infty), \quad \sigma \in [0, \infty), \quad Q \in C[[t_0, \infty), \mathbb{R}^+] \quad \text{and} \quad \int_{t_0}^{\infty} Q(s)\, ds = \infty.$$

Does every non-oscillatory solution of this equation tends to zero as $t \to \infty$?

6.12.10   Consider the neutral delay differential equation

$$\frac{d}{dt}[x(t) + px(t - \tau)] + Q(t)x(t - \sigma) = 0, \qquad t \geqslant 0$$

where

$$p \in \mathbb{R} - \{0\}, \qquad \tau \in (0, \infty), \qquad \sigma \in [0, \infty), \qquad \text{and} \qquad Q \in C[[0, \infty), \mathbb{R}^+].$$

In each of the following cases, find sufficient conditions for the oscillation of all solutions under the indicated restrictions on the function Q and the delays τ and σ.

(a) $-1 \leqslant p < 0$. Do not assume that

$$\int_0^{\infty} Q(t)\, dt = \infty.$$

(b) $p > 0$. Do not assume that $Q(t)$ is τ-periodic.

(c) $p > 0$ and $\sigma \leqslant \tau$.

7

OSCILLATIONS OF DELAY DIFFERENCE EQUATIONS

Our primary goal in this chapter is to develop an oscillation theory for delay (and also for advanced) difference equations which parallels the oscillation theory of delay differential equations.

Let k and r be positive integers and for each $i = 1, 2, \ldots, k$ let $\{P_i(n)\}_{n=0}^{\infty}$ be a sequence of real $r \times r$ matrices. We are interested in the oscillations of all solutions of the difference equation

$$a_{n+k} + P_1(n)a_{n+k-1} + \cdots + P_k(n)a_n = 0, \qquad n = 0, 1, 2, \ldots \quad (7.0.1)$$

For the most part our results will be about scalar difference equations. A notable exception is Section 7.1, where we establish a general necessary and sufficient condition for the oscillation of all solutions of eqn (7.0.1) when the coefficient matrices $P_i(n)$ are all constants.

By a *solution* of eqn (7.0.1) we mean a sequence $\{a_n\}$ of points $a_n \in \mathbb{R}^r$ for $n = 0, 1, 2, \ldots$ which satisfies eqn (7.0.1). A sequence of real numbers $\{x_n\}$ is said to *oscillate* if the terms x_n are not all eventually positive or eventually negative. Let $\{a_n\}$ be a solution of eqn (7.0.1) with $a_n = [a_n^1, \ldots, a_n^r]^T$ for $n = 0, 1, 2, \ldots$. We say that the solution $\{a_n\}$ oscillates *componentwise* or simply *oscillates* if each component $\{a_n^j\}$ oscillates. Otherwise the solution $\{a_n\}$ is called *non-oscillatory*. Therefore a solution of eqn (7.0.1) is non-oscillatory if it has a component $\{a_n^j\}$ which is eventually positive or eventually negative.

For the purpose of obtaining oscillation results for difference equations it is preferable, for the most part, that the equation be written in the form

$$a_{n+1} - a_n + \sum_{i=1}^{m} P_i(n)a_{n-k_i} = 0, \qquad n = 0, 1, 2, \ldots \quad (7.0.2)$$

where $k_i \in \mathbb{Z} = \{\ldots, -1, 0, 1, \ldots\}$ for $i = 1, 2, \ldots, m$. One may think of eqn (7.0.2) as being a discrete analogue of the differential equation

$$\dot{y}(t) + \sum_{i=1}^{m} P_i(t)y(t - \tau_i) = 0, \qquad t \geq 0 \quad (7.0.3)$$

where $\tau_i \in \mathbb{R}$ for $i = 1, 2, \ldots, m$.

By analogy with eqn (7.0.3), eqn (7.0.2) is said to be of the *delay, advanced,* or *mixed* type depending on whether the k_i's are all non-negative, all non-positive, or some non-negative and some non-positive, respectively.

More precisely, let

$$k = \max\{0, k_1, \ldots, k_m\} \quad \text{and} \quad l = \max\{1, -k_1, \ldots, -k_m\}. \quad (7.0.4)$$

Then eqn (7.0.2) is a difference equation of order $(k + l)$. If $k \geqslant 0$ and $l = 1$, we say that eqn (7.0.2) is a *delay difference equation*. When $k = 0$ and $l \geqslant 2$, eqn (7.0.2) is called an *advanced difference equation*. When $k \geqslant 1$ and $l \geqslant 2$, then eqn (7.0.2) is of the *mixed* type.

By a *solution* of eqn (7.0.2) we mean a sequence $\{a_n\}$ which is defined for $n \geqslant -k$ and which satisfies eqn (7.0.2) for $n \geqslant 0$. With the k and l as defined in (7.0.4), eqn (7.0.2) can be rearranged into the form

$$a_{n+1} - a_n + \sum_{j=-k}^{l} Q_j(n) a_{n+j} = 0, \quad n = 0, 1, 2, \ldots \quad (7.0.5)$$

Throughout this chapter we assume that

$$\left. \begin{array}{l} |l - 1| + |\det(Q_1 + I)| \neq 0 \text{ and} \\ \text{if } k = 0 \text{ and } l \geqslant 2 \text{ then } \det Q_l \neq 0 \end{array} \right\}. \quad (7.0.6)$$

Let $A_{-k}, \ldots, A_{l-1}$ be $(k + l)$ given vectors in $\mathbb{R}^r$. Then under the assumption (7.0.6), eqn (7.0.2) has a unique solution $\{a_n\}$ which satisfies the initial conditions

$$a_i = A_i \quad \text{for } i = -k, \ldots, l - 1. \quad (7.0.7)$$

On the other hand, when $l = 1$ and $\det(Q_1 + I) = 0$ or when $k = 0$, $l \geqslant 2$ and $\det Q_l \neq 0$, the initial value problem (7.0.5) and (7.0.7) has in general no solution.

7.1 Necessary and sufficient conditions for the oscillation of linear systems of difference equations

Let k and r be positive integers and for each $i = 1, 2, \ldots, k$ let P_i be an $r \times r$ matrix with real entries. Consider the difference equation

$$a_{n+k} + P_1 a_{n+k-1} + \cdots + P_k a_n = 0, \quad n = 0, 1, 2, \ldots \quad (7.1.1)$$

and its associated characteristic equation

$$\det(\lambda^k I + \lambda^{k-1} P_1 + \cdots + \lambda P_{k-1} + P_k) = 0 \quad (7.1.2)$$

where I is the $r \times r$ identity matrix. Our aim in this section is to establish the following necessary and sufficient condition for the oscillation component-wise of all solutions of eqn (7.1.1).

Theorem 7.1.1. *Let k and r be positive integers and let $P_1, \ldots, P_k$ be real $r \times r$ matrices. Then the following statements are equivalent.*

(a) *Every solution $\{a_n\}_{n=0}^{\infty}$ of eqn (7.1.1) oscillates componentwise.*

(b) *The characteristic equation (7.1.2) has no positive roots.*

Proof. The proof that (a) $\Rightarrow$ (b) is simple. If (a) holds, (7.1.2) cannot have a positive root, since if λ_0 were such a root then there would be a non-zero vector $\zeta \in \mathbb{R}^r$ such that

$$(\lambda^k I + \lambda^{k-1} P_1 + \cdots + P_k)\zeta = 0.$$

But then $a_n = \lambda_0^n \zeta$ is a solution of (7.1.1) with at least one non-oscillatory component.

The proof that (b) $\Rightarrow$ (a) uses the z-transform (see Section 1.4). Assume that (b) holds and, for the sake of contradiction, assume that (7.1.1) has a solution $a_n = [a_n^1, \ldots, a_n^r]^T$ with at least one non-oscillatory component. With no loss of generality we assume that $\{a_n^1\}$ is eventually positive. As (7.1.1) is autonomous, we assume in fact that $a_n^1 > 0$ for $n \geqslant 0$. By Lemma 1.4.2 there exist constants $b, c \in (0, \infty)$ such that $\|a_n\| < bc^n$. Then, as we saw in Section 1.4.1, the z-transform of $\{a_n\}$,

$$A(z) = \sum_{n=0}^{\infty} a_n z^{-n}$$

exists for $|z| > c$. By taking the z-transform of both sides of eqn (7.1.1) and by using Lemma 1.4.1 we find that

$$F(z)A(z) = \Phi(z)$$

holds for $|z| > c$ where (with $P_0 = I$),

$$F(z) = \sum_{i=0}^{k} P_i z^{k-i}$$

and

$$\Phi(z) = \sum_{i=0}^{k} P_i \sum_{j=0}^{k-i-1} z^{k-i-j} a_j.$$

By hypothesis, $\det(F(z)) \neq 0$ for $z \in (0, \infty)$. Furthermore, $\det(F(z)) \to \infty$ for real $z \to \infty$, so

$$\det(F(z)) > 0$$

for $z \in (0, \infty)$. Let $A_1(z)$ be the z-transform of the first component $\{a_n^1\}$ of the solution $\{a_n\}$ and let M be the modulus of the largest zero of $\det(F(z))$.

Then by Cramer's rule, for $|z| > \max\{c, M\}$,

$$\det(F(z))A_1(z) = \det(D(z)) \qquad (7.1.3)$$

where $D(z)$ has components of $F(z)$ and $\Phi(z)$ as its entries. Clearly, the determinants in (7.1.3) are polynomials in z.

Suppose

$$A_1(z) = \sum_{n=0}^{\infty} a_n^1 z^{-n}$$

and let $W(z) = A_1(1/z)$ so that $W(z)$ is a power series with positive coefficients having radius of convergence $\rho > 0$. Equation (7.1.3) holds for $|z| > 1/\rho$; equivalently, $\det[F(1/z)]W(z) = \det[D(1/z)]$ for $0 < |z| < \rho$. Now it is known (see Hille 1963, p. 133) that a power series with positive coefficients having radius of convergence $\rho < \infty$ has a singularity (in the sense of analytic continuation) at $z = \rho$. But since $\det[F(1/z)] \neq 0$ we see that $\det[D(1/z)]/\det[F(1/z)]$ is analytic in a disc centred at ρ and agrees with $W(z)$ on that part of the disc where $|z| < \rho$. This contradiction shows that we must have $\rho = \infty$, and it follows that (7.1.3) holds for $|z| > 0$. But then $a_n^1 = 0$ for all sufficiently large n, since otherwise the left-hand side of (7.1.3), but not the right-hand side, would have an essential singularity at $z = 0$. This contradicts the assumption that $\{a_n^1\}$ is non-oscillatory and the proof is complete.

Remark 7.1.1. *One should observe that we proved a little more than we stated in Theorem 7.1.1. Namely, if (7.1.1) has a solution $\{a_n\}$ with one component eventually non-negative but not eventually identically zero, then the characteristic equation (7.1.2) has a real root.*

Remark 7.1.2. *One can easily see that the conclusion of Theorem 7.1.1 is also true when k is a negative integer and $\det P_k \neq 0$.*

Consider the difference equation

$$a_{n+1} - a_n + \sum_{j=-k}^{l} Q_j a_{n+j} = 0, \qquad n = 0, 1, 2, \ldots \qquad (7.1.4)$$

where

$$k, l \in N \quad \text{and} \quad Q_j \in \mathbb{R}^{r \times r} \quad \text{for } j = -k, \ldots, l. \qquad (7.1.5)$$

In agreement with the discussion at the end of Section 7.0 we assume that

$$\left. \begin{array}{l} |l - 1| + |\det(Q_1 + I)| \neq 0 \text{ and} \\ \text{if } k = 0 \text{ and } l \geqslant 2 \text{ then } \det Q_l \neq 0 \end{array} \right\}. \qquad (7.1.6)$$

Clearly the characteristic equation associated with (7.1.4) is

$$\det\left(\lambda I - I + \sum_{j=-k}^{l} Q_j \lambda^j\right) = 0. \tag{7.1.7}$$

The proof of the following theorem is an immediate consequence of Theorem 7.1.1 and Remark 7.1.2 and is therefore omitted.

Theorem 7.1.2. *Assume that (7.1.5) and (7.1.6) hold. Then the following statements are equivalent.*

(a) *Every solution of eqn (7.1.4) oscillates componentwise.*

(b) *The characteristic equation (7.1.7) has no positive roots.*

It is an elementary observation that for the scalar difference equation

$$a_{n+1} - a_n + \sum_{j=-k}^{l} q_j a_{n+j} = 0 \tag{7.1.8}$$

the conclusion of Theorem 7.1.2 is true without the assumption that (7.1.6) holds. That is, the following result is true.

Corollary 7.1.1. *Assume that*

$$k, l \in \mathbb{N} \quad and \quad q_j \in \mathbb{R} \quad for \ j = -k, \ldots, l. \tag{7.1.9}$$

Then the following statements are equivalent.

(a) *Every solution of eqn (7.1.8) oscillates.*

(b) *The characteristic equation*

$$\lambda - 1 + \sum_{j=-k}^{l} q_j \lambda^j = 0 \tag{7.1.10}$$

has no positive roots.

7.2 Explicit necessary and sufficient conditions for oscillations

In this section we obtain necessary and sufficient conditions for the oscillation of all solutions of the difference equation

$$a_{n+1} - a_n + p a_{n-k} + q a_{n-l} = 0, \quad n = 0, 1, 2, \ldots \tag{7.2.1}$$

where

$$p, q \in \mathbb{R}, \quad k \in \mathbb{Z} \quad and \quad l \in \{-1, 0\}.$$

The conditions will be given explicitly in terms of p, q, k, and l.
The following result corresponds to the case $q = 0$.

Theorem 7.2.1. *Consider the difference equation*

$$a_{n+1} - a_n + pa_{n-k} = 0, \qquad n = 0, 1, 2, \ldots \tag{7.2.2}$$

where

$$p \in \mathbb{R} \quad and \quad k \in \mathbb{Z}. \tag{7.2.3}$$

Then every solution of eqn (7.2.2) oscillates if and only if one of the following conditions holds:

(a) $k = -1$ *and* $p \leqslant -1$;

(b) $k = 0$ *and* $p \geqslant 1$;

(c) $k \in \{\ldots, -3, -2\} \cup \{1, 2, \ldots\}$ *and* $p\dfrac{(k+1)^{k+1}}{k^k} > 1$.

For the proof of Theorem 7.2.1 see Remark 7.2.1 at the end of this section.
One should recall (see Theorem 2.2.3) that every solution of the differential equation

$$\dot{x}(t) + px(t - \tau) = 0 \tag{7.2.4}$$

where

$$p, \tau \in \mathbb{R}$$

oscillates if and only if

$$p\tau \, e > 1. \tag{7.2.5}$$

Now observe that condition (c) of Theorem 7.2.1 can be written in the form

$$p(k + 1)\frac{(k+1)^k}{k^k} > 1 \tag{7.2.6}$$

and that

$$\frac{(k+1)^k}{k^k} = \left(1 + \frac{1}{k}\right)^k \uparrow e \qquad \text{as } k \to \infty.$$

Therefore one can think of (7.2.6) as being the discrete analogue of (7.2.5) with the 'delay' of (7.2.2) being $(k + 1)$.

Table 7.1

	p	q	k	l	Necessary and sufficient conditions for oscillations
1	+	+	+	-1	$p(1 + q)^k K > 1$
2	+	+	0	-1	$p \geqslant 1$
3	+	+	$\leqslant -2$	-1	There exist non-oscillatory solutions
4	+	+	+	0	$q \leqslant -1$ or $q \in (0, 1)$ and $pK > (1 - q)^{k+1}$
5	+	+	$\leqslant -2$	0	$q \geqslant 1$
6	+	−	+	-1	$q > -1$ and $p(1 + q)^k K > 1$
7	+	−	0	-1	$p = -1$ or $q = 1$ or $(p + 1)(q - 1) > 0$
8	+	−	$\leqslant -2$	$-1, 0$	There exist non-oscillatory solutions
9	+	−	+	0	$pK > (1 + q)^{k+1}$
10	−	+	$+, 0$	-1	There exist non-oscillatory solutions
11	−	+	$\leqslant -2$	-1	$p(1 + q)^k K < 1$
12	−	+	+	0	There exist non-oscillatory solutions
13	−	$+ -$	$\leqslant -2$	0	$pK < (1 - q)^{k+1}$
14	−	−	$+, 0$	-1	$q \leqslant -1$
15	−	−	$\leqslant -2$	-1	$q \leqslant -1$ or $q > -1$ and $p(1 + q)^k K < 1$
16	−	−	+	0	There exist non-oscillatory solutions

We now state necessary and sufficient conditions for the oscillation of all solutions of eqn (7.2.1) where

$$p, q \in \mathbb{R} - \{0\}, \qquad k \in \mathbb{Z}, \qquad l \in \{-1, 0\} \qquad \text{and} \qquad k \neq l. \quad (7.2.7)$$

These conditions are given, whenever they exist, in Table 7.1. In this table, K denotes the constant $(k + 1)^{k+1}/k^k$.

As the table shows, in cases 3, 8, 10, 12, and 16 there exist non-oscillatory solutions. This is because in each of these cases the characteristic equation

$$F(\lambda) \equiv \lambda - 1 + p\lambda^{-k} + q\lambda^{-l} = 0 \quad (7.2.8)$$

has a positive root. This can be easily seen by computing $F(0+)$, $F(1)$, and $F(\infty)$ and by using the intermediate value theorem.

In each of the remaining cases the given condition is necessary and sufficient for the oscillation of all solutions. The proof can be obtained by computing the extreme value of $F(\lambda)$ as given by (7.2.8) and by applying Corollary 7.1.1. As an example we give the details in Case 1, where

$$p > 0, \qquad q > 0, \qquad k \in \{1, 2, \ldots\} \qquad \text{and} \qquad l = -1. \quad (7.2.9)$$

In this case (7.2.8) becomes

$$F(\lambda) = (1 + q)\lambda - 1 + p\lambda^{-k} = 0.$$

We have

$$F'(\lambda) = (1 + q) - pk\lambda^{-(k+1)}$$

$$F''(\lambda) = pk(k + 1)\lambda^{-(k+2)} > 0 \qquad \text{for } \lambda > 0.$$

The only critical point of $F(\lambda)$ in $(0, \infty)$ is $\lambda_0 = (pk/(1 + q))^{1/(k+1)}$ and $F(\lambda)$ has a minimum at $\lambda = \lambda_0$. Also $F(0+) = \infty$ and $F(\infty) = \infty$. Therefore $F(\lambda)$ has a global minimum in $(0, \infty)$ at the point λ_0.

In view of Theorem 7.1.1, if (7.2.9) holds, eqn (7.2.1) oscillates if and only if

$$F(\lambda_0) > 0.$$

But

$$F(\lambda_0) = (1 + q)\lambda_0 - 1 + p\lambda_0^{-k} = \lambda_0\left[(1 + q) - \frac{1}{\lambda_0} + p\lambda_0^{-(k+1)}\right]$$

$$= \lambda_0\left[(1 + q) - \frac{1}{\lambda_0} + \frac{1 + q}{k}\right] = \lambda_0\left[(1 + q)\frac{k + 1}{k} - \frac{1}{\lambda_0}\right].$$

Hence $F(\lambda_0) > 0$ if and only if $\lambda_0 > \dfrac{k}{k + 1}\dfrac{1}{1 + q}$ if and only if

$$\frac{pk}{1 + q} = \lambda_0^{k+1} > \left(\frac{k}{k + 1}\right)^{k+1}\frac{1}{(1 + q)^{k+1}}$$

if and only if $p(1 + q)^k K > 1$, which completes the proof in Case 1. The proofs in the other cases are similar and are omitted.

Remark 7.2.1. *The proof of Theorem 7.2.1 when $k \notin \{-1, 0\}$ follows from the above proof with $q = 0$. On the other hand, when $k \in \{-1, 0\}$, eqn (7.2.2) can be solved explicitly and the result is a simple consequence of the form of the solution.*

7.3 Sufficient conditions for oscillations

In this section we obtain sufficient conditions for the oscillation of all solutions of the difference equation

$$a_{n+1} - a_n + \sum_{i=1}^{m} p_i a_{n-k_i} = 0, \qquad n = 0, 1, 2, \ldots \qquad (7.3.1)$$

where either

$$p_i \in (0, \infty) \qquad \text{and} \qquad k_i \in \{0, 1, 2, \ldots\} \qquad \text{for } i = 1, 2, \ldots, m \qquad (7.3.2)$$

or

$$p_i \in (-\infty, 0) \quad \text{and} \quad k_i \in \{\ldots, -3, -2, -1\} \quad \text{for } i = 1, 2, \ldots, m$$

$$(7.3.3)$$

Throughout this chapter we use the convention that $0^0 = 1$.

Theorem 7.3.1. *Assume that either* (7.3.2) *or* (7.3.3) *holds and suppose that*

$$\sum_{i=1}^{m} p_i \frac{(k_i + 1)^{k_i+1}}{k_i^{k_i}} > 1. \tag{7.3.4}$$

Then every solution of eqn (7.3.1) *oscillates.*

Proof. In view of Corollary 7.1.1, it suffices to prove that the characteristic equation

$$F(\lambda) = \lambda - 1 + \sum_{i=1}^{m} p_i \lambda^{-k_i} = 0 \tag{7.3.5}$$

has no positive roots. First, assume that (7.3.2) holds. Then eqn (7.3.5) has no roots in $[1, \infty)$. Observe that for $i = 1, 2, \ldots, m$,

$$\min_{0 < \lambda < 1} \left(\frac{\lambda^{-k_i}}{1 - \lambda} \right) = \frac{(k_i + 1)^{k_i+1}}{k_i^{k_i}}$$

while

$$\inf_{0 < \lambda < 1} \left(\frac{1}{1 - \lambda} \right) = 1.$$

Hence for $0 < \lambda < 1$ we have

$$F(\lambda) = (1 - \lambda)\left(-1 + \sum_{i=1}^{m} p_i \frac{\lambda^{-k_i}}{1 - \lambda} \right)$$

$$\geqslant (1 - \lambda)\left[-1 + \sum_{i=1}^{m} p_i \frac{(k_i + 1)^{k_i+1}}{k_i^{k_i}} \right] > 0$$

which completes the proof when (7.3.2) holds.

Next, assume that (7.3.3) holds. Then eqn (7.3.5) has no roots in $(0, 1]$. Observe that for $i = 1, 2, \ldots, m$

$$\inf_{\lambda > 1} \left(\frac{\lambda^{-k_i}}{\lambda - 1} \right) = -\frac{(k_i + 1)^{k_i+1}}{k_i^{k_i}}.$$

Hence for $\lambda > 1$ we have

$$F(\lambda) = (\lambda - 1)\left(1 + \sum_{i=1}^{m} p_i \frac{\lambda^{-k_i}}{\lambda - 1}\right) \leqslant (\lambda - 1)\left[1 - \sum_{i=1}^{m} p_i \frac{(k_i + 1)^{k_i + 1}}{k_i^{k_i}}\right] < 0$$

which completes the proof of the theorem.

For delay differential equations of the form

$$\dot{x}(t) + \sum_{i=1}^{m} p_i x(t - \tau_i) = 0 \tag{7.3.6}$$

where

$$p_i, \tau_i \in (0, \infty) \qquad \text{for } i = 1, 2, \ldots, m$$

it has been shown in Theorem 2.2.1(a) that every solution oscillates provided that 'the sum of the torques $p_i \tau_i$ is greater than $1/e$', that is,

$$\sum_{i=1}^{m} p_i \tau_i > \frac{1}{e}. \tag{7.3.7}$$

If we rewrite condition (7.3.4) in the form

$$\sum_{i=1}^{m} p_i(k_i + 1) \frac{(k_i + 1)^{k_i}}{k_i^{k_i}} > 1$$

and if we observe that

$$\frac{(k_i + 1)^{k_i}}{k_i^{k_i}} \uparrow e \qquad \text{as } k_i \to \infty,$$

then we can interpret (7.3.4) as the discrete analogue of (7.3.7).

Another condition, independent of (7.3.7), which implies that every solution of eqn (7.3.6) oscillates is (see Theorem 2.2.1(b))

$$\left(\prod_{i=1}^{m} p_i\right)^{1/m} \sum_{i=1}^{m} \tau_i > \frac{1}{e}. \tag{7.3.8}$$

This has motivated the following result.

Theorem 7.3.2. *Assume that either (7.3.2) or (7.3.3) holds and suppose that*

$$m\left(\prod_{i=1}^{m} |p_i|\right)^{1/m} \left|\frac{(k + 1)^{k+1}}{k^k}\right| > 1 \tag{7.3.9}$$

where

$$k = \frac{1}{m} \sum_{i=1}^{m} k_i.$$

Then every solution of eqn (7.3.1) oscillates.

Proof. We prove the theorem when (7.3.2) holds. The proof when (7.3.3) holds is similar and is omitted. In view of Corollary 7.1.1 it suffices to prove that the characteristic equation (7.3.5) has no positive roots. Clearly eqn (7.3.5) has no roots in $[1, \infty)$. On the other hand, for $0 < \lambda < 1$, by employing the arithmetic mean–geometric mean inequality we find

$$F(\lambda) = (1 - \lambda)\left(-1 + \sum_{i=1}^{m} p_i \frac{\lambda^{-k_i}}{1 - \lambda}\right) \geq (1 - \lambda)\left[-1 + m\left(\prod_{i=1}^{m} p_i\right)^{1/m} \frac{\lambda^{-k}}{1 - \lambda}\right]$$

$$\geq (1 - \lambda)\left[-1 + m\left(\prod_{i=1}^{m} p_i\right)^{1/m} \frac{(k + 1)^{k+1}}{k^k}\right] > 0$$

and the proof is complete.

7.4 Linearized oscillations for delay difference equations

In this section we develop a linearized oscillation result for delay difference equations. Roughly speaking, we prove that certain non-linear difference equations have the same oscillatory character as an associated linear equation. The statements in this section (but not their proofs) are the discrete analogues of those in Section 4.1.

Consider the non-linear delay difference equation

$$a_{n+1} - a_n + \sum_{i=1}^{m} p_i f_i(a_{n-k_i}) = 0, \qquad n = 0, 1, 2, \ldots \qquad (7.4.1)$$

where for $i = 1, 2, \ldots, m$,

$$p_i \in (0, \infty) \quad \text{and} \quad k_i \in \mathbb{N} \quad \text{with } \sum_{i=1}^{m} (p_i + k_i) \neq 1, \qquad (7.4.2)$$

$$f_i \in C[\mathbb{R}, \mathbb{R}] \quad \text{and} \quad u f_i(u) > 0 \quad \text{for } u \neq 0. \qquad (7.4.3)$$

For easy reference in the sequel, we also list the following hypotheses on f, which will be assumed only when explicitly indicated.

(H$_1$) $$\liminf_{u \to 0} \frac{f_i(u)}{u} \geq 1 \quad \text{for } i = 1, 2, \ldots, m. \qquad (7.4.4)$$

(H$_2$) $$\lim_{u \to 0} \frac{f_i(u)}{u} = 1 \quad \text{for } i = 1, 2, \ldots, m. \qquad (7.4.5)$$

(H$_3$) There exists a positive constant δ such that

$$\left.\begin{array}{llll} \text{either} & f_i(u) \leq u & \text{for } o \leq u \leq \delta & \text{and} & i = 1, 2, \ldots, m \\ \text{or} & f_i(u) \geq u & \text{for } -\delta \leq u \leq 0 & \text{and} & i = 1, 2, \ldots, m \end{array}\right\}.$$

$$(7.4.6)$$

Whenever condition (7.4.4) or (7.4.5) is satisfied, the linear equation

$$b_{n+1} - b_n + \sum_{i=1}^{m} p_i b_{n-k_i} = 0, \qquad n = 0, 1, 2, \ldots \tag{7.4.7}$$

will be called the *linearized equation* associated with eqn (7.4.1).
The main results in this section are the following.

Theorem 7.4.1. *Assume that (7.4.2), (7.4.3), and (7.4.4) hold and suppose that every solution of the linearized equation (7.4.7) is oscillatory. Then every solution of eqn (7.4.1) also oscillates.*

Theorem 7.4.2. *Assume that (7.4.2), (7.4.3), and (7.4.6) hold and suppose that every solution of eqn (7.4.1) is oscillatory. Then every solution of the linearized equation (7.4.7) also oscillates.*

Theorem 7.4.2 is a partial converse of Theorem 7.4.1. By combining both of these theorems we obtain the following powerful linearized oscillation result.

Corollary 7.4.1. *Assume that (7.4.2), (7.4.3), (7.4.5), and (7.4.6) hold. Then every solution of eqn (7.4.1) oscillates if and only if every solution of its linearized equation (7.4.7) oscillates.*

Before we can establish Theorems 7.4.1 and 7.4.2 we need to prove two lemmas which are interesting in their own right. The first lemma shows that, under appropriate hypotheses, if a linear inequality has an eventually positive solution, so does the corresponding 'limiting' equation.

Lemma 7.4.1. *For each $i = 1, 2, \ldots, m$, assume that (7.4.2) holds and let $\{P_i(n)\}$ be sequences of real numbers such that*

$$\liminf_{n \to \infty} P_i(n) \geq p_i \qquad for \; i = 1, 2, \ldots, m. \tag{7.4.8}$$

Suppose that the linear difference inequality

$$x_{n+1} - x_n + \sum_{i=1}^{m} P_i(n) x_{n-k_i} \leq 0, \qquad n = 0, 1, 2, \ldots \tag{7.4.9}$$

has an eventually positive solution $\{x_n\}$. Then the corresponding limiting equation (7.4.7) also has an eventually positive solution.

Proof. First assume that $k = \max\{k_1, k_2, \ldots, k_m\}$ is zero. Then (7.4.9) and

(7.4.7) reduce to

$$x_{n+1} \leqslant \left(1 - \sum_{i=1}^{m} P_i(n)\right) x_n \tag{7.4.10}$$

$$b_{n+1} = \left(1 - \sum_{i=1}^{m} p_i\right) b_n \tag{7.4.11}$$

respectively. As $\{x_n\}$ is eventually positive, it follows that for n sufficiently large,

$$\sum_{i=1}^{m} P_i(n) < 1. \tag{7.4.12}$$

From (7.4.8) it follows that for any $\varepsilon > 0$, there exists an n_0 such that (7.4.12) holds for $n \geqslant n_0$ and also

$$0 < p_i \leqslant P_i(n) + \varepsilon/m \qquad \text{for } n \geqslant n_0.$$

Hence

$$0 < \sum_{i=1}^{m} p_i \leqslant \sum_{i=1}^{m} P_i(n) + \varepsilon < 1 + \varepsilon \qquad \text{for } n \geqslant n_0.$$

As $\varepsilon > 0$ is arbitrary, this implies that

$$0 < \sum_{i=1}^{m} p_i \leqslant 1.$$

But (7.4.2), together with the hypothesis that $k_1 = k_2 = \cdots = k_m = 0$, implies that

$$\sum_{i=1}^{m} p_i \neq 1.$$

Therefore

$$0 < \sum_{i=1}^{m} p_i < 1.$$

Clearly, under this condition, the solution of eqn (7.4.11) with $b_0 = 1$ is positive.

Next, assume that $k \geqslant 1$. For n sufficiently large, set

$$\beta_n = x_n/x_{n-1}.$$

Then eventually $0 < \beta_n < 1$ and (7.4.9) yields

$$\beta_{n+1} - 1 + \sum_{i=1}^{m} P_i(n) \left[\prod_{j=0}^{k_i-1} \frac{1}{\beta_{n-j}} \right] \leqslant 0. \tag{7.4.13}$$

Set

$$\beta = \limsup_{n \to \infty} \beta_n.$$

It follows from (7.4.13) that $0 < \beta_n < 1$ and that $0 < \beta < 1$. We also claim that

$$\beta - 1 + \sum_{i=1}^{m} p_i \beta^{-k_i} \leqslant 0.$$

Indeed, from (7.4.2) and (7.4.8), for every $\varepsilon \in (0, 1)$ there exists an $N_\varepsilon > 0$ such that $P_i(n) \geqslant (1 - \varepsilon) p_i$ for $i = 1, 2, \ldots, m$ and $n \geqslant N_\varepsilon$. Therefore,

$$\beta_{n+1} \leqslant 1 - (1 - \varepsilon) \sum_{i=1}^{m} p_i \left[\prod_{j=0}^{k_i-1} \frac{1}{\beta_{n-j}} \right] \qquad \text{for } n \geqslant N_\varepsilon.$$

Let $\tilde{N}_\varepsilon \geqslant N_\varepsilon + k$ be such that

$$\beta_n \leqslant (1 + \varepsilon)\beta \qquad \text{for } n \geqslant \tilde{N}_\varepsilon - k.$$

Then for $n \geqslant \tilde{N}_\varepsilon - k$,

$$\beta_{n+1} \leqslant 1 - (1 - \varepsilon) \sum_{i=1}^{m} p_i \beta^{-k_i}(1 + \varepsilon)^{-k_i}$$

and so

$$\beta \leqslant 1 - (1 - \varepsilon) \sum_{i=1}^{m} p_i \beta^{-k_i}(1 + \varepsilon)^{-k_i}.$$

As this is true for every $\varepsilon \in (0, 1)$, it follows that

$$\beta \leqslant 1 - \sum_{i=1}^{m} p_i \beta^{-k_i}$$

which proves our claim. Set

$$F(\lambda) = \lambda - 1 + \sum_{i=1}^{m} p_i \lambda^{-k_i}.$$

Then $F(0+) = \infty$, while $F(\beta) \leqslant 0$. It follows that the characteristic equation of eqn (7.4.7) has a positive root λ_0. Then $b_n = \lambda_0^n$ is a positive solution of eqn (7.4.7). The proof is complete.

The following corollary of Lemma 7.4.1 and its 'dual' for advanced inequalities is interesting in its own right.

Corollary 7.4.1. *Assume that*

$$p_i \in (0, \infty) \quad \text{and} \quad k_i \in \mathbb{N} \quad \text{for } i = 1, 2, \ldots, m.$$

Then the following statements are true.

(a) *The difference inequality*

$$x_{n+1} - x_n + \sum_{i=1}^{m} p_i x_{n-k_i} \leqslant 0, \qquad n = 0, 1, 2, \ldots$$

has an eventually positive solution if and only if the difference equation

$$y_{n+1} - y_n + \sum_{i=1}^{m} p_i y_{n-k_i} = 0, \qquad n = 0, 1, 2, \ldots$$

has an eventually positive solution.

(b) *The difference inequality*

$$x_{n+1} - x_n - \sum_{i=1}^{m} p_i x_{n+k_i} \geqslant 0, \qquad n = 0, 1, 2, \ldots$$

has an eventually positive solution if and only if the difference equation

$$y_{n+1} - y_n - \sum_{i=1}^{m} p_i y_{n+k_i} = 0, \qquad n = 0, 1, 2, \ldots$$

has an eventually positive solution.

The following lemma is a useful comparison result for difference inequalities.

Lemma 7.4.2. *For each $i = 1, 2, \ldots, m$, assume that* (7.4.2) *holds and let λ_0 be a positive root of the characteristic equation*

$$\lambda - 1 + \sum_{i=1}^{m} p_i \lambda^{-k_i} = 0$$

of eqn (7.4.7). *Let $N_1 \in N$, $N_1 \geqslant 1$ and $\Theta \in (0, \infty)$. Set $k = \max\{k_1, k_2, \ldots, k_m\}$ and assume that $\{c_n\}$ is a solution of the difference inequality*

$$c_{n+1} - c_n + \sum_{i=1}^{m} p_i c_{n-k_i} \geqslant 0, \qquad n = 0, 1, \ldots, N_1 - 1 \qquad (7.4.14)$$

with initial conditions

$$c_n = \Theta \lambda_0^n, \qquad n = -k, \ldots, 0.$$

Then

$$c_n \geqslant \Theta \lambda_0^n, \qquad n = 1, 2, \ldots, N_1.$$

Proof. The case where $k = 0$ is simple and is omitted. So assume that $k > 0$. Set

$$\gamma_n = c_n / c_{n-1} \qquad \text{for } n = -k + 1, \ldots, -1, 0, 1, \ldots$$

provided that $c_{n-1} \neq 0$. Then, from (7.4.14), we see that

$$0 \leqslant \gamma_1 - 1 + \sum_{i=1}^{m} p_i \left(\prod_{j=0}^{k_i - 1} \frac{1}{\gamma - j} \right) = \gamma_1 - 1 + \sum_{i=1}^{m} p_i \lambda_0^{-k_i} = \gamma_1 - \lambda_0$$

and so $\gamma_1 \geqslant \lambda_0$ or, equivalently, $c_1 \geqslant \Theta \lambda_0$. In a similar manner, (7.4.14) yields

$$0 \leqslant \gamma_2 - 1 + \sum_{i=1}^{m} p_i \left(\prod_{j=0}^{k_i - 1} \frac{1}{\gamma_1 - j} \right) \leqslant \gamma_2 - 1 + \sum_{i=1}^{m} p_i \lambda_0^{-k_i} = \gamma_2 - \lambda_0$$

and so $\gamma_2 \geqslant \lambda_0$ or $c_2 \geqslant \Theta \lambda_0^n$. The proof follows by induction and is omitted.

Proof of Theorem 7.4.1. Assume, for the sake of contradiction, that eqn (7.4.1) has a non-oscillatory solution $\{a_n\}$. We assume that $\{a_n\}$ is eventually positive. The case where $\{a_n\}$ is eventually negative is similar and is omitted. It is not difficult to see that

$$\lim_{n \to \infty} a_n = 0.$$

We can now rewrite eqn (7.4.1) in the form

$$a_{n+1} - a_n + \sum_{i=1}^{m} P_i(n) a_{n-k_i} = 0$$

where

$$P_i(n) = p_i \frac{f(a_{n-k_i})}{a_{n-k_i}}.$$

By (7.4.4),

$$\liminf_{n \to \infty} P_i(n) \geqslant p_i$$

and so the hypotheses of Lemma 7.4.1 are satisfied. Hence the limiting equation (7.1.7) also has an eventually positive solution. This contradicts the hypothesis and the proof is complete.

Proof of Theorem 7.4.2. Assume that (7.4.6) holds with $f_i(u) \leqslant u$ for $0 \leqslant u \leqslant \delta$ and $i = 1, 2, \ldots, m$. The case where $f_i(u) \geqslant u$ for $-\delta \leqslant u \leqslant 0$ and $i = 1, 2, \ldots, m$ is similar and is omitted. Now, assume, for the sake of contradiction, that eqn (7.4.7) has an eventually positive solution $\{b_n\}$. Then, by Corollary 7.1.1, the characteristic equation of eqn (7.4.7),

$$\lambda - 1 + \sum_{i=1}^{m} p_i \lambda^{-k_i} = 0,$$

has a positive root λ_0. Also, as $p_i > 0$, $\lambda_0 \in (0, 1)$. Let $\{a_n\}$ be the solution of eqn (7.4.1) with initial conditions

$$a_n = \Theta \lambda_0^n \qquad \text{for } n = -k, \ldots, -1, 0$$

where

$$k = \max\{k_1, k_2, \ldots, k_m\} \qquad \text{and} \qquad \Theta = \delta \lambda_0^k.$$

To complete the proof it suffices to show that

$$a_n > 0 \qquad \text{for } n = 1, 2, \ldots$$

Otherwise, there exists an integer $N_1 \geqslant 1$ such that

$$a_n > 0 \qquad \text{for } -k \leqslant n < N_1 \text{ and } a_{N_1} \leqslant 0.$$

It follows from eqn (7.4.1) that

$$a_{n+1} < a_n \qquad \text{for } 0 \leqslant n \leqslant N_1 - 1$$

and in particular

$$0 < a_n < a_0 = \Theta = \delta \lambda_0^k < \delta \qquad \text{for } n = 1, 2, \ldots, N_1 - 1.$$

Then, by using (7.4.6), eqn (7.4.1) yields the inequality

$$a_{n+1} - a_n + \sum_{i=1}^{m} p_i a_{n-k_i} \geqslant 0 \qquad \text{for } n = 0, 1, \ldots, N_1 - 1.$$

In view of Lemma 7.4.2, this implies that

$$a_{N_1} \geqslant \Theta \lambda_0^{N_1} > 0$$

and this contradiction completes the proof of the theorem.

7.5 A sharp condition for the oscillation of difference equations

Consider the difference equation

$$a_{n+1} - a_n + p_n a_{n-k} = 0, \qquad n = 0, 1, 2, \ldots \tag{7.5.1}$$

where $\{p_n\}$ is a sequence of non-negative real numbers and k is a positive integer.

Our main goal in this section is to establish the following result.

Theorem 7.5.1. *Suppose that $\{p_n\}$ is a non-negative sequence of real numbers and let k be a positive integer. Then*

$$\liminf_{n \to \infty} \left[\frac{1}{k} \sum_{i=n-k}^{n-1} p_i \right] > \frac{k^k}{(k+1)^{k+1}} \tag{7.5.2}$$

is a sufficient condition for every solution of eqn (7.5.1) to be oscillatory.

The theorem is sharp in that the lower bound $k^k/(k+1)^{k+1}$ cannot be improved. Moreover, when

$$p_n = p \qquad \text{for } n = 0, 1, 2, \ldots,$$

condition (7.5.2) reduces to

$$p > k^k/(k+1)^{k+1}, \tag{7.5.3}$$

which, as we saw in Theorem 7.2.1, is a necessary and sufficient condition for the oscillation of all solutions of the difference equation

$$a_{n+1} - a_n + p a_{n-k} = 0, \qquad n = 0, 1, 2, \ldots \tag{7.5.4}$$

If

$$\liminf_{n \to \infty} p_n > k^k/(k+1)^{k+1} \tag{7.5.5}$$

then it follows from Theorem 7.2.1 and Lemma 7.4.1 that every solution of eqn (7.5.1) oscillates. Clearly (7.5.2) is a substantial improvement over (7.5.5), replacing the p_n of (7.5.5) by the arithmetic mean of the terms $p_{n-k}, \ldots, p_{n-1}$ in (7.5.2).

Theorem 7.5.1 should be looked upon as a discrete analogue of Theorem 2.3.1 about the oscillation of the delay differential equation

$$\dot{x}(t) + p(t)x(t - \tau) = 0, \qquad t \geqslant t_0 \tag{7.5.6}$$

where

$$p \in C[[t_0, \infty), \mathbb{R}^+] \qquad \text{and} \qquad \tau \in (0, \infty),$$

which states that

$$\liminf_{t \to \infty} \int_{t-\tau}^{t} p(s) \, ds > 1/e$$

is a sufficient condition for the oscillation of all solutions of eqn (7.5.6). One should notice that condition (7.5.2) can be written in the form

$$\liminf_{n \to \infty} \left[\sum_{i=n-k}^{n-1} p_i \right] > \left(\frac{k}{k+1} \right)^{k+1} \tag{7.5.2'}$$

and that

$$\lim_{k \to \infty} \left(\frac{k}{k+1} \right)^{k+1} = \lim_{k \to \infty} \left[\frac{1}{(1+1/k)^k} \cdot \frac{1}{1+1/k} \right] = \frac{1}{e}.$$

Proof of Theorem 7.5.1. Assume, for the sake of contradiction, that eqn (7.5.1) has an eventually positive solution $\{a_n\}$. Then eventually

$$a_{n+1} - a_n = -p_n a_{n-k} \leqslant 0$$

and so $\{a_n\}$ is an eventually decreasing sequence of positive numbers. It follows from eqn (7.5.1) that eventually

$$a_{n+1} - a_n + p_n a_n \leqslant 0$$

or

$$p_n \leqslant 1 - \frac{a_{n+1}}{a_n}$$

and so eventually,

$$\frac{1}{k} \sum_{i=n-k}^{n-1} p_i \leqslant \frac{1}{k} \sum_{i=n-k}^{n-1} \left(1 - \frac{a_{i+1}}{a_i} \right). \tag{7.5.7}$$

Set

$$\alpha = k^k / (k+1)^{k+1}. \tag{7.5.8}$$

Then, from (7.5.2), it is clear that we can choose a constant β such that for n sufficiently large,

$$\alpha < \beta \leqslant \frac{1}{k} \sum_{i=n-k}^{n-1} p_i. \tag{7.5.9}$$

Thus, in view of (7.5.7),

$$\beta \leqslant \frac{1}{k} \sum_{i=n-k}^{n-1} \left(1 - \frac{a_{i+1}}{a_i} \right) \qquad \text{for all large } n. \tag{7.5.10}$$

By using (7.5.10) and the well-known inequality between the arithmetic and geometric mean we find that for n sufficiently large,

$$\beta \leqslant \frac{1}{k} \sum_{i=n-k}^{n-1} \left(1 - \frac{a_{i+1}}{a_i}\right) = 1 - \frac{1}{k} \sum_{i=n-k}^{n-1} \frac{a_{i+1}}{a_i}$$

$$\leqslant 1 - \left[\sum_{i=n-k}^{n-1} \frac{a_{i+1}}{a_i}\right]^{1/k} = 1 - \left(\frac{a_n}{a_{n-k}}\right)^{1/k},$$

that is,

$$\left(\frac{a_n}{a_{n-k}}\right)^{1/k} \leqslant 1 - \beta \qquad \text{for all large } n. \tag{7.5.11}$$

In particular, this implies that $0 < \beta < 1$.

Now observe that

$$\max_{0 \leqslant \lambda \leqslant 1} \left[(1 - \lambda)\lambda^{1/k}\right] = k/(k+1)^{1+1/k} = \alpha^{1/k}$$

where α is the positive constant defined by (7.5.8). Therefore

$$1 - \lambda \leqslant \alpha^{1/k}\lambda^{-1/k} \qquad \text{for } 0 < \lambda \leqslant 1$$

and (7.5.11) yields

$$\frac{\beta}{\alpha} a_n \leqslant a_{n-k} \qquad \text{for all large } n. \tag{7.5.12}$$

By using (7.5.12) in eqn (7.5.1) and then by repeating the above arguments we find that

$$(\beta/\alpha)^2 a_n \leqslant a_{n-k} \qquad \text{for all large } n$$

and, by induction, for every $m = 1, 2, \ldots$ there exists an integer n_m such that

$$(\beta/\alpha)^m a_n \leqslant a_{n-k} \qquad \text{for } n \geqslant n_m \text{ and } m = 1, 2, \ldots \tag{7.5.13}$$

Next, observe that because of (7.5.9), for n sufficiently large,

$$\sum_{i=n-k}^{n} p_i \geqslant \sum_{i=n-k}^{n-1} p_i \geqslant k\beta.$$

Hence, for all sufficiently large n,

$$\sum_{i=n-k}^{n} p_i \geqslant M, \tag{7.5.14}$$

where $M = k\beta > 0$. Choose m such that

$$(\beta/\alpha)^m > (2/M)^2, \tag{7.5.15}$$

which is possible because, from (7.5.9), $\beta > \alpha$. Then for n sufficiently large, say for $n \geq n_0$, (7.5.13) is satisfied for the specific m that was chosen in (7.5.15), and also (7.5.9) and (7.5.14) hold, and $\{a_n\}$ is decreasing for $n \geq n_0$. Now in view of (7.5.14), and for $n \geq n_0 + k$, there exists an integer n^* with $n - k \leq n^* \leq n$ such that

$$\sum_{i=n-k}^{n^*} p_i \geq M/2 \qquad \text{and} \qquad \sum_{i=n^*}^{n} p_i \geq M/2.$$

From eqn (7.5.1) and the decreasing nature of $\{a_n\}$ we have

$$a_{n^*+1} - a_{n-k} = \sum_{i=n-k}^{n^*} (a_{i+1} - a_i) = - \sum_{i=n-k}^{n^*} p_i a_{i-k} \leq - \left(\sum_{i=n-k}^{n^*} p_i \right) a_{n^*-k}$$

$$\leq -\frac{M}{2} a_{n^*-k}.$$

Hence

$$\frac{M}{2} a_{n^*-k} \leq a_{n-k}. \tag{7.5.16}$$

Similarly,

$$a_{n+1} - a_{n^*} = \sum_{i=n^*}^{n} (a_{i+1} - a_i) = - \sum_{i=n^*}^{n} p_i a_{i-k} \leq - \left(\sum_{i=n^*}^{n} p_i \right) a_{n-k} \leq -\frac{M}{2} a_{n-k}.$$

and so

$$\frac{M}{2} a_{n-k} \leq a_{n^*}. \tag{7.5.17}$$

From (7.5.16) and (7.5.17) we find

$$(M/2)^2 a_{n^*-k} \leq a_{n^*}$$

which in view of (7.5.13) yields

$$(\beta/\alpha)^m \leq \frac{a_{n^*-k}}{a_{n^*}} \leq (2/M)^2.$$

This contradicts (7.5.15) and so the proof of the theorem is complete.

In Theorem 7.5.1 we assumed that k is a positive integer. When $k = 0$, it is not difficult to see that every solution of the difference equation

$$a_{n+1} - a_n + p_n a_n = 0, \qquad n = 0, 1, 2, \ldots \tag{7.5.18}$$

oscillates provided that the sequence $\{1 - p_n\}$ is not eventually positive. Theorem 7.5.1 can also be extended to the case where k is any integer. The proof is an easy modification of the proof of Theorem 7.5.1 and is omitted.

Theorem 7.5.2. *Let* $\{p_n\}$ *be a sequence of real numbers and let* k *be an integer. Then in each of the following cases, every solution of the difference equation (7.5.1) is oscillatory.*

(a) $k \geqslant 1$, p_n *is eventually non-negative and*

$$\liminf_{n \to \infty} \sum_{i=n-k}^{n-1} p_i > \left(\frac{k}{k+1}\right)^{k+1}.$$

(b) $k = 0$ *and* $(1 - p_n)$ *is not eventually positive.*

(c) $k = -1$ *and* $(1 + p_n)$ *is not eventually positive.*

(d) $k \leqslant -2$, p_n *is eventually non-positive and*

$$\liminf_{n \to \infty} \sum_{i=n+k+1}^{n-1} (-p_i) > \left(\frac{k}{k+1}\right)^{k}.$$

We close this section with a simple discrete analogue of Theorem 3.4.3, which, although it does not possess the sharpness of Theorem 7.5.1, nevertheless gives additional sufficient conditions for the oscillation of all solutions of eqn (7.5.1).

Theorem 7.5.3. *Let* $\{p_n\}$ *be a sequence of real numbers and let* k *be an integer. Then in each of the following two cases, every solution of eqn (7.5.1) oscillates.*

(a) $k \geqslant 0$, p_n *is eventually non-negative and*

$$\limsup_{n \to \infty} \sum_{i=n}^{n+k} p_i > 1.$$

(b) $k \leqslant -1$, p_n *is eventually non-positive and*

$$\limsup_{n \to \infty} \sum_{i=n+k+1}^{n} (-p_i) > 1.$$

Proof. We prove (a). The proof of (b) is similar and is omitted. Assume, for the sake of contradiction, that eqn (7.5.1) has an eventually positive solution $\{a_n\}$. Clearly $\{a_n\}$ is non-increasing and so for n sufficiently large,

$$0 = a_{n+1} - a_n + p_n a_{n-k} \geqslant a_{n+1} - a_n + p_n a_n,$$

$$0 = a_{n+2} - a_{n+1} + p_{n+1} a_{n+1-k} \geqslant a_{n+2} - a_{n+1} + p_{n+1} a_n,$$

$$\cdots$$

$$0 = a_{n+1+k} - a_{n+k} + p_{n+k} a_n \geqslant a_{n+1+k} - a_{n+k} + p_{n+k} a_n.$$

Summing up, we obtain

$$a_{n+1+k} + \left(-1 + \sum_{i=n}^{n+k} p_i \right) a_n \leqslant 0$$

which is a contradiction and the proof is complete.

7.6 Difference inequalities

A slight modification in the proof of Theorem 7.5.1 leads to the following result about the difference inequalities

$$x_{n+1} - x_n + p_n x_{n-k} \leqslant 0, \qquad n = 0, 1, 2, \ldots \qquad (7.6.1)$$

$$y_{n+1} - y_n - p_n y_{n+k} \geqslant 0, \qquad n = 0, 1, 2, \ldots \qquad (7.6.2)$$

where $\{p_n\}$ is a sequence of non-negative real numbers and k is a positive integer.

Theorem 7.6.1. *Assume that $\{p_n\}$ is a non-negative sequence of real numbers and let k be a positive integer. Suppose that (7.5.2) holds. Then the difference inequalities (7.6.1) and (7.6.2) cannot have eventually positive solutions.*

The next result is an extension of Theorem 7.6.1 to the difference inequality (7.6.1). A similar result is also true for (7.6.2).

Theorem 7.6.2. *Let k be a positive integer and let $\{p_n\}$ be a sequence of non-negative real numbers such that*

$$\sum_{j=0}^{k-1} p_{n+j} > 0 \qquad \text{for all large } n. \qquad (7.6.3)$$

Assume that $\{x_n\}$ is a solution of (7.6.1) such that

$$x_n > 0 \qquad \text{for } n \geqslant -k.$$

Then the difference equation

$$a_{n+1} - a_n + p_n a_{n-k} = 0, \qquad n = 0, 1, 2, \ldots \qquad (7.6.4)$$

has a solution $\{a_n\}$ such that

$$0 < a_n \leqslant x_n \qquad \text{for } n \geqslant -k$$

and

$$\lim_{n \to \infty} a_n = 0.$$

Proof. For $\tilde{n} > n$, we have,

$$x_{\tilde{n}+1} - x_n + \sum_{i=n}^{\tilde{n}} p_i x_{i-k} \leq 0$$

and so

$$\sum_{i=n}^{\infty} p_i x_{i-k} \leq x_n \qquad \text{for } n = 0, 1, 2, \ldots \qquad (7.6.5)$$

Consider the space S of all sequences $\{a_n\}$ for $n \geq -k$ that are such that

$$a_n = x_n \qquad \text{for } n \leq 0$$

and

$$0 \leq a_n \leq x_n \qquad \text{for } n \geq 1.$$

Define the operator T on S as follows. For every $a = \{a_n\} \in S$, set $TA = B = \{b_n\}$ where

$$b_n = x_n \qquad \text{for } n \leq 0$$

and

$$b_n = \sum_{i=n}^{\infty} p_i a_{i-k} \qquad \text{for } n \geq 1. \qquad (7.6.6)$$

It follows from (7.6.5) that

$$b_n \leq \sum_{i=n}^{\infty} p_i x_{i-k} \leq x_n \qquad \text{for } n \geq 1$$

and so T is well-defined and $T: S \to S$. If $a^1 = \{a_n^1\}$ and $a^2 = \{a_n^2\}$ are two sequences in S, we say that $a^1 \leq a^2$ if and only if $a_n^1 \leq a_n^2$ for $n \geq -k$. With this definition the operator T is monotonic in the sense that if $a^1, a^2 \in S$ with $a^1 \leq a^2$, then $Ta^1 \leq Ta^2$.

Next, we define the sequence $\{a^r\}$ for $r = 0, 1, \ldots$ of points $a^r \in S$ as follows:

$$a^0 = \{x_n\}$$

$$a^{r+1} = Ta^r \qquad \text{for } r = 0, 1, 2, \ldots \qquad (7.6.7)$$

It follows by induction that

$$\cdots \leq a^{r+1} \leq a^r \leq \cdots \leq a^1 \leq a^0.$$

Set

$$a^r = \{a_n^r\} \qquad \text{and} \qquad a_n = \lim_{r \to \infty} a_n^r.$$

Then one can see that

$$a_n = \sum_{i=n}^{\infty} p_i a_{i-k}$$

and so

$$a_{n+1} - a_n = -p_n a_{n-k}, \qquad (7.6.8)$$

that is, $\{a_n\}$ is a solution of eqn (7.6.4). It is also clear that

$$0 \leqslant a_n \leqslant x_n \qquad \text{and} \qquad \lim_{n \to \infty} a_n = 0.$$

Finally, we claim that $a_n > 0$ for $n \geqslant -k$. Otherwise, there exists $n_0 \geqslant 1$ such that

$$a_n > 0 \qquad \text{for } n = n_0 - k, \ldots, n_0 - 1 \text{ and } a_{n_0} = 0.$$

Then, by summing up both sides of (7.6.8) from $n = n_0$ to $n = n_0 + k - 1$, we find

$$0 \leqslant a_{n_0} + 1 = - \sum_{j=n_0}^{n_0+k-1} p_j a_{j-k} < 0.$$

This is a contradiction and the proof is complete.

The following result is an obvious corollary of Theorem 7.6.2 and its version for eqn (7.6.2).

Corollary 7.6.1.
   (a) *Let $\{p_n\}$ be a sequence of non-negative real numbers and let k be a positive integer such that*

$$\sum_{i=n-k}^{n-1} p_i > 0 \qquad \text{for all large } n.$$

Then the difference inequality

$$x_{n+1} - x_n + p_n x_{n-k} \leqslant 0, \qquad n = 0, 1, 2, \ldots$$

has an eventually positive solution if and only if the difference equation

$$z_{n+1} - z_n + p_n z_{n-k} = 0, \qquad n = 0, 1, 2, \ldots$$

has an eventually positive solution.
   (b) *Let $\{p_n\}$ be a sequence of non-negative real numbers and let $k \in \{2, 3, \ldots\}$ be such that*

$$\sum_{i=n-k+1}^{n-1} p_i > 0 \qquad \text{for all large } n.$$

Then the difference inequality

$$y_{n+1} - y_n - p_n y_{n+k} \geqslant 0, \qquad n = 0, 1, 2, \ldots$$

has an eventually positive solution if and only if the difference equation

$$w_{n+1} - w_n - p_n w_{n+k} = 0, \qquad n = 0, 1, 2, \ldots$$

has an eventually positive solution.

7.7 Equations with positive and negative coefficients

Consider the difference equations with positive and negative coefficients

$$a_{n+1} - a_n + p a_{n-k} - q a_{n-l} = 0, \qquad n = 0, 1, 2, \ldots \tag{7.7.1}$$

$$a_{n+1} - a_n + p a_{n+k} - q a_{n+l} = 0, \qquad n = 0, 1, 2, \ldots \tag{7.7.2}$$

where

$$p, q \in \mathbb{R}^+ \qquad \text{and} \qquad k, l \in \mathbb{N}.$$

Our aim in this section is to establish the following theorems.

Theorem 7.7.1. *Assume that*

$$p > q \geqslant 0, \qquad k \geqslant l \geqslant 0, \qquad q(k - l) \leqslant 1 \tag{7.7.3}$$

and that

$$\left.\begin{array}{ll} p - q > k^k/(k + 1)^{k+1} & \text{if } k \geqslant 1 \quad \text{and} \\ p - q \geqslant 1 & \text{if } k = 0 \end{array}\right\}. \tag{7.7.4}$$

Then every solution of eqn (7.7.1) oscillates.

Theorem 7.7.2. *Assume that*

$$0 \leqslant p < q, \qquad 1 \leqslant k \leqslant l, \qquad p(l - k) \leqslant 1 \tag{7.7.5}$$

and that

$$\left.\begin{array}{ll} q - p > (l - 1)^{l-1}/l^l & \text{if } l \geqslant 2 \\ q - p \geqslant 1 & \text{if } l = 1 \end{array}\right\}. \tag{7.7.6}$$

Then every solution of eqn (7.4.2) oscillates.

Proof of Theorem 7.7.1. The case $k = l$ reduces to Theorem 7.2.1. So suppose $k > l$. Assume, for the sake of contradiction, that eqn (7.7.1) has an eventually positive solution $\{a_n\}$. Then there exists $n_0 \in \mathbb{N}$ such that $a_n > 0$ for $n \geqslant n_0$.

Set

$$c_n = a_n - q \sum_{j=l+1}^{k} a_{n-j}, \qquad n \geq n_0 + k. \tag{7.7.7}$$

Then

$$c_{n+1} - c_n = a_{n+1} - a_n - q(a_{n-l} - a_{n-k}) = -(p - q)a_{n-k} < 0, \qquad n \geq n_0 + k. \tag{7.7.8}$$

Thus c_n is a strictly decreasing sequence for $n \geq n_0 + k$. We claim that

$$L \equiv \lim_{n \to \infty} c_n \in \mathbb{R}. \tag{7.7.9}$$

Otherwise, $L = -\infty$ and $\{a_n\}$ must be unbounded. Hence there exists $n_1 \geq n_0 + k$ such that $a_{n_1} = \max\{a_n : n \leq n_1\}$ and $c_{n_1} < 0$. Then

$$0 > c_{n_1} = a_{n_1} - q \sum_{j=l+1}^{k} a_{n_1-j} \geq a_{n_1}[1 - q(k - l)] \geq 0,$$

which is a contradiction. Thus (7.7.9) holds. It now follows, by taking limits in (7.7.8), that $\lim_{n \to \infty} a_n = 0$. As the sequence $\{c_n\}$ decreases to zero, we conclude that

$$c_n > 0 \qquad \text{for } n \geq n_0 + k. \tag{7.7.10}$$

Also, from (7.7.7) we see that $c_n < a_n$ for $n \geq n_0 + k$, so (7.7.8) yields the inequality

$$c_{n+1} - c_n + (p - q)c_{n-k} < 0 \qquad \text{for } n \geq n_0 + 2k. \tag{7.7.11}$$

But in view of Theorem 7.2.1, Corollary 7.4.1 and the hypothesis (7.7.4), the difference inequality (7.7.11) cannot have an eventually positive solution. This contradicts (7.7.10) and completes the proof of the theorem.

Proof of Theorem 7.7.2. The case $k = l$ reduces to Theorem 7.2.1. So suppose $k < l$. Assume, for the sake of contradiction, that eqn (7.7.2) has an eventually positive solution $\{a_n\}$. Set

$$c_n = a_n - p \sum_{j=n+k}^{n+l-1} a_j \tag{7.7.12}$$

Then

$$c_{n+1} - c_n = (a_{n+1} - a_n) - p(a_{n+l} - a_{n+k}) = (q - p)a_{n+l} > 0. \tag{7.7.13}$$

Thus $\{c_n\}$ is eventually strictly increasing and either

$$\lim_{n \to \infty} c_n = \infty \tag{7.7.14}$$

or

$$\lim_{n \to \infty} c_n = c \in \mathbb{R}. \tag{7.7.15}$$

Assume that (7.7.15) holds. Then from (7.7.13) and (7.7.12) we see that

$$\lim_{n \to \infty} a_n = 0 = \lim_{n \to \infty} c_n.$$

Hence there exists an index n such that

$$c_{n_1} < 0 \quad \text{and} \quad a_{n_1} \geq a_n > 0 \quad \text{for } n \geq n_1.$$

Then (7.7.12) yields

$$0 > c_{n_1} = a_{n_1} - p \sum_{j=n_1+k}^{n_1+l-1} a_j \geq a_{n_1}[1 - p(l - k)] \geq 0,$$

which is a contradiction. Therefore (7.7.14) holds. From (7.7.13) and (7.7.12) we find $c_{n+1} - c_n - (p - q)c_{n+l} \geq 0$. Also $c_n > 0$. In view of Corollary 7.4.1 this implies that the difference equation

$$y_{n+1} - y_n - (p - q)y_{n+l} = 0$$

has an eventually positive solution. This contradicts (7.7.6) and completes the proof.

By combining the results in Theorems 7.3.1, 7.7.1, and 7.7.2 we obtain Table 7.2 which gives sufficient conditions for the oscillation of all solutions

Table 7.2

	p	q	k	l	Sufficient conditions for oscillation
1	$+$	$+$	$+$	$+$	$pK + qL > 1$
2	$+$	$+$	$+$	$-$	$pK > 1$
3	$+$	$+$	$-$	$-$	There exist non-oscillatory solutions
4	$+$	$-$	$+$	$+$	$1 + q(k - l) \geq 0$ and $(p + q)K > 1$
5	$+$	$-$	$+$	$-$	There exist non-oscillatory solutions
6	$+$	$-$	$-$	$-$	$1 - p(k - l) \geq 0$ and $(p + k)L > 1$
7	$-$	$+$	$+$	$+$	There exist non-oscillatory solutions
8	$-$	$+$	$+$	$-$	There exist non-oscillatory solutions
9	$-$	$+$	$-$	$-$	There exist non-oscillatory solutions
10	$-$	$-$	$+$	$+$	There exist non-oscillatory solutions
11	$-$	$-$	$+$	$-$	$qL > 1$
12	$-$	$-$	$-$	$-$	$pK + qL > 1$

of the difference equation

$$a_{n+1} - a_n + pa_{n-k} + qa_{n-l} = 0, \qquad n = 0, 1, 2, \ldots$$

where

$$p, q \in \mathbb{R} - \{0\}, \qquad k, l \in \mathbb{Z} - \{0, 1\}, \qquad \text{and} \qquad k > l.$$

In Table 7.2, K and L denote the constants $(k+1)^{k+1}/k^k$ and $(l+1)^{l+1}/l^l$ respectively.

7.8 Existence of positive solutions

In this section we establish sufficient conditions for the existence of positive solutions of the difference equation

$$a_{n+1} - a_n + \sum_{i=1}^{m} P_i(n)a_{n-k_i} = 0, \qquad n = 0, 1, 2, \ldots \qquad (7.8.1)$$

where

$$P_i(n) \in \mathbb{R} \qquad \text{and} \qquad k_i \in \mathbb{N} \qquad \text{for } i = 1, 2, \ldots, m. \qquad (7.8.2)$$

The first result in this section is the following theorem.

Theorem 7.8.1. *Assume that (7.8.2) holds and that there exists a $\mu \in (0, 1]$ such that*

$$\mu - 1 + \sum_{i=1}^{m} P_i^+(n)\mu^{-k_i} \leqslant 0 \qquad (7.8.3)$$

where

$$P_n^+(n) = \max\{0, P_i(n)\} \qquad for \ i = 1, 2, \ldots, m \ and \ n = 0, 1, \ldots$$

Set

$$k = \max_{1 \leqslant i \leqslant m} k_i \qquad (7.8.4)$$

and let $\alpha_j \in (0, \infty)$ for $j = -k, \ldots, 0$ be chosen in such a way that

$$\alpha_{j+1} \geqslant \mu\alpha_j > 0 \qquad for \ j = -k, \ldots, -1. \qquad (7.8.5)$$

Then the unique solution $\{a_n\}$ of eqn (7.8.1) which satisfies the initial conditions

$$a_j = \alpha_j \qquad for \ j = -k, \ldots, 0 \qquad (7.8.6)$$

is positive for all n.

Before we prove Theorem 7.8.1 we need the following lemma which is interesting in its own right.

Lemma 7.8.1. *Assume that (7.8.2) holds and let k be as given by (7.8.4). Then a non-trivial sequence $\{a_n\}$ is a solution of eqn (7.8.1) with constant sign for all $n \geqslant -k$ if and only if it can be written in the form*

$$a_{n+1} = \left(\prod_{j=-k}^{n} \lambda_j \right) a_{-k} \qquad for \ n \geqslant -k \qquad (7.8.7)$$

where $\{\lambda_n\}$ is a sequence of positive numbers, for $n \geqslant -k$, such that

$$\lambda_n = 1 - \sum_{i=1}^{m} P_i(n) \prod_{j=1}^{k_i} 1/\lambda_{n-j}, \qquad n = 0, 1, \ldots \qquad (7.8.8)$$

Proof. Let $\{a_n\}$ be a solution of eqn (7.8.1) with constant sign, that is,

$$\operatorname{sgn} a_n = \operatorname{sgn} a_{-k} \neq 0 \qquad for \ n \geqslant -k.$$

Set

$$\lambda_n = a_{n+1}/a_n \qquad for \ n \geqslant -k. \qquad (7.8.9)$$

Then $\lambda_n > 0$, (7.8.7) is clearly satisfied and for all $l = 1, \ldots, k$ and $n = 0, 1, \ldots,$

$$a_{n-l}/a_n = \prod_{j=1}^{l} a_{n-j}/a_{n-j+1} = \prod_{j=1}^{l} 1/\lambda_{n-j}. \qquad (7.8.10)$$

Finally, it follows from eqn (7.8.1) that for $n = 0, 1, \ldots,$

$$\lambda_n = a_{n+1}/a_n = 1 - \sum_{i=1}^{m} P_i(n) \frac{a_{n-k_i}}{a_n} = 1 - \sum_{i=1}^{m} P_i(n) \prod_{j=1}^{k_i} 1/\lambda_{n-j} \qquad (7.8.11)$$

and (7.8.8) holds.

Conversely, assume that $\{\lambda_n\}$, for $n \geqslant -k$, is a sequence of positive numbers which satisfies (7.8.8) and let $\{a_n\}$ be a non-trivial sequence given by (7.8.7). Clearly, $\{a_n\}$ has constant sign for $n \geqslant -k$. Also observe that (7.8.9) and (7.8.10) are satisfied, so (7.8.8) yields

$$a_{n+1}/a_n = 1 - \sum_{i=1}^{m} P_i(a) a_{n-k_i}/a_n.$$

This shows that $\{a_n\}$ is a solution of eqn (7.8.1) and the proof is complete.

Proof of Theorem 7.8.1. Set

$$\lambda_n = a_{n+1}/a_n \qquad (7.8.12)$$

for $n = -k, \ldots, -1$ and for as long as $a_n \neq 0$. Then (7.8.5) implies that

$$\lambda_n \geqslant \mu > 0 \qquad (7.8.13)$$

for $n = -k, \ldots, -1$. Now one can show, as in the proof of Lemma 7.8.1, that (7.8.8) is true for all values of $n \geqslant 0$ for which $a_0 \cdots a_n \neq 0$. Thus for $n = 0$, (7.8.8) yields

$$\lambda_0 = 1 - \sum_{i=1}^{m} P_i(0) \prod_{j=1}^{k_i} 1/\lambda_{-j},$$

which in view of (7.8.13) and (7.8.3) implies that

$$\lambda_0 \geqslant 1 - \sum_{i=1}^{m} P_i(0)\mu^{-k_i} \geqslant \mu.$$

Hence (7.8.13) is true for $n = 0$ and (7.8.12) implies that $a_1 = \lambda_0 a_0 > 0$. It follows, by induction, that λ_n is well-defined for all $n \geqslant -k$ and satisfies the hypotheses of Lemma 7.8.1. Since $a_j = \alpha_j > 0$ for $j = -k, \ldots, 0$, it follows that $a_n > 0$ for $n \geqslant -k$ and the proof is complete.

Corollary 7.8.1. *Assume that* $k \in \{1, 2, \ldots\}$, $P(n) > 0$ *for* $n \geqslant 0$, *and that*

$$P(n) \leqslant k^k/(k+1)^{k+1}. \qquad (7.8.14)$$

Then the difference equation

$$a_{n+1} - a_n + P(n)a_{n-k} = 0, \qquad n = 0, 1, 2, \ldots \qquad (7.8.15)$$

has a positive solution $\{a_n\}$.

Proof. It suffices to prove that there exists a $\mu \in (0, 1]$ such that

$$\mu - 1 + P(n)\mu^{-k} \leqslant 0. \qquad (7.8.16)$$

In view of (7.8.14), (7.8.16) is true with $\mu = k/(k+1)$. The proof is complete.

It should be noticed that (7.8.14) is a sharp condition for eqn (7.8.15) to have a positive solution. This is because, by Theorem 7.8.1, when $P(n)$ is a positive constant and $k \geqslant 1$, then (7.8.14) is a necessary and sufficient condition for eqn (7.8.15) to have a positive solution.

The next theorem provides a sufficient condition for the existence of a positive and increasing solution of eqn (7.8.1).

Theorem 7.8.2. *Assume that* (7.8.2) *holds,*

$$0 \leqslant k_1 \leqslant k_2 \leqslant \cdots \leqslant k_m, \qquad (7.8.17)$$

and

$$\sum_{i=1}^{s} P_i(n) \leqslant 0 \qquad \text{for } s = 1, 2, \ldots, m \text{ and } n = 0, 1, \ldots \qquad (7.8.18)$$

Then eqn (7.8.1) has a positive and increasing solution.

Proof. Let $\alpha_j \in (0, \infty)$ for $j = -k, \ldots, 0$ be chosen in such a way that $\alpha_0 > 0$ and

$$\alpha_{j+1} - \alpha_j \geqslant 0 \qquad \text{for } j = -k, \ldots, -1. \qquad (7.8.19)$$

Consider the unique solution $\{a_n\}$ of eqn (7.8.1) which satisfies the initial condition (7.8.6). We will show that for all $n \geqslant 0$,

$$a_{n+1} \geqslant a_n > 0. \qquad (7.8.20)$$

If $m = 1$, then $P_1(n) \leqslant 0$ for $n = 0, 1, \ldots$ Hence

$$a_1 - a_0 = -P_1(0)\alpha_0 \geqslant 0$$

and (7.8.18) follows by induction. If $m > 1$, then from (7.8.1) it follows that

$$a_{n+1} - a_n = -\sum_{i=1}^{m-1} \left(\sum_{j=1}^{i} P_j(n) \right)(a_{n-k_i} - a_{n-k_{i+1}}) \qquad (7.8.21)$$

for all $n = 0, 1, \ldots$ Thus (7.8.17), (7.8.18), and (7.8.21) yield that (7.8.20) is true for $n = 0$. Now it follows, by induction, that (7.8.20) is also true for all $n = 1, 2, \ldots$ and the proof is complete.

The following result is an immediate consequence of Theorem 7.8.2.

Corollary 7.8.2. *Consider the delay difference equation*

$$a_{n+1} - a_n + \sum_{i=1}^{m} P_i(n)a_{n-k_i} - P(n)a_{n-k} = 0 \qquad (7.8.22)$$

where

$$P_i(n), P(n) \in \mathbb{R}, \qquad \text{and} \qquad k_i, k \in \mathbb{N} \qquad \text{for } i = 1, 2, \ldots, m \text{ and } n \in \mathbb{N},$$

$$k \leqslant \max\{k_1, \ldots, k_m\} \qquad \text{and} \qquad \sum_{i=1}^{m} P_i(n) \leqslant P(n) \qquad \text{for } n \in \mathbb{N}.$$

Then eqn (7.8.22) has a positive increasing solution.

7.9 Oscillations in a discrete delay logistic model

Pielou (1969, p. 22 and 1974, p. 79) considered the delay difference equation

$$N_{t+1} = \frac{\alpha N_t}{1 + \beta N_{t-k}} \qquad (7.9.1)$$

where

$$\alpha \in (1, \infty), \qquad \beta \in (0, \infty), \qquad \text{and} \qquad k \in \mathbb{N} \qquad (7.9.2)$$

as the discrete analogue of the delay logistic equation

$$\dot{N}(t) = rN(t)[1 - N(t - \tau)/K] \qquad (7.9.3)$$

where r and K are the growth rate and the carrying capacity of the population respectively.

One can derive eqn (7.9.1) as follows. When $\tau = 0$, the solution of eqn (7.9.3) with $N(0) = N_0$ is given by

$$N(t) = \frac{K}{1 + [K/N_0 - 1] e^{-rt}}.$$

An elementary calculation shows that

$$N(t + 1) = \frac{\alpha N(t)}{1 + \beta N(t)} \qquad (7.9.4)$$

where

$$\alpha = e^r \qquad \text{and} \qquad \beta = (e^r - 1)/K.$$

Pielou arrived at eqn (7.9.1) by assuming that there is a delay k in the response of the growth rate per individual to density changes. This is also how one arrives at the delay logistic equation (7.9.3) from the logistic equation

$$\dot{N}(t) = rN(t)[1 - N(t)/K].$$

Pielou's interest in eqn (7.9.1) was in showing that 'the tendency to oscillate is a property of the populations themselves and is independent of any extrinsic factors'. That is, population sizes oscillate 'even though the environment remains constant'. According to Pielou, 'oscillations can be set up in a population if its growth rate is governed by a density dependent mechanism and if there is a delay in the response of the growth rate to density changes. When this happens the size of the population alternately overshoots and undershoots its equilibrium level'.

The blowfly (*Lucilia cuprina*) studied by Nicholson (1954) is an example of a laboratory population which behaves in the manner described above.

Our aim in this section is to obtain necessary and sufficient conditions for the oscillation of all positive solutions of the delay difference equation

$$x_{n+1} = \frac{\alpha x_n}{1 + \beta x_{n-k}}, \qquad n = 0, 1, 2, \ldots \tag{7.9.5}$$

If $\alpha_{-k}, \ldots, \alpha_0$ are $(k + 1)$ given constants, then eqn (7.9.5) has a unique solution satisfying the initial conditions

$$x_i = \alpha_i \qquad \text{for } i = -k, \ldots, 0. \tag{7.9.6}$$

If the initial values are such that

$$\alpha_i \geqslant 0 \qquad \text{for } i = -k, \ldots, -1 \text{ and } \alpha_0 > 0, \tag{7.9.7}$$

then the unique solution of the IVP (7.9.5) and (7.9.6) is positive for $n \geqslant 0$. In the sequel, we will investigate only those solutions of eqn (7.9.5) whose initial values satisfy condition (7.9.7).

A sequence $\{x_n\}$ is said to oscillate about zero, or simply oscillate, if the terms x_n are not all eventually positive or eventually negative. A sequence $\{x_n\}$ is said to oscillate about x^* if the sequence $\{x_n - x^*\}$ oscillates. The main result in this section is the following theorem.

Theorem 7.9.1. *Assume that (7.9.2) holds. Then every positive solution of eqn (7.9.5) oscillates about its positive equilibrium $(\alpha - 1)/\beta$ if and only if*

$$(\alpha - 1)/\alpha > k^k/(k + 1)^{k+1}. \tag{7.9.8}$$

Proof. The change of variables

$$x_n = \frac{\alpha - 1}{\beta} e^{y_n}$$

transforms eqn (7.9.5) to the difference equation

$$y_{n+1} - y_n + \frac{\alpha - 1}{\alpha} f(y_{n-k}) = 0, \qquad n = 0, 1, 2, \ldots \tag{7.9.9}$$

where

$$f(u) = \frac{\alpha}{\alpha - 1} \ln \frac{(\alpha - 1) e^u + 1}{\alpha}.$$

Clearly, every solution of eqn (7.9.9) oscillates about zero if and only if every solution of eqn (7.9.5) oscillates about $(\alpha - 1)/\beta$. One can easily see now that all the hypotheses of Corollary 7.4.1 are satisfied for the difference equation (7.9.9). In particular, note that (7.4.6) is satisfied because

$$f(u) \geqslant u \qquad \text{for } u < 0.$$

The linearized equation associated with eqn (7.9.9) is

$$z_{n+1} - z_n + \frac{\alpha - 1}{\alpha} z_{n-k} = 0, \qquad (7.9.10)$$

and by Theorem 7.2.1, every solution of eqn (7.9.10) oscillates if and only if (7.9.8) holds. The proof is now an elementary consequence of Corollary 7.4.1.

7.10 Notes

The book of Pielou (1974, p. 79) and the papers of Levin and May (1976) and Clark (1976) seem to be the earliest sources which make use of the term delay difference equation (see also Myskis 1972). Our interest in delay difference equations was initially stimulated by our study of equations with piecewise constant argument (see Chapter 8) and by the results of Erbe and Zhang (1988).

The results of Section 7.1 are from Györi *et al.* (1990a); see also Györi and Ladas (1989), Ladas *et al.* (1989g), and Partheniadis (1988). The results of Sections 7.2 and 7.3 are extracted from Ladas (1990a); see also Erbe and Zhang (1989) and Gopalsamy *et al.* (1989a). The results of Section 7.4 are adapted from Györi and Ladas (1989). Theorem 7.5.1 is due to Ladas *et al.* (1989f). Theorem 7.5.3(a) is given in Erbe and Zhang (1989). The results of Section 7.6 are adapted from Ladas *et al.* (1989f) and the results of Section 7.7 are from Ladas (1988b). Section 7.8 is from Györi *et al.* (1990c). Theorem 7.9.1 is extracted from Kuruklis and Ladas (1990). For oscillation results in difference equations with periodic coefficients, see Philos (1990). For some interesting results on the existence of positive solutions for higher order difference equations, see Philos and Sficas (1990).

In this chapter we developed an oscillation theory for delay difference equations which parallels the oscillation theory of delay differential equations. However, there is a vast amount of interesting research on the oscillation of difference equations that has been developed independently of our framework of delay equations. See for example Hooker *et al.* (1987), Peterson and Ridenhour (1989), Hall and Trimble (1990), Mingarelli (1983), and the book of Kelly and Peterson (1991).

7.11 Open problems

7.11.1   Obtain explicit conditions in terms of the matrices Q and the indices k and l for the oscillation of all solutions of eqn (7.1.4).

7.11.2   Extend the linearized oscillation results of Section 7.4 to systems of difference equations and to higher-order difference equations.

7.11.3   Extend Theorem 7.5.1 to difference equations with oscillating coefficients p_n.

7.11.4   Extend Theorem 7.5.1 to systems of difference equations.

7.11.5   Obtain some discrete analogues of the results in Chapter 3.

7.11.6   Extend Theorem 7.9.1 to delay difference equations of the form

$$x_{n+1} = \frac{\alpha x_n}{1 + \beta x_{n-k} + \gamma x_{n-l}}, \qquad n = 0, 1, 2, \ldots$$

where

$$\alpha \in (1, \infty), \qquad \beta, \gamma \in (0, \infty), \qquad \text{and} \qquad k, l \in \mathbb{N}.$$

7.11.7   Study the oscillatory character and the periodic nature of all positive solutions of the delay difference equation

$$x_{n+1} = \frac{\alpha + \beta x_n + \gamma x_{n-k}}{A + B x_n + C x_{n-l}}, \qquad n = 0, 1, 2, \ldots$$

where

$$\alpha, \beta, \gamma, A, B, C \in [0, \infty) \qquad \text{with } B + C > 0 \text{ and } k, l \in \mathbb{N}.$$

7.11.8   Obtain discrete analogues of the results in Section 2.5, 4.2, 4.3, and 4.4.

8

OSCILLATIONS OF EQUATIONS WITH PIECEWISE CONSTANT ARGUMENTS

Let $[\cdot]$ denote the greatest-integer function. Then

$$\dot{x}(t) + px([t - 1]) = 0 \qquad (8.0.1)$$

$$\dot{x}(t) + px(2[(t + 1)/2]) = 0 \qquad (8.0.2)$$

$$\dot{N}(t) = pN(t)(1 - N([t])/K) = 0 = 0 \qquad (8.0.3)$$

are examples of equations with piecewise constant arguments. The precise meaning of a solution of such an equation will be given in Sections 8.1, 8.2, and 8.3. These equations have been the subject of intense investigations; see for example Aftabizadeh *et al.* (1987), Cooke and Wiener (1984, 1987*b*), Györi and Ladas (1990*b*), Gopalsamy *et al.* (1990*b*) and the references cited therein.

As we shall see, within intervals of certain length, each of the above equations has the structure of a continuous dynamical system, while the continuity of the solution at the points that join consecutive intervals leads to difference equations. It is interesting to note that equations with piecewise constant arguments provide the simplest examples of differential equations capable of displaying chaotic behaviour. For example, one can easily see that the unique solution of the initial value problem

$$\begin{cases} \dot{y}(t) = 3y([t]) - y^2([t]) \\ y(0) = A_0 \end{cases}$$

has the property that

$$y(n + 1) = 4y(n) - y^2(n), \qquad n = 0, 1, 2, \ldots$$

If we choose $A_0 = 4 \sin^2(\pi/9)$ then the unique solution of this difference equations is

$$y(n) = 4 \sin^2(2^n \pi/9), \qquad n = 0, 1, 2, \ldots$$

which has period three. By the well-known result which states that 'period three implies chaos' (see Li and Yorke 1975), the solution of the above differential equation exhibits chaos. The chaotic behaviour of solutions of eqn (8.0.3) is described in Section 8.2.

Fractals and Chaos are everywhere and different equations are responsible for that. It should be mentioned that the study of equations with piecewise constant argument of the form of eqn (8.0.1) was our initial motivation for most of the results on difference equations in Chapter 7.

Concerning eqn (8.0.2), it is interesting to observe that its argument deviation

$$\tau(t) = t - 2[(t+1)/2]$$

is a periodic function of period two. Furthermore, for every integer n, $\tau(t)$ is negative for $2n - 1 \leqslant t < 2n$ and positive for $2n < t \leqslant 2n + 1$. Therefore, on each interval $[2n - 1, 2n + 1)$, eqn (8.0.2) is of alternately advanced and retarded type.

In summary, equations with piecewise constant arguments are interesting in their own right, have some curious and unpredictable properties and also can be used to inspire and motivate new results on difference equations.

8.1 Necessary and sufficient conditions for the oscillation of linear systems with piecewise constant arguments

Let $[\cdot]$ denote the greatest-integer function, $\mathbb{N}$ the set of non-negative integers and $\mathbb{R}^{r \times r}$ the set of all $r \times r$ matrices with real components. In this section we apply Theorem 7.1.1 to obtain necessary and sufficient conditions for the oscillation of all solutions of the system of linear differential equations with piecewise constant arguments

$$\dot{x}(t) + \sum_{j=-k}^{l} Q_j x([t + j]) = 0, \qquad t \geqslant 0, \tag{8.1.1}$$

where

$$k, l \in \mathbb{N} \quad \text{and} \quad Q_j \in \mathbb{R}^{r \times r} \quad \text{for } j = -k, \ldots, l. \tag{8.1.2}$$

With eqn (8.1.1) one associates initial conditions of the form

$$x(j) = a_j \in \mathbb{R}^r \quad \text{for } j \in \{-k, \ldots, 0\} \cup \{0, \ldots, l - 1\} \tag{8.1.3}$$

with the convention that if $l = 0$, the set $\{0, \ldots, l - 1\}$ is empty.

By a solution of eqn (8.1.1) we mean a function x which is defined on the set $\{-k, \ldots, 0\} \cup (0, \infty)$ with values in $\mathbb{R}^r$ and which satisfies the following properties:

(a) x is continuous on $[0, \infty)$.

(b) The derivative $\dot{x}(t)$ exists at each point $t \in (0, \infty)$ with the possible exception of the points $t \in \mathbb{N}$ where finite one-sided derivatives exist.

(c) Equation (8.1.1) is satisfied on each interval $[n, n + 1]$ for $n \in \mathbb{N}$.

Let $a_j \in \mathbb{R}^r$ for $j \in \{-k, \dots, l = 1\}$ be given. Then, as we show in the next lemma, the initial value problem (8.1.1) and (8.1.3) has a unique solution provided that

$$\left.\begin{array}{c} |l - 1| + |\det(Q_1 + I)| \neq 0 \\ \text{and if } k = 0 \quad \text{and} \quad l \geqslant 2 \quad \text{then } \det Q_l \neq 0 \end{array}\right\}, \qquad (8.1.4)$$

where I is the $r \times r$ identity matrix. When

$$l = 1 \quad \text{and} \quad \det(Q_1 + I) = 0 \quad \text{or when } k = 0, \quad l \geqslant 2 \quad \text{and} \quad \det Q_l = 0,$$

the initial value problem (8.1.1) and (8.1.3) has in general no solution.

Lemma 8.1.1. *Assume that (8.1.2) and (8.1.4) hold. Then the initial value problem (8.1.1) and (8.1.3) has a unique solution $x(t)$. Furthermore, $x(t)$ is given by*

$$x(t) = a_n - \left(\sum_{j=-k}^{l} Q_j a_{n+j}\right)(t - n) \qquad \text{for } n \leqslant t < n + 1 \text{ and } n \in \mathbb{N} \tag{8.1.5}$$

where (a_n) is the unique sequence of vectors in $\mathbb{R}^r$ which satisfies the difference equation

$$a_{n+1} - a_n + \sum_{j=-k}^{l} Q_j a_{n+j} = 0 \qquad \text{for } n \geqslant 0. \tag{8.1.6}$$

Proof. Let $x(t)$ be a solution of (8.1.1) and (8.1.3). Then in the interval $n \leqslant t < n + 1$, for any $n \in N$, eqn (8.1.1) becomes

$$\dot{x}(t) + \sum_{j=-k}^{l} Q_j a_{n+j} = 0 \tag{8.1.7}$$

where we use the notation

$$a_n = x(n) \qquad \text{for } n \in \{-k, \dots, 0, 1, 2, \dots\}.$$

Clearly, the solution of (8.1.7) is given by (8.1.5). By continuity of the solution, as $t \to n + 1$, (8.1.5) yields (8.1.6). So far we have proved that if $x(t)$ is a solution of (8.1.1) and (8.1.3) then $x(t)$ is given by (8.1.5) where the sequence $\{a_n\}$ satisfies (8.1.6).

Conversely, let $\{a_n\}$ be the solution of (8.1.6) with initial values $a_{-k}, \dots, a_0, \dots, a_{l-1}$. Note that this solution exists and is unique provided that (8.1.4) holds. Now define $x(t)$ by (8.1.3) and (8.1.5). Then one can easily show by direct substitution that $x(t)$ satisfies (8.1.1). The proof is complete.

We are now ready to present the main result in this section, which states that every solution of eqn (8.1.1) oscillates componentwise if and only if every solution of eqn (8.1.6) oscillates componentwise if and only if the characteristic equation of eqn (8.1.6),

$$\det\left(\lambda I - I + \sum_{j=-k}^{l} Q_j \lambda^j\right) = 0, \tag{8.1.8}$$

has no positive roots.

Theorem 8.1.1. *Assume that conditions (8.1.2) and (8.1.4) hold. Then the following statements are equivalent.*

(a) *Every solution of eqn (8.1.1) oscillates componentwise.*

(b) *Every solution of eqn (8.1.6) oscillates componentwise.*

(c) *The characteristic equation (8.1.8) has no positive roots.*

Proof. The fact that (b) is equivalent to (c) follows from Theorem 7.1.1. The proof that (b) $\Rightarrow$ (a) in an obvious consequence of the fact that $x(n) = a_n$. It remains to show that (a) $\Rightarrow$ (b). To this end, assume that every component of every solution of eqn (8.1.1) oscillates, that is, the solutions of eqn (8.1.1) are oscillatory componentwise, and that eqn (8.1.6) has a solution $\{a_n\}$ which is non-oscillatory. Let

$$a_n = [a_n^1, a_n^2, \ldots, a_n^r]^T \qquad \text{for } n \geqslant 0.$$

Then one of the components of a_n is eventually positive or eventually negative. Without loss of generality we assume that the first component a_n^1 is eventually positive, that is, there exists an $n_0 \geqslant 0$ such that

$$a_n^1 > 0 \qquad \text{for } n \geqslant n_0.$$

From (8.1.5) and (8.1.6) it follows that for $n \in N$ and $t \in [n, n+1)$,

$$x(t) = a_n - (a_{n+1} - a_n)(t - n) = (1 + (t - n))a_n + (n - t)a_{n+1}.$$

Hence the first component $x^1(t)$ of $x(t) = [x^1(t), \ldots, x^r(t)]^T$ is such that

$$x^1(t) = 1 + (t - n)a_n^1 + (n - t)a_{n+1}^1 > 0$$

for $n \geqslant n_0$ and $t \in [n, n+1)$. This contradicts the hypothesis that every component of $x(t)$ oscillates and the proof is complete.

Now consider the scalar differential equation with piecewise constant arguments of the form

$$\dot{x}(t) + px(t) + \sum_{j=-k}^{l} q_j x([t+j]) = 0, \qquad t \geqslant 0 \tag{8.1.9}$$

where

$$k, l \in N \quad \text{and} \quad p, q_j \in \mathbb{R} \quad \text{for } j = -k, \ldots, l. \quad (8.1.10)$$

First assume that

$$l = 1 \quad \text{and} \quad q_1 = p(e^{-p} - 1)^{-1}$$

where, for $p = 0$, the indeterminate quantity $p(e^{-p} - 1)^{-1}$ should be understood to be its limit as $p \to 0$, namely, equal to 1. In this case one can see that the initial value problem consisting of eqn (8.1.9) and the initial condition

$$x(j) = a_j \in \mathbb{R} \quad \text{for } j \in \{-k, \ldots, 0\} \cup \{0, \ldots, l-1\} \quad (8.1.11)$$

has in general no solution. For example, the equation

$$\dot{x}(t) + q_0 x([t]) - x([t+1]) = 0 \quad \text{with } q_0 \neq 1$$

has only the trivial solution, while the equation

$$\dot{x}(t) + x([t]) - x([t+1]) = 0$$

has infinitely many solutions. Hence in the sequel we shall assume that

$$\left. \begin{array}{c} |l - 1| + |q_1 - p(e^{-p} - 1)^{-1}| \neq 0 \\ \text{and if } k = 0 \quad \text{and} \quad l \geq 2 \quad \text{then } q_l \neq 0 \end{array} \right\}. \quad (8.1.12)$$

The following result is the analogue of Lemma 8.1.1 for the scalar equation (8.1.9).

Lemma 8.1.2. *Assume that (8.1.10) and (8.1.12) hold. Then the initial value problem (8.1.9) and (8.1.11) has a unique solution $x(t)$ given by*

$$x(t) = e^{-p(t-n)} a_n - \frac{1 - e^{-p(t-n)}}{p} \sum_{j=-k}^{l} q_j a_{n+j} \quad (8.1.13)$$

for $n \leq t < n+1$ and $n \in N$, where the sequence $\{a_n\}$ satisfies the difference equation

$$a_{n+1} - e^{-p} a_n + \frac{1 - e^{-p}}{p} \sum_{j=-k}^{l} q_j z a_{n+j} = 0 \quad \text{for } n \geq 0. \quad (8.1.14)$$

Proof. Let $x(t)$ be a solution of (8.1.9) and (8.1.11). Then in the interval $n \leq t < n+1$ for any $n \in N$, eqn (8.1.9) becomes

$$\dot{x}(t) + px(t) + \sum_{j=-k}^{l} q_j a_{n+j} = 0 \quad (8.1.15)$$

where we use the notation

$$a_n = x(n) \qquad \text{for } n \in \{-k, \ldots, 0, 1, 1, \ldots\}.$$

The proof is now a consequence of (8.1.15) and an argument similar to that given in Lemma 8.1.1.

The characteristic equation associated with eqn (8.1.14) is

$$\lambda - e^{-p} + \frac{1 - e^{-p}}{p} \sum_{j=-k}^{l} q_j \lambda^j = 0 \tag{8.1.16}$$

and the next theorem is the analogue of Theorem 8.1.1 for the scalar equation (8.1.9).

Theorem 8.1.2. *Assume that (8.1.10) and (8.1.11) hold. Then the following statements are equivalent.*

(a) *Every solution of eqn (8.1.9) oscillates.*

(b) *Every solution of eqn (8.1.14) oscillates.*

(c) *The characteristic equation (8.1.16) has no positive roots.*

Proof. The fact that (b) $\Leftrightarrow$ (c) follows from Theorem 7.1.1. The statement that (b) $\Rightarrow$ (a) is obvious. So it remains to establish that (a) $\Rightarrow$ (b). From (8.1.14) we find

$$\sum_{j=-k}^{l} q_j a_n^{+j} = (e^{-p} a_n - a_{n+1}) \frac{p}{1 - e^{-p}}.$$

By substituting this into (8.1.13) and simplifying, we obtain

$$x(t) = \frac{e^{p(n+1-t)} - 1}{e^p - 1} a_n + \frac{e^{-p(t-n)} - 1}{e^{-p} - 1} a_{n+1} \tag{8.1.17}$$

for $n \leqslant t < n + 1$ and $n \geqslant 0$. Now assume, for the sake of contradiction, that eqn (8.1.14) has an eventually positive solution $\{a_n\}$. As the coefficients a_n and a_{n+1} in (8.1.17) are non-negative and they are not both zero for any $p \in \mathbb{R}$, it follows that $x(t)$ is eventually positive. This contradicts the hypothesis that $x(t)$ oscillates and the proof is complete.

By applying Theorem 8.2.1 we obtain the following corollaries.

Corollary 8.1.1. *Consider the equation*

$$\dot{x}(t) + px(t) + q_o x([t]) = 0 \tag{8.1.18}$$

where p and q_0 are real numbers. Then every solution of (8.1.18) oscillates if and only if

$$q_0 \geqslant p/(e^p - 1). \tag{8.1.19}$$

Proof. From Theorem 8.1.2 we know that every solution of eqn (8.1.18) oscillates if and only if the associated characteristic equation

$$\lambda - e^{-p} + \frac{1 - e^{-p}}{p} q_0 = 0$$

has no positive roots. But this is clearly equivalent to (8.1.19).

Corollary 8.1.2. *Consider the equation*

$$\dot{x}(t) + px(t) + q_{-1}x([t-1]) = 0 \tag{8.1.20}$$

where p and q_{-1} are real numbers. Then every solution of eqn (8.1.20) oscillates if and only if

$$q_{-1} > \frac{p\,e^{-p}}{4(e^p - 1)}. \tag{8.1.21}$$

Proof. The characteristic equation associated with eqn (8.1.20) is

$$\lambda - e^{-p} + \frac{1 - e^{-p}}{p} q_{-1} \lambda^{-1} = 0. \tag{8.1.22}$$

Clearly (8.1.22) has no positive roots if and only if the equation

$$f(\lambda) = \lambda^2 - e^{-p}\lambda + \frac{1 - e^{-p}}{p} q_{-1} = 0$$

has no positive roots. By evaluating the minimum of $f(\lambda)$ one can easily see that eqn (8.1.22) has no positive roots if and only if (8.2.21) holds. The proof is now a consequence of Theorem 8.1.2.

Corollary 8.1.3. *Consider the equation*

$$\dot{x}(t) + px(t) + q_0 x([t]) + q_1 x([t+1]) = 0 \tag{8.1.23}$$

where p, q_0 and q_1 are real numbers and

$$q_1 \neq p(e^{-p} - 1)^{-1}.$$

Then every solution of eqn (8.1.22) oscillates if and only if

$$\left(q_1 + \frac{p}{1 - e^{-p}} \right) \left(q_0 - \frac{p\,e^{-p}}{1 - e^{-p}} \right) \geqslant 0. \tag{8.1.24}$$

Proof. The characteristic equation of (8.1.23) is

$$\left(1 + \frac{1 - e^{-p}}{p} q_1\right)\lambda = e^{-p} - \frac{1 - e^{-p}}{p} q_0$$

and has no positive roots if and only if (8.1.24) is satisfied.

Corollary 8.1.4. *Assume that $q \in \mathbb{R}$ and $k, l \in N$ with $l \geq 1$. Then the following statements hold.*

(a) *Every solution of*

$$\dot{x}(t) + qx([t - k]) = 0$$

oscillates if and only if

$$q > k^k/(k + 1)^{k+1} \qquad if\ k \geq 1$$

and

$$q \geq 1 \qquad if\ k = 0.$$

(b) *Every solution of*

$$\dot{x}(t) - qx([t+l]) = 0$$

oscillates if and only if

$$q > (l - 1)^{l-1}/l^l \qquad if\ l \geq 2$$

and

$$q \geq 1 \qquad if\ l = 1.$$

Proof. This corollary is a simple consequence of Theorems 8.1.2 and 7.1.1.

8.2 Linearized oscillations and the logistic equation with piecewise constant arguments

In this section we study the oscillatory properties of the logistic equation with piecewise constant arguments

$$\frac{dN(t)}{dt} = rN(t)\left\{1 - \sum_{j=0}^{m} p_j N([t - j])\right\}, \qquad t \geq 0 \qquad (8.2.1)$$

where

$$r, p_m \in (0, \infty), \qquad p_0, p_1, \ldots, p_{m-1} \in [0, \infty), \qquad m + r \neq 1 \qquad (8.2.2)$$

$[\cdot]$ denotes the greatest-integer function.

The possible complex behaviour of the solutions of eqn (8.2.1) can be demonstrated by looking at a simple special case of eqn (8.2.1), namely,

$$\frac{dy(t)}{dt} = ry(t)\{1 - y([t])/K\}, \qquad t \geq 0. \tag{8.2.3}$$

On any interval of the form $[n, n + 1)$ for $n = 0, 1, 2, \ldots,$, by integrating (8.2.3) we obtain

$$y(t) = y(n) \exp\{r(1 - y(n)/K)(t - n)\} \tag{8.2.4}$$

for $n \leq t < n + 1$ and $n = 0, 1, 2, \ldots$ by taking limits, $t \to n + 1$ in eqn (8.2.4), we find

$$y(n + 1) = y(n) \exp\{r(1 - y(n)/K)\}, \qquad n = 0, 1, 2, \ldots \tag{8.2.5}$$

The first-order difference equation (8.2.5) has been considered in its own right, with no reference to eqn (8.2.3), as a discrete population model of single species with non-overlapping generations. It is known from the works of May (1975) and May and Oster (1976) that for certain parameter values of r, the asymptotic behaviour of the solutions of eqn (8.2.5) is 'chaotic'. Since (8.2.3) inherits all the behaviour of (8.2.5) in a natural way through (8.2.4), eqn (8.2.3) [and also (8.2.2)] provides a simple example of a continuous one-dimensional dynamical system capable of displaying chaotic behaviour.

Clearly eqn (8.2.1) has a unique positive steady state

$$N^* = \left(\sum_{j=0}^{m} p_j \right)^{-1}. \tag{8.2.6}$$

In this section we give necessary and sufficient conditions for the oscillation of all positive solutions of eqn (8.2.1) about the positive steady state N^*.

The definition of a solution of eqn (8.2.1) is similar to the definition of solution of eqn (8.1.1) in Section 8.1. Throughout this section we assume that eqn (8.2.1) is supplemented with initial conditions of the form

$$N(0) = N_0 > 0 \qquad \text{and} \qquad N(-j) = N_{-j} \geq 0, \qquad j = 1, 2, \ldots, m. \tag{8.2.6}$$

The following lemma implies that (8.2.1) and (8.2.6) has a unique positive solution.

Lemma 8.2.1. *Let $N_0 > 0$ and $N_{-j} \geq 0$ for $j = 1, 2, \ldots, m$ be given. The initial value problem* (8.2.1) *and* (8.2.6) *has a unique positive solution $N(t)$ given by*

$$N(t) = N_n \exp\left\{ r\left(1 - \sum_{j=0}^{m} p_j N_{n-j}\right)(t - n)\right\} \qquad \text{for } n \leq t < n + 1 \tag{8.2.7}$$

and $n = 0, 1, 2, \ldots$ *where the sequence* $\{N_n\}$ *satisfies the difference equation*

$$N_{n+1} = N_n \exp\left\{ r\left(1 - \sum_{j=0}^{m} p_j N_{n-j} \right) \right\} \qquad \textit{for } n = 0, 1, 2, \ldots \quad (8.2.8)$$

Proof. For every $n = 0, 1, 2, \ldots$ and for $n \leqslant t < n + 1$, eqn (8.2.1) becomes

$$\frac{dN(t)}{dt} = rN(t)\left(1 - \sum_{j=0}^{m} p_j N_{n-j} \right), \qquad n \leqslant t < n + 1 \qquad (8.2.9)$$

where we use the notation $N_n = N(n)$ for $n \in \{-m, \ldots, 0, 1, \ldots\}$. By using (8.2.9) and an argument similar to that given in the proof of Lemma 8.1.1 we see that (8.2.7) is the unique solutions of (8.2.1) and (8.2.6). It is also clear from (8.2.7) that for as long as $N_0 > 0$, $N(t) > 0$ for $t \geqslant t_0$.

Remark 8.2.1. *One should note from the proof of Lemma 8.2.1 that* $N_0 > 0$ *implies* $N(t) > 0$ *for* $t \geqslant 0$ *for any* $N_{-j} \in \mathbb{R}$ *for* $j = 1, 2, \ldots, m$. *However, we assume in (8.2.6) that* $N_{-j} \geqslant 0$ *only for 'biological reasons'.*

Let $N(t)$ be the positive solution of (8.2.1) and (8.2.6) and set

$$N(t) = N^* \exp(x(t)), \qquad t \geqslant 0.$$

Then $x(t)$ satisfies the equation

$$\dot{x}(t) + \sum_{j=0}^{m} q_j f(x([t - j])) = 0, \qquad t \geqslant m, \qquad (8.2.10)$$

where

$$f(u) = e^u - 1 \qquad \text{and} \qquad q_j = rN^* p_j = \frac{r}{\sum_{i=0}^{m} p_i} p_j. \qquad (8.2.11)$$

Also,

$$x(j) = \ln \frac{N(j)}{N^*} \qquad \text{for } j = 1, 2, \ldots, m.$$

Clearly, $N(t)$ oscillates about N^* if and only if $x(t)$ oscillates about zero. Observe that the function f satisfies the following properties:

$$f \in C[\mathbb{R}, \mathbb{R}], \qquad uf(u) > 0 \qquad \text{for } u \neq 0, \qquad (8.2.12)$$

$$f(u) \geqslant u \qquad \text{for } u \leqslant 0 \qquad \text{and} \qquad \lim_{u \to 0} \frac{f(u)}{u} = 1. \qquad (8.2.13)$$

The linearized equation associated with eqn (8.2.10) is

$$\dot{y}(t) + \sum_{j=0}^{m} q_j y([t - j]) = 0. \qquad (8.2.14)$$

We now establish a linearized oscillation result for the equation with piecewise constant arguments (8.2.10), which we use to obtain necessary and sufficient conditions for the oscillaton of all positive solutions of eqn (8.2.1) about the positive equilibrium N^*.

Theorem 8.2.1. *Consider the equation*

$$\dot{x}(t) + \sum_{j=0}^{m} q_j f(x([t-j])) = 0, \qquad t \geq t_0 \tag{8.2.15}$$

where $q_0, \ldots, q_m \in [0, \infty)$, $m + q_0 \neq 1$ *and the function* f *satisfies* (8.2.12) *and* (8.2.13). *Then every solution of* (8.2.15) *oscillates if and only if every solution of the linear equation* (8.2.14) *oscillates, that is, if and only if the equation*

$$\lambda - 1 + \sum_{j=0}^{m} q_j \lambda^{-j} = 0 \tag{8.2.16}$$

has no positive roots.

Proof. By an argument similar to that given in the proof of Lemma 8.1.1 we see that $x(t)$ is a solution of (8.2.15) if and only if

$$x(t) = x_n - \sum_{j=0}^{m} q_j f(x_{n-j})(t-n) \tag{8.2.17}$$

for $t \in [n, n+1)$ and $n = 0, 1, 2, \ldots$, where the sequence $\{x_n\}$ satisfies the difference equation

$$x_{n+1} - x_n + \sum_{j=0}^{m} q_j j(x_{n-j}) = 0 \qquad \text{for } n \geq 0. \tag{8.2.18}$$

On the other hand, from (8.1.17) and (8.4.18) it follows that the solution $x(t)$ is given by

$$x(t) = (1 + (t-n))x_n + (n-t)x_{n+1} \text{ for } n \geq 0 \text{ and } t \in [n, n+1).$$

Hence one can easily see that $x(t)$ is an oscillatory solution of eqn (8.2.15) if and only if the sequence $\{x_n\}$ is an oscillatory solution of (8.2.18). By applying Corollary 7.4.1 to eqn (8.2.18) we see that every solution of eqn (8.2.18) oscillates if and only if every solution of its linearized equation

$$y_{n+1} - y_n + \sum_{j=0}^{m} q_j y_{n-j} = 0 \qquad \text{for } n \geq 0 \tag{8.2.19}$$

oscillates. But by Theorem 8.1.2 we know that every solution of eqn (8.2.19) oscillates if and only if its characteristic equation (8.2.16) has no positive roots. The proof of the theorem is complete.

The following result is a consequence of Theorem 8.2.1 and the above discussion about the logistic equation (8.2.1).

Theorem 8.2.2. *Assume that (8.2.2) is satisfied. Then every positive solution of eqn (8.2.1) oscillates about its positive equilibrium N* if and only if the equation*

$$\lambda - 1 + \frac{r}{\sum_{i=0}^{m} p_i} \sum_{j=0}^{m} p_j \lambda^{-j} = 0 \qquad (8.2.20)$$

has no positive roots.

By using the explicit conditions of Section 7.2 for the oscillation of all solutions of linear difference equations and by applying Theorem 8.2.2, one can obtain several explicit conditions for the oscillation of all solutions of eqn (8.2.1). For example, we have the following results.

Corollary 8.2.1. *Assume that r and K are positive constants. Then every positive solution of the logistic equation (8.2.4) oscillates about K if and only if r > 1.*

Remark 8.2.1. *When r = 1, the linearized oscillation Theorem 8.2.1 does not apply. However, one can easily see that in this case the solutions of eqn (8.2.5) with y(0) > 0 and y(0) ≠ K converge monotonically to K.*

Corollary 8.2.2. *Assume that r and K are positive real numbers and l ≥ 1 is an integer. Then every positive solution of the logistic equation*

$$\frac{dN(t)}{dt} = rN(t)\left[1 - \frac{N([t - l])}{K}\right], \qquad t \geq 0$$

oscillates about K if and only if

$$r > \frac{l^l}{(l + 1)^{l+1}}.$$

8.3 Oscillations in non-autonomous equations with piecewise constant arguments

In this section we establish sufficient conditions for the oscillation of all solutions of some non-autonomous equations with piecewise constant arguments. The first two theorems deal with the equation

$$\dot{x}(t) + p(t)x(t) + q(t)x([t - 1]) = 0, \qquad t \geq 0 \qquad (8.3.1)$$

where

$$p \in C[[0, \infty), \mathbb{R}] \quad \text{and} \quad q \in C[[0, \infty), \mathbb{R}^+] \tag{8.3.2}$$

and $[\cdot]$ denotes the greatest-integer function. In this section the definition of oscillation is similar to that given in Section 8.1 for eqn (8.1.1).

The first result is in the spirit of Theorem 3.4.3.

Theorem 8.3.1. *Assume that (8.3.2) holds and that*

$$\limsup_{n \to \infty} \int_n^{n+1} q(s) \exp\left(\int_{n-1}^s p(\zeta)\, d\zeta\right) ds > 1.$$

$$\tag{8.3.3}$$

Then every solution of eqn (8.3.1) oscillates.

Proof. Assume, for the sake of contradiction, that eqn (8.3.1) has an eventually positive solution $x(t)$. Then there exists an integer $N \geqslant 0$ such that

$$x(t) > 0 \quad \text{for } t \geqslant N.$$

Therefore for $n \geqslant N + 1$ and for $t \in [n, n+1)$ we have

$$\left.\begin{array}{c} \dot{x}(t) + p(t)x(t) = -a_{n-1}q(t) \\ x(n) = a_n \end{array}\right\} \tag{8.3.4.}$$

where we use the notation

$$x(n) = a_n \quad \text{for all } n \geqslant N.$$

The unique solution of the IVP (8.3.4) satisfies the equation

$$x(t) \exp\left(\int_n^t p(\zeta)\, d\zeta\right) = a_n - a_{n-1} \int_n^t q(s) \exp\left(\int_n^s p(\zeta)\, d\zeta\right) ds.$$

By using the continuity of $x(t)$, as $t \to n + 1$ we find

$$a_{n+1} \exp\left(\int_n^{n+1} p(\zeta)\, d\zeta\right) = a_n - a_{n-1} \int_n^{n+1} q(s) \exp\left(\int_n^s p(\zeta)\, d\zeta\right) ds. \tag{8.3.5}$$

It follows from (8.3.5) that

$$a_{n+1} \exp\left(\int_n^{n+1} p(\zeta)\, d\zeta\right) < a_n \quad \text{for } n \geqslant N+1$$

or

$$a_n \exp\left(\int_{n-1}^n p(\zeta)\, d\zeta\right) < a_{n-1} \quad \text{for } n \geqslant N + 2. \tag{8.3.6}$$

By using (8.3.6) in (8.3.5) we obtain

$$a_{n+1} \exp\left(\int_n^{n+1} p(\zeta)\,d\zeta\right) < a_n\left[1 - \int_n^{n+1} q(s) \exp\left(\int_{n-1}^s p(\zeta)\,d\zeta\right)ds\right]$$

$$\text{for } n \geqslant N + 2$$

and since $a_n, a_{n+1} \in (0, \infty)$ for $n \geqslant N$ we conclude that

$$\int_n^{n+1} q(s) \exp\left(\int_{n-1}^s p(\zeta)\,d\zeta\right) < 1.$$

Hence

$$\limsup_{n \to \infty} \int_n^{n+1} q(s) \exp\left(\int_{n-1}^s p(\zeta)\,d\zeta\right) \leqslant 1$$

which contradicts (8.3.3) and completes the proof of the theorem.

The next theorem gives the 'best-possible' condition for the oscillation of all solutions of eqn (8.3.1) in the sense that when $p(t)$ and $q(t)$ are constants the condition is necessary and sufficient.

Theorem 8.3.2. *Assume that* (8.3.2) *holds,*

$$\limsup_{n \to \infty} \int_n^{n+1} p(s)\,ds > -\infty \tag{8.3.7}$$

and

$$\liminf_{n \to \infty} \left(\exp\left(\int_n^{n+1} p(s)\,ds\right)\right) \cdot \liminf_{n \to \infty} \int_n^{n+1} q(s) \exp\left(\int_n^s p(\zeta)\,d\zeta\right) ds > \tfrac{1}{4}. \tag{8.3.8}$$

Then every solution of eqn (8.3.1) *oscillates.*

Proof. Assume, for the sake of contradiction, that eqn (8.3.1) has an eventually positive solution $x(t)$. Then there exists an integer $N \geqslant 0$ such that

$$x(t) > 0 \qquad \text{for } t \geqslant N.$$

Therefore, as in the proof of Theorem 8.3.1. (8.3.5) holds and so for $n \geqslant N + 1$,

$$\frac{a_{n+1}}{a_n} \frac{a_n}{a_{n-1}} \exp\left(\int_n^{n+1} p(\zeta)\,d\zeta\right) + \int_n^{n+1} q(s) \exp\left(\int_n^s p(\zeta)\,d\zeta\right) ds = \frac{a_n}{a_{n-1}}. \tag{8.3.9}$$

Set $w_n = a_n/a_{n-1}$. Thus $w_n > 0$ for $n \geqslant N + 1$. We have two cases to consider.
  Case 1.

$$\rho \equiv \lim_{n \to \infty} \inf w_n < \infty.$$

Then (8.3.9) yields

$$\left(\lim_{n \to \infty} \inf w_{n+1} \right) \left(\lim_{n \to \infty} \inf w_n \right) \left(\lim_{n \to \infty} \inf \exp \left(\int_n^{n+1} p(\zeta) \, d\zeta \right) \right)$$

$$+ \lim_{n \to \infty} \inf \int_n^{n+1} q(s) \exp \left(\int_n^s p(\zeta) \, d\zeta \right) ds \leqslant \lim_{n \to \infty} \inf w_n. \quad (8.3.10)$$

Set

$$A = \lim_{n \to \infty} \inf \exp \left(\int_n^{n+1} p(\zeta) \, d\zeta \right)$$

and

$$B = \lim_{n \to \infty} \inf \int_n^{n+1} q(s) \exp \left(\int_n^s p(\zeta) \, d\zeta \right) ds.$$

Then (8.3.10) implies that

$$A\rho^2 + B \leqslant \rho$$

and by completing the square we find

$$A \left[\left(\rho - \frac{1}{2A} \right)^2 + \frac{4AB - 1}{4A^2} \right] \leqslant 0.$$

Hence $AB \leqslant \frac{1}{4}$, which contradicts the hypothesis (8.3.8) and completes the proof in this case.
  Case 2.

$$\lim_{n \to \infty} w_n = \infty.$$

Then (8.3.5) implies that

$$0 < \exp \left(\int_n^{n+1} p(\zeta) \, d\zeta \right) < 1/w_{n+1}$$

and so

$$\lim_{n \to \infty} \exp \left(\int_n^{n+1} p(\zeta) \, d\zeta \right) = 0,$$

which contradicts the hypothesis (8.3.7) and completes the proof of the theorem.

Remark 8.3.1. *Assume the functions $p(t)$ and $q(t)$ of Theorem 8.3.2 are identically equal to the constants p and q respectively. Then the condition (8.3.7) is automatically satisfied, while the condition (8.3.8) reduces to*

$$q > \frac{p\,e^{-p}}{4(e^p - 1)}. \tag{8.3.11}$$

As we saw in Corollary 8.1.2, (8.3.11) is a necessary and sufficient condition for the oscillation of all solutions of the equation

$$\dot{x}(t) + px(t) + qx([t-1]) = 0.$$

The final result in this section deals with the oscillation of all solutions of the equation with piecewise constant arguments

$$\dot{x}(t) + \sum_{i=0}^{m} q_i(t)x([t-k_i]) = 0, \qquad t \geqslant 0 \tag{8.3.12}$$

where

$$q_i \in C[\mathbb{R}^+, \mathbb{R}^+] \qquad \text{and} \qquad k_i \in \mathbb{N} \qquad \text{for } i = 1, 2, \ldots, m. \tag{8.3.13}$$

Let $k = \max\{k_1, \ldots, k_m\}$ and assume that $a_{-k}, \ldots, a_{-1}, a_0$ are given real numbers. Then as we shall see in the next lemma, eqn (8.3.12) has a unique solution x satisfying the initial conditions

$$x(-k) = a_{-k}, \ldots, x(0) = a_0. \tag{8.3.14}$$

Lemma 8.3.1. *Suppose that (8.3.13) holds and assume that $a_{-k}, \ldots, a_{-1}, a_0 \in \mathbb{R}$. Then the initial value problem (8.3.12) and (8.3.14) has a unique solution $x(t)$. Furthermore, $x(t)$ is given by*

$$x(t) = a_n - \sum_{i=0}^{m} a_{n-k_i} \int_n^t q_i(s)\,ds \qquad \text{for } n \leqslant t < n+1 \qquad \text{and} \qquad n \in \mathbb{N}$$

$$\tag{8.3.15}$$

where the sequence $\{a_n\}$ satisfies the difference equation

$$a_{n+1} - a_n + \sum_{i=0}^{m} a_{n-k_i} \int_n^{n+1} q_i(s)\,ds = 0 \qquad \text{for } n = 0, 1, 2, \ldots \tag{8.3.16}$$

The proof of Lemma 8.3.1 is similar to that of Lemma 8.1.1 and is omitted.

Theorem 8.3.3. *Assume that (8.3.13) holds. Then every solution of (8.3.12) oscillates if and only if every solution of the difference equation (8.3.16) oscillates.*

Proof. If $x(t)$ is a solution of (8.3.12), then from (8.3.15) and (8.3.16) we see that $x(n) = a_n$ for all n. Hence if every solution of (8.3.16) oscillates, so does every solution of eqn (8.3.12). Conversely, assume that every solution (8.3.12) oscillates and for the sake of contradiction assume that (8.3.16) has an eventually positive solution $\{a_n\}$. Then for n sufficiently large and for all $t \in [n, n+1)$,

$$x(t) \geqslant a_n - \sum_{i=0}^{m} a_{n-k_i} \int_n^{n+1} q_i(s) \, ds = a_{n+1} > 0,$$

which is a contradiction and the proof is complete.

8.4 Oscillations in equation of alternately retarded and advanced type

In this section we obtain necessary and sufficient conditions for the oscillation of all solutions of the following two equations with piecewise constant argument:

$$\dot{x}(t) + px\left(2\left[\frac{t+1}{2}\right]\right) = 0, \qquad t \geqslant 0 \tag{8.4.1}$$

$$\dot{x}(t) + px([t + \tfrac{1}{2}]) = 0, \qquad t \geqslant 0 \tag{8.4.2}$$

where p is a real number and $[\cdot]$ denotes the greatest integer function.

One can look on equations (8.4.1) and (8.4.2) as equations of the form

$$\dot{x}(t) + px(t - \tau(t)) = 0, \qquad t \geqslant 0 \tag{8.4.1$'$}$$

and

$$\dot{x}(t) + px(t - \sigma(t)) = 0, \qquad t \geqslant 0 \tag{8.4.2$'$}$$

where the argument deviation is given by

$$\tau(t) = t - 2\left[\frac{t+1}{2}\right] \quad \text{and} \quad \sigma(t) = t - [t + \tfrac{1}{2}]$$

respectively. The arguments $\tau(t)$ and $\sigma(t)$ are piecewise linear periodic functions with period 2 and 1 respectively. More precisely, for every integer n,

$$\tau(t) = t - 2n \qquad \text{for } 2n - 1 \leqslant t < 2n + 1$$

$$\sigma(t) = t - n \qquad \text{for } n - \tfrac{1}{2} \leqslant t < n + \tfrac{1}{2}.$$

Also,

$$-1 \leqslant \tau(t) < 1 \qquad \text{for } 2n - 1 \leqslant t < 2n + 1$$

$$-\tfrac{1}{2} \leqslant \sigma(t) < \tfrac{1}{2} \qquad \text{for } n - \tfrac{1}{2} \leqslant t < n + \tfrac{1}{2}.$$

Therefore, in each interval $[2n - 1, 2n + 1)$, eqn (8.4.1) is of alternately advanced and retarded type. It is of advanced type in $[2n - 1, 2n]$ and of retarded type in $(2n, 2n + 1)$. Similarly, in each interval $[n - \tfrac{1}{2}, n + \tfrac{1}{2})$, eqn (8.4.2) is of alternately advanced and retarded type. It is of advanced type in $[n - \tfrac{1}{2}, n)$ and of retarded type in $(n, n + \tfrac{1}{2})$.

By a *solution* of eqn (8.4.1) [respectively, eqn (8.4.2)] we mean a function $x(t)$ which satisfies the following properties:

(a) $x(t)$ is continuous on $[0, \infty)$.

(b) The derivative $\dot{x}(t)$ exists at each point $t \in [0, \infty)$, with the possible exception of the points $t = 2n + 1$ (respectively, $t = n + \tfrac{1}{2}$) for $n \in \mathbb{N}$ where one-sided derivatives exist.

(c) Equation (8.4.1) (respectively, eqn (8.4.2)) is satisfied on each interval of the form $[2n - 1, 2n + 1) \cap \mathbb{R}^+$ (respectively, $[n - \tfrac{1}{2}, n + \tfrac{1}{2}) \cap \mathbb{R}^+$) for $n \in \mathbb{N}$.

With each of these two equations we associate an initial condition of the form

$$x(0) = a_0 \tag{8.4.3}$$

where a_0 is a given real number.

The following two lemmas deal with existence and uniqueness of solutions.

Lemma 8.4.1. *Assume that $p, a_0 \in \mathbb{R}$ and $p \neq 1$. Then the initial value problem* (8.4.1) *and* (8.4.3) *has a unique solution $x(t)$. Furthermore, $x(t)$ is given by*

$$x(t) = [1 - p(t - 2n)]a_n \quad \text{for } t \in [2n - 1, 2n + 1) \cap \mathbb{R}^+ \quad \text{and} \quad n \in \mathbb{N},$$
$$\tag{8.4.4}$$

where the sequence $\{a_n\}$ satisfies the difference equation

$$a_{2n+1} = (1 - p)a_{2n} \quad \text{for } n = 0, 1, 2, \ldots$$

and

$$a_{2n-1} = (1 + p)a_{2n} \quad \text{for } n = 1, 2, \ldots$$
$$\left.\phantom{\begin{matrix} a \\ a \end{matrix}}\right\} . \tag{8.4.5}$$

Proof. Let $x(t)$ be a solution of (8.4.1) and (8.4.3). Then in the interval $[2n - 1, 2n + 1) \cap \mathbb{R}^+$ and for any $n \in \mathbb{N}$, eqn (8.4.1) becomes

$$\dot{x}(t) + pa_{2n} = 0, \tag{8.4.6}$$

where we have used the notation $a_n = x(n)$ for $n \in \mathbb{N}$. Clearly the solution of (8.4.6) with initial condition $x(n) = a_{2n}$ is given by (8.4.4). By the continuity of the solution as $t \to 2n + 1$ and for $t = 2n - 1$, (8.4.4) yields (8.4.5) and (8.4.6). So far we have proved that if $x(t)$ is a solution of (8.4.1) and (8.4.3) then $x(t)$ is given by (8.4.4) where the sequence $\{a_n\}$ satisfies (8.4.5). Conversely, given $a_0 \in \mathbb{R}$ and because $p \neq -1$, the difference equation (8.4.5) has a unique solution $\{a_n\}$. Now by direct substitution into eqn (8.4.1) one can see that the function $x(t)$ as defined by (8.4.4) is a solution. The proof is complete.

Lemma 8.4.2. *Assume that* p, $a_0 \in \mathbb{R}$ *and* $p \neq -2$. *Then the initial value problem* (8.4.2) *and* (8.4.3) *has a unique solution* $x(t)$. *Furthermore,* $x(t)$ *is given by*

$$x(t) = [1 - p(t - n)]a_n \quad \text{for } t \in [n - \tfrac{1}{2}, n + \tfrac{1}{2}) \cap \mathbb{R}^+ \quad \text{and} \quad n \in \mathbb{N}$$
(8.4.7)

where the sequence $\{a_n\}$ *satisfies the difference equation*

$$a_{n+1} = \frac{2 - p}{2 + p} a_n \quad \text{for } n = 0, 1, 2, \ldots$$
(8.4.8)

Proof. Let $x(t)$ be a solution of (8.4.2) and (8.4.3). Then in the interval $[n - \tfrac{1}{2}, n + \tfrac{1}{2}) \cap \mathbb{R}^+$ for any $n \in \mathbb{N}$, eqn (8.4.2) becomes

$$\dot{x}(t) + pa_n = 0$$
(8.4.9)

where we have used the notation $a_n = x(n)$ for $n \in \mathbb{N}$. The solution of (8.4.9) with initial condition $x(n) = a_n$ is given by (8.4.7). By the continuity of the solutions as $t \to n + \tfrac{1}{2}$ and for $t = n - \tfrac{1}{2}$, (8.4.7) yields

$$x(n + \tfrac{1}{2}) = (1 - \tfrac{1}{2}p)a_n \quad \text{and} \quad x(n - \tfrac{1}{2}) = (1 + \tfrac{1}{2}p)a_n,$$

from which (8.4.8) follows. The remaining part of the proof is similar to that of Lemma 8.4.1 and is omitted. The proof is complete.

The following two theorems provide necessary and sufficient conditions for the oscillation of all solutions of eqns (8.4.1) and (8.4.2) respectively.

Theorem 8.4.1. *Assume that* $p \in \mathbb{R}$ *and* $p \neq -1$. *Then every solution of eqn* (8.4.1) *oscillates if and only if*

$$p \in (-\infty, -1) \cup [1, \infty).$$
(8.4.10)

Proof. Assume that (8.4.10) holds. Then either $p < -1$ or $p \geq 1$ and in either case it follows from (8.4.5) that the sequence $\{a_n\}$ oscillates. As $x(n) = a_n$ for

$n \in \mathbb{N}$, $x(t)$ also oscillates. Conversely, assume that every solution $x(t)$ of eqn (8.4.1) oscillates and, for the sake of contradiction, assume that

$$|p| < 1. \tag{8.4.11}$$

Let $x(t)$ be the solution of (8.4.1) with $x(0) = a_0 = 1$. Then, from (8.4.5) and because of (8.4.11),

$$a_n > 0 \qquad \text{for } n = 0, 1, 2, \ldots.$$

Hence for $t \in [2n - 1, 2n + 1)$ and $n \in \mathbb{N}$, $|2n - t| \leqslant 1$, so (8.4.4) yields

$$x(t) = [1 - p(t - 2n)]a_{2n} \geqslant [1 - |p|\,|t - 2n|]a_{2n} \geqslant (1 - |p|)a_{2n} > 0.$$

This contradicts the assumption that $x(t)$ oscillates and the proof is complete.

Theorem 8.4.2. *Assume that $p \in \mathbb{R}$ and $p \neq -2$. Then every solution of eqn* (8.4.2) *oscillates if and only if*

$$p \in (-\infty, -2) \cup [2, \infty). \tag{8.4.12}$$

Proof. Assume that (8.4.12) holds. Then either $p < -2$ or $p \geqslant 2$ and in either case it follows from (8.4.8) that the sequence $\{a_n\}$ oscillates. As $x(n) = a_n$ for $n \in \mathbb{N}$, $x(t)$ also oscillates. Conversely, assume that every solution $x(t)$ of eqn (8.4.2) oscillates and, for the sake of contradiction, assume that

$$|p| < 2. \tag{8.4.13}$$

Let $x(t)$ be the solution of (8.4.2) with $x(0) = a_0 = 1$. Then from (8.4.8), $a_n > 0$ for $n \in \mathbb{N}$. Hence for $t \in [n - \frac{1}{2}, n + \frac{1}{2})$ and $n \in \mathbb{N}$, $|t - n| \leqslant \frac{1}{2}$, so (8.4.7) yields

$$x(t) = [1 - p(t - n)]a_n \geqslant [1 - |p|\,|t - n|]a_n \geqslant (1 - |p|)a_n > 0.$$

This contradicts the assumption that $x(t)$ oscillates and the proof is complete.

8.5 Oscillations in equations with continuous and piecewise constant arguments

In this section we consider the oscillations of the equation with continuous and piecewise constant arguments

$$\dot{x}(t) + px(t - \tau) + qx([t - k]) = 0 \tag{8.5.1}$$

where

$$p, \tau, q \in \mathbb{R}^+ \qquad \text{and} \qquad k \in \mathbb{N}. \tag{8.5.2}$$

By a solution of eqn (8.5.1) we mean a function x which is defined on the set

$$\{-k, \ldots, -1, 0\} \cup [-\tau, \infty)$$

and which satisfies the following properties:

(a) x is continuous on $[-\tau, \infty)$.

(b) The derivative $\dot{x}(t)$ exists at each point $t \in (0, \infty)$ with the possible exception of the points $t \in \mathbb{N}$ where one-sided derivatives exist.

(c) Equation (8.5.1) is satisfied on each interval $[n, n + 1)$ for $n \in \mathbb{N}$.

Let $\phi \in C([-\tau, 0], \mathbb{R})$ and let $a_{-k}, \ldots, a_{-1}, a_0$ be given real numbers such that

$$a_{-j} = \phi(-j) \quad \text{for } j \leqslant \tau \quad \text{and} \quad j = 0, 1, 2, \ldots, k.$$

Then one can show that eqn (8.5.1) has a unique solution x satisfying the initial conditions

$$x(t) = \phi(t) \quad \text{for } -\tau \leqslant t \leqslant 0 \quad \text{and} \quad x(-j) = a_{-j} \quad \text{for } j = 0, 1, 2, \ldots, k.$$

When $q = 0$, eqn (8.5.1) reduces to the delay equation

$$\dot{u}(t) + pu(t - \tau) = 0 \tag{8.5.3}$$

and by Theorem 2.1.2 we know that every solution of eqn (8.5.3) oscillates if and only if its characteristic equation

$$F(\lambda) \equiv \lambda + p\,e^{-\lambda\tau} = 0 \tag{8.5.4}$$

has no real roots, or equivalently,

$$p\tau > 1/e. \tag{8.5.5}$$

On the other hand, when $p = 0$, eqn (8.5.1) reduces to the equation with piecewise constant argument

$$\dot{v}(t) + qv([t - k]) = 0 \tag{8.5.6}$$

and by Corollary 8.1.4 we know that every solution of eqn (8.5.6) oscillates if and only if

$$q > k^k/(k + 1)^{k+1} \quad \text{if } k \geqslant 1 \tag{8.5.7}$$

$$q \geqslant 1 \quad \text{if } k = 0. \tag{8.5.8}$$

When $p\tau\, e > 1$, one can easily see that every solution of eqn (8.5.1) oscillates. Otherwise, eqn (8.5.1) has an eventually positive solution $x(t)$. Then as $q \geqslant 0$, eqn (8.5.1) implies that eventually

$$\dot{x}(t) + px(t - \tau) \leqslant 0.$$

By Corollary 2.4.1 it follows that eqn (8.5.3) has an eventually positive solution, or equivalently, $p\tau\, e \leqslant 1$, which is a contradiction.

From this observation it follows that the case $p\tau\, e \leqslant 1$ is critical. In this case we have the following lemma.

Lemma 8.5.1. *Assume that*

$$p\tau\,e \leqslant 1.$$

Then eqn (8.5.4) *has a unique root* $\lambda_0 \in [-1/\tau, 0]$. *(When* $\tau = 0$, *we use the convention that* $1/\tau$ *stands for* ∞).

Proof. If $\tau = 0$, then $\lambda_0 = -p \leqslant 0$ is a root of (8.5.4). If $\tau > 0$, then

$$F\left(-\frac{1}{\tau}\right)F(0) = \frac{p\tau\,e - 1}{\tau}\,p \leqslant 0$$

and so eqn (8.5.4) has a root $\lambda_0 \in [/1/\tau, 0]$. If $p = 0$ then $\lambda_0 = 0$, while if $p\tau = 1/e$ then $\lambda_0 = -1/\tau$. Observe also that when $p\tau\,e \leqslant 1$,

$$F'(\lambda) = 1 - p\tau\,e^{-\lambda\tau} \geqslant 0 \qquad \text{for } -1/\tau < \lambda \leqslant 0,$$

from which the result follows.

When $p\tau\,e \leqslant 1$, let λ_0 be as defined in Lemma 8.5.1 and set

$$A = \lambda_0(1 + \lambda_0\tau) \qquad \text{and} \qquad B = e^{-\lambda_0\tau}(1 = e^{-\lambda_0}\tau). \tag{8.5.9}$$

The following theorem is the main result in this section.

Theorem 8.5.1. *Consider eqn* (8.5.1) *and assume that condition* (8.5.2) *is satisfied. Let A and B be as defined by* (8.5.9). *Then each of the following four conditions implies that every solution of eqn* (8.5.1) *oscillates:*

(a) $p\tau\,e > 1$ *and* $q \geqslant 0$;

(b) $p\tau\,e = 1$ *and* $q > 0$;

(c) $p\tau\,e < 1$, $p > 0$, $q > 0$ *and the equation*

$$A(\lambda - 1) + B\lambda^{-k} = 0 \tag{8.5.10}$$

has no roots in $[0, 1)$;

(d) $p \geqslant 0$ *and* $q > k/(k + 1)^{k+1}$ *if* $k \geqslant 1$

$$p \geqslant 0 \qquad \text{and} \qquad q \geqslant 1 \qquad \text{if } k = 0.$$

Proof. Case (a) was established in the discussion preceding Lemma 8.5.1. Next, assume that (b) or (c) holds. Clearly, in these cases $-1/\tau \leqslant \lambda_0 < 0$, where λ_0 is as defined in Lemma 8.5.1. Assume for the sake of contradiction, that eqn (8.5.1) has an eventually positive solution $x(t)$. Set

$$y(t) = e^{-\lambda_0 t}x(t)$$

$$u(t) = y(t) + \lambda_0 \int_{t-\tau}^{t} y(s)\,ds. \tag{8.5.11}$$

Then for $n \leqslant t < n + 1$ and $n = 0, 1, 2, \ldots,$

$$
\begin{aligned}
\dot{u}(t) &= \dot{y}(t) + \lambda_0 y(t) - \lambda_0 y(t - \tau) \\
&= -\lambda_0 e^{-\lambda_0 t} x(t) + e^{-\lambda_0 t} \dot{x}(t) + \lambda_0 e^{-\lambda_0 t} x(t) - \lambda_0 e^{-\lambda_0(t-\tau)} x(t - \tau) \\
&= e^{-\lambda_0 t}[\dot{x}(t) - \lambda_0 e^{-\lambda_0 \tau} x(t - \tau)] \\
&= e^{-\lambda_0 t}(\dot{x}(t) + p x(t - \tau)) \\
&= -e^{-\lambda_0 t} q x([t - k]).
\end{aligned}
$$

Therefore, for $n \leqslant t < n + 1$ and $n = 0, 1, 2, \ldots,$

$$
\dot{u}(t) = -e^{-\lambda_0 t} q x([t - k]) < 0 \tag{8.5.12}
$$

and so $u(t)$ is strictly decreasing function. We claim that $u(t)$ is a bounded function. Otherwise there exists a t_1 such that $u(t_1) < 0$ and $y(t_1) = \max_{s \leqslant t_1} y(s)$. Then (8.5.11) implies that

$$
0 > u(t_1) = y(t_1) + \lambda_0 \int_{t_1 - \tau}^{t_1} y(s)\, ds \geqslant y(t_1)(1 + \lambda_0 \tau) \geqslant 0,
$$

which is a contradiction. Thus $u(t)$ is bounded and decreasing, so

$$
L = \lim_{t \to +\infty} u(t) \tag{8.5.13}
$$

exists and is finite. We claim that $L = 0$. Indeed, by integrating (8.5.12) from t to $+\infty$ we see that

$$
\int_t^\infty e^{-\lambda_0 s} q x([s - k])\, ds = u(t) - L,
$$

which implies that $y \in L^1[0, \infty)$. Hence $u \in L^1[0, \infty)$, so L has to be zero. As $u(t)$ decreases to zero, it follows that eventually, $u(t) > 0$. Hence from (8.5.11) we find

$$
y(t) = u(t) - \lambda_0 \int_{t-\tau}^t y(s)\, ds > u(t) - \lambda_0 \tau u(t) = (1 - \lambda_0 \tau) u(t).
$$

One can show by induction that for every $N \in \mathbf{N}$ there exists a $t_N > 0$ such that

$$
y(t) \geqslant c_N u(t), \qquad t \geqslant t_N \tag{8.5.14}
$$

where

$$
c_N = \sum_{i=0}^{N-1} = \begin{cases} N & \text{if } -\lambda_0 \tau = 1 \\[2ex] \dfrac{1 - (-\lambda_0 \tau)^N}{1 + \lambda_0 \tau} & \text{if } -\lambda_0 \tau < 1. \end{cases}
$$

By using (8.5.12) and (8.5.13) and the definition of y we find that for $t \geqslant t_N$,

$$\dot{u}(t) + q\, e^{-\lambda_0 t}\, e^{\lambda_0(n-k)} c_N u(n-k) \leqslant 0, \qquad n \leqslant t < n+1. \qquad (8.5.15)$$

By integrating (8.5.15) from n to t and by letting $t \to n+1$ we obtain

$$u(n+1) - u(n) + q c_N\, e^{-\lambda_0 k}\, \frac{1 - e^{-\lambda_0}}{\lambda_0}\, u(n-k) \leqslant 0, \qquad n \geqslant t_N.$$

That is,

$$B_{n+1} - B_n + qQ B_n^{-k} \leqslant 0, \qquad n \geqslant t_N \qquad (8.5.16)$$

where

$$B_n = u(n) \qquad \text{and} \qquad Q = e^{-\lambda_0 k}\, \frac{1 - e^{-\lambda_0}}{\lambda_0}\, c_N.$$

Since B_n is eventually positive, it follows from Corollary 7.4.1 that the equation

$$\lambda - 1 + qQ\lambda^{-k} = 0 \qquad (8.5.17)$$

has a root in $(0, 1]$.

Now assume that (b) holds. Choose N so large that

$$qQ \geqslant 1.$$

Then for $0 < \lambda \leqslant 1$,

$$\lambda - 1 + qQ\lambda^{-k} > -1 + \lambda^{-k} \geqslant 0,$$

which contradicts the fact that eqn (8.5.17) has a root in $(0, 1]$.

Next, assume that (c) holds. As eqn (8.5.10) has no roots in $(0, 1]$, $q > 0$. By using an elementary calculation, one can easily show that for all small $\varepsilon \in (0, q)$,

$$\lambda - 1 + \frac{B}{A}(q - \varepsilon)\lambda^{-k} > 0 \qquad \text{for } 0 < \lambda \leqslant 1. \qquad (8.5.18)$$

Now observe that the sequence $\{c_N\}$ is strictly increasing and

$$\lim_{N \to \infty} c_N = \frac{1}{1 + \lambda_0 \tau} > 0.$$

Choose N so large that

$$q c_N\, e^{-\lambda_0 k}\, \frac{1 - e^{-\lambda_0}}{\lambda_0} > (q - \varepsilon)\frac{B}{A},$$

that is,

$$c_N > \frac{q - \varepsilon}{q} \frac{1}{1 + \lambda_0 \tau}.$$

Then

$$qQ > (q - \varepsilon) \frac{B}{A}$$

and so

$$\lambda - 1 + qQ\lambda^{-k} > \lambda - 1 + (q - \varepsilon) \frac{B}{A} \lambda^{-k} > 0 \qquad \text{for } 0 < \lambda \leqslant 1.$$

This contradicts the fact that eqn (8.5.17) has a root in $(0, 1]$ and completes the proof in this case.

Finally, assume that (d) is satisfied and that, for the sake of contradiction, eqn (8.5.1) has an eventually positive solution $x(t)$. Then

$$\dot{x}(t) + qx([t - k]) \leqslant 0. \tag{8.5.19}$$

By integrating (8.5.19) from n to t for $t \in [n, n + 1)$ and then by letting $t \rightarrow n + 1$, we obtain

$$a_{n+1} - a_n + qa_{n-k} \leqslant 0, \tag{8.5.20}$$

where $a_n = x(n)$ for $n = 0, 1, 2, \ldots$. Thus (8.5.20) has an eventually positive solution. However, in view of Corollary 7.4.1, this implies that the equation

$$b_{n+1} - b_n + qb_{n-k} = 0$$

has an eventually positive solution. In view of the hypothesis and Theorem 7.2.1, this is impossible and the proof is complete.

Remark 8.5.1. *Case (d) of Theorem* 8.5.1 *can be considered as a limiting case of case (c) of the same theorem. In fact, when* $p = 0$ *then* $\lambda_0 = 0$ *and*

$$\lim_{\lambda_0 \rightarrow 0} \frac{B}{A} = 1.$$

therefore eqn (8.5.10) as $\lambda_0 \rightarrow 0$ *yields*

$$\lambda - 1 + q\lambda^{-k} = 0.$$

8.6 Notes

The study of equations with piecewise constant argument was originated by the works of Wiener and his collaborators. See for example Cooke and

Wiener (1984, 1987a,b), Aftabizadeh and Wiener (1985, 1988), and Aftabizadeh et al. (1987). In addition to its own interest, this area has stimulated much activity in the study of delay difference equations.

Theorem 8.1.1 is from Györi et al. (1990a). Theorem 8.1.2 and Corollary 8.1.4 are from Ladas (1988a). Corollaries 8.1.1 and 8.1.3 are extracted from Aftabizadeh and Wiener (1985) and Corollary 8.1.2 is from Aftabizadeh et al. (1987).

Theorem 8.2.1 is from Györi and Ladas (1989) and Theorem 8.2.2 from Gopalsamy et al. (1990a).

The results of Section 8.3 are from Aftabizadeh et al. (1987).

Theorem 8.4.1 is extracted from Cooke and Wiener (1987a) and Theorem 8.4.2 from Aftabizadeh and Wiener (1988).

Section 8.5 is from Gopalsamy et al. (1989a).

For some additional results on this subject, see Carvalho and Cooke (1988), Huang (1988, 1989), Shah and Wiener (1983), and the results of Wiener (1990) on partial differential equations with piecewise constant argument.

8.7 Open problems

8.7.1   Obtain a 'characteristic equation' for an equation with continuous and piecewise constant argument of the form

$$\dot{x}(t) + px(t - \tau) + qx([t - k]) = 0$$

which when $q = 0$ reduces to

$$\lambda + p\,e^{-\lambda\tau} = 0$$

and when $p = 0$ reduces to

$$\lambda - 1 + q\lambda^{-k} = 0.$$

(See Grove et al. 1990a.)

8.7.2   Obtain necessary and sufficient conditions for the oscillation of all solutions of the equation

$$\dot{x}(t) + px(t - 1) + qx([t - 1]) = 0$$

where $p, q \in \mathbb{R}$.

8.7.3   Obtain explicit conditions in terms of the matrices Q_j and the integers k and l for the oscillation of all solutions of eqn (8.1.1).

8.7.4   Obtain necessary and sufficient conditions for the oscillation of all solutions of the equations

$$\dot{x}(t) + Px\left(2\left[\frac{t+1}{2}\right]\right) = 0$$

$$\dot{x}(t) + Px([t + \tfrac{1}{2}]) = 0$$

$$\dot{x}(t) + Px\left(\left[\frac{t+1}{2}\right]\right) = 0$$

where P is a real $r \times r$ matrix.

8.7.5   Extend theorem 8.2.1 to systems and to higher-order equations.

9

OSCILLATIONS OF INTEGRODIFFERENTIAL EQUATIONS

In Sections 9.1 and 9.3 we obtain necessary and sufficient conditions for the existence of positive solutions of the linear integrodifferential equations

$$\dot{x}(t) + \int_0^t k(t-s)x(s)\, ds = 0, \qquad t \geqslant T \tag{9.0.1}$$

and

$$\dot{x}(t) + bx(t) + \int_{-\infty}^t k(t-s)x(s)\, ds = 0, \qquad t \geqslant 0 \tag{9.0.2}$$

respectively. In Section 9.2 we obtain a comparison result for the positive solutions $x(t)$ and $y(t)$ of the integrodifferential inequalities

$$\dot{x}(t) + p(t)x(t) + \int_0^t k(t-s)x(s)\, ds < 0, \qquad 0 \leqslant t < T \tag{9.0.3}$$

and

$$\dot{y}(t) + p(t)y(t) + \int_0^t k(t-s)y(s)\, ds \geqslant 0, \qquad 0 \leqslant t < T. \tag{9.0.4}$$

In Section 9.4 we present a 'linearized' oscillation result for the Volterra-type integrodifferential equation of population dynamics:

$$\dot{N}(t) = N(t)\left[a - bN(t) - \int_{-\infty}^t k(t-s)N(s)\, ds\right], \qquad t \geqslant 0 \tag{9.0.5}$$

where

$$a \in (0, \infty), \qquad b \in [0, \infty), \qquad k \in C[[0, \infty), \mathbb{R}^+]$$

and

$$0 < \int_0^\infty k(s)\, ds < \infty.$$

We also obtain in Section 9.4 a sufficient condition for eqn (9.0.5) to have a solution $N(t)$ such that

$$N(t) > N^* \qquad \text{for } -\infty < t < \infty$$

where

$$N^* = \frac{a}{b + \int_0^\infty K(s)\,\mathrm{d}s}$$

is the positive equilibrium of eqn (9.0.5).

9.1 Oscillations of linear integrodifferential equations

Consider the linear Volterra-type integrodifferential equation

$$\dot{x}(t) + \int_0^t k(t - s)x(s)\,\mathrm{d}s = 0, \qquad t \geqslant T \qquad (9.1.1)$$

where $T \geqslant 0$ and $k \in C[\mathbb{R}^+, \mathbb{R}^+]$. By a solution of (9.1.1), we mean a continuous function x which is defined for $t \geqslant 0$ and which satisfies (9.1.1) for $t \geqslant T$.

We prove in this section that under the condition

$$-\lambda + \int_0^\infty e^{\lambda s}k(s)\,\mathrm{d}s > 0 \qquad \text{for all } \lambda > 0, \qquad (9.1.2)$$

eqn (9.1.1) cannot have a solution x which is positive on $[0, \infty)$, while if $T > 0$ and k is not identically zero on $[0, T]$ and if

$$-\lambda + \int_0^\infty e^{\lambda s}k(s)\,\mathrm{d}s \leqslant 0 \qquad \text{for some } \lambda > 0, \qquad (9.1.3)$$

then eqn (9.1.1) has a solution x which is positive on $[0, \infty)$.

The main result in this section is the following theorem.

Theorem 9.1.1. *Suppose that $k \in C[\mathbb{R}^+, \mathbb{R}^+]$ and let $T > 0$ be such that k is not identically zero on $[0, T]$. Then (9.1.3) is a necessary and sufficient condition for eqn (9.1.1) to have a solution x which is positive on $[0, \infty)$.*

This theorem is an immediate consequence of the following two results which are interesting in their own right.

Theorem 9.1.2. *Suppose that $k \in C[\mathbb{R}^+, \mathbb{R}^+]$ and let $T \geqslant 0$. Assume that condition (9.1.2) is satisfied. Then eqn (9.1.1) cannot have a solution x which is positive (or negative) on $[0, \infty)$.*

Theorem 9.1.3. *Suppose that $k \in C[\mathbb{R}^+, \mathbb{R}^+]$ and let $T > 0$ be such that k is not identically zero on $[0, T]$. Assume that condition (9.1.3) is satisfied. Then there exists a solution x of eqn (9.1.1) which is positive on $[0, \infty)$ and which* *satisfies*

$$\dot{x}(t) + \int_0^t k(t-s)x(s)\, ds \leqslant 0 \qquad \text{for } 0 \leqslant t \leqslant T \tag{9.1.4}$$

$$0 < x(t) \leqslant e^{-\lambda t} \qquad \text{for } t \geqslant 0. \tag{9.1.5}$$

Proof of Theorem 9.1.2. Assume, for the sake of contradiction, that there exists a solution x of (9.1.1) which is positive on $[0, \infty)$. Set

$$\Lambda = \{\lambda \geqslant 0 : \dot{x}(t) + \lambda x(t) \leqslant 0 \text{ for all large } t\}.$$

Clearly, $0 \in \Lambda$ and Λ is a subinterval of $\mathbb{R}^+$. First we establish that there exists a positive number λ_1 such that $\lambda_1 \in \Lambda$. To this end, (9.1.2) implies that k is not identically zero on $[0, \infty)$ and so there exists $\varepsilon > 0$ and $t_1 > \varepsilon$ such that

$$A \equiv \int_\varepsilon^{t_1} k(s)\, ds > 0.$$

By using the decreasing nature of x on $[T, \infty)$, it follows from (9.1.1) that for $t \geqslant t_1 + T$,

$$0 \geqslant \dot{x}(t) + \int_T^{t-\varepsilon} k(t-s)x(s)\, ds \geqslant \dot{x}(t) + \int_T^{t-\varepsilon} k(t-s)\, ds\, x(t-\varepsilon)$$

$$= \dot{x}(t) + \int_\varepsilon^{t-T} k(s)\, ds\, x(t-\varepsilon) \geqslant \dot{x}(t) + Ax(t-\varepsilon).$$

Thus

$$\dot{x}(t) + Ax(t-\varepsilon) \leqslant 0 \qquad \text{for } t \geqslant t_1 + T \tag{9.1.6}$$

and so also

$$\dot{x}(t) + Ax(t) \leqslant 0 \qquad \text{for } t \geqslant t_1 + T,$$

which shows that $\lambda_1 \equiv A \in \Lambda$. Next we prove that $\sup \Lambda < \infty$. In view of (9.1.6), it follows from Lemma 1.6.1 that

$$x(t-\varepsilon) < Bx(t) \qquad \text{for sufficiently large } t, \tag{9.1.7}$$

where $B = 4/(\varepsilon A)^2$. Clearly, B has to be greater than one, or else (9.1.7) is already a contradiction. We now claim that $\sup \Lambda \leqslant \ln B/\varepsilon$. Otherwise, $\Theta \equiv \ln(B/\varepsilon) \in \Lambda$. Hence there exists a $t_2 \geqslant T$ such that for $t \geqslant t_2$,

$$\frac{d}{dt}[e^{\Theta t}x(t)] = e^{\Theta t}[\dot{x}(t) + \Theta x(t)] \leqslant 0,$$

which implies that the function $e^{\Theta t}x(t)$ is decreasing on $[t_2, \infty)$. Hence for

$t \geqslant t_2 + \varepsilon$,

$$e^{\Theta(t-\varepsilon)}x(t - \varepsilon) \geqslant e^{\Theta t}x(t) \qquad \text{or} \qquad x(t - \varepsilon) \geqslant e^{\Theta \varepsilon}x(t) = Bx(t).$$

This contradicts (9.1.7) and proves our claim that sup $\Lambda < \infty$.

Set $\lambda^* = \sup \Lambda$ and let μ be an arbitrary positive number in the interval $(0, \lambda^*]$. Then $\lambda^* - \mu \equiv r \in \Lambda$ and so there exists a $t_3 \geqslant T$ such that $\dot{x}(t) + rx(t) \leqslant 0$ for $t \geqslant t_3$. Thus for any t, s with $t \geqslant t_3$ and $0 \leqslant s \leqslant t - t_3$, we have

$$\frac{x(t - s)}{x(t)} = \exp\left(-\ln \frac{x(t)}{x(t - s)}\right) = \exp\left(-\int_{t-s}^{t} \frac{\dot{x}(\xi)}{x(\xi)}\, d\xi\right) \geqslant e^{rs}.$$

Hence

$$x(t - s) \geqslant e^{rs}x(t) \qquad \text{for } t \geqslant t_3 \text{ and } 0 \leqslant s \leqslant t - t_3. \tag{9.1.8}$$

From (9.1.1) and (9.1.8), it follows that for $t \geqslant t_3$,

$$0 = \dot{x}(t) + \int_0^t k(s)x(t - s)\, ds \geqslant \dot{x}(t) + \int_0^{t-t_3} k(s)\, e^{rs}x(t)\, ds,$$

that is,

$$\dot{x}(t) + \left(\int_0^{t-t_3} e^{rs}k(s)\, ds\right)x(t) \leqslant 0 \qquad \text{for } t \geqslant t_3. \tag{9.1.9}$$

We now claim that

$$\int_0^{t-t_3} e^{rs}k(s) \leqslant \lambda^* \qquad \text{for every } t \geqslant t_3. \tag{9.1.10}$$

Otherwise, there exists a $\tilde{t}_3 > t_3$ such that

$$\tilde{\lambda} \equiv \int_0^{\tilde{t}_3 - t_3} e^{rs}k(s)\, ds > \lambda^*$$

and therefore (9.1.9) yields for $\tilde{t} \geqslant t_3$

$$0 \geqslant \dot{x}(t) + \left(\int_0^{\tilde{t}_3 - t_3} e^{rs}k(s)\, ds\right)x(t) = \dot{x}(t) + \tilde{\lambda}x(t).$$

Hence $\tilde{\lambda} \in \Lambda$, which contradicts the hypothesis that $\tilde{\lambda} > \lambda^*$. Thus (9.1.10) has been established. Finally, (9.1.10) implies that

$$\int_0^\infty e^{rs}k(s)\, ds \leqslant \lambda^* \qquad \text{or} \qquad \int_0^\infty e^{(\lambda^* - \mu)s}k(s)\, ds \leqslant \lambda^*.$$

As $\mu \in (0, \lambda^*]$ is arbitrary, it follows that

$$\int_0^\infty e^{\lambda^* s} k(s)\, ds \leqslant \lambda^*,$$

which contradicts (9.1.2) and completes the proof of the theorem.

Proof of Theorem 9.1.3. Let λ be such that (9.1.3) is satisfied, and define

$$y(t) = e^{-\lambda t}, \qquad t \geqslant 0.$$

Then one can easily verify that

$$\dot{y}(t) + \int_0^t k(t - s)\, y(s)\, ds \leqslant 0, \qquad t \geqslant 0. \tag{9.1.11}$$

By integrating (9.1.11) from t to $\tilde{t}$, with $\tilde{t} \geqslant t \geqslant 0$, we find

$$y(\tilde{t}) + \int_t^{\tilde{t}} \int_0^u k(u - s)\, y(s)\, ds\, du \leqslant y(t)$$

from which it follows that

$$\int_0^\infty \int_0^u k(u - s)\, y(s)\, ds\, du \leqslant y(t), \qquad t \geqslant 0.$$

Let

$$X = \{x \in C[\mathbb{R}^+, \mathbb{R}^+]: x(t) \leqslant y(t) \text{ for } t \geqslant 0\}$$

and define the monotone operator $S: X \to X$ as follows:

$$(Sx)(t) = \begin{cases} \displaystyle\int_t^\infty \int_0^u k(u - s)\, x(s)\, ds\, du, & t \geqslant T \\[2ex] \displaystyle\int_T^\infty \int_0^u k(u - s)\, x(s)\, ds\, du \\[2ex] \quad + \displaystyle\int_t^T \int_0^u k(u - s)\, y(s)\, ds\, du, & 0 \leqslant t < T. \end{cases}$$

Consider the sequence $\{x_n\}$ defined by

$$x_0 = y \qquad \text{and} \qquad x_n = Sx_{n-1} \qquad \text{for } n = 1, 2, \ldots$$

Clearly, for every $t \geqslant 0$, $y(t) = x_0(t) \geqslant x_1(t) \geqslant \cdots \geqslant x_n(t) \geqslant \cdots \geqslant 0$. Set $x(t) = \lim_{n \to +\infty} x_n(t)$ for each $t \geqslant 0$. Then by Lebesgue's dominated convergence theorem

$$
x(t) = \begin{cases} \displaystyle\int_t^\infty \int_0^u k(u - s)x(s)\, ds\, du, & t \leqslant T \\[2.5ex] \displaystyle\int_T^\infty \int_0^u k(u - s)x(s)\, ds\, du \\[2.5ex] \quad + \displaystyle\int_t^T \int_0^u k(u - s)y(s)\, ds\, du, & 0 \leqslant t < T. \end{cases} \tag{9.1.12}
$$

Clearly, $x \in C[\mathbb{R}^+, \mathbb{R}^+]$, $x(t) \leqslant y(t) = e^{-\lambda t}$, $(t \geqslant 0)$ holds and from (9.1.12) we see that x is a solution of eqn (9.1.1). For $0 \leqslant t < T$, (9.1.12) yields

$$
0 = \dot{x}(t) + \int_0^t k(t - s)y(s)\, ds \geqslant \dot{x}(t) + \int_0^t k(t - s)x(s)\, ds
$$

and so (9.1.4) holds. It remains to show that $x(t) > 0$ for $t \geqslant 0$. Clearly for $0 \leqslant t < T$,

$$
x(t) \geqslant \int_t^T \int_0^u k(u - s)y(s)\, ds\, du = \int_t^T \int_0^u k(s)y(u - s)\, ds\, du > 0
$$

because $y(t) > 0$ on $[0, T]$ and k is not identically zero on $[0, T]$. Next, we prove that x is also positive on $[T, \infty)$. Assume, for the sake of contradiction, that $\tau \geqslant T$ is the first zero of $x(t)$ to the right of 0. That is,

$$
x(t) > 0 \qquad \text{for } 0 \leqslant t < \tau, \qquad x(\tau) = 0 \qquad \text{and} \qquad \tau \geqslant T.
$$

Then from (9.1.12),

$$
0 = x(\tau) = \int_\tau^\infty \int_0^u k(u - s)x(s)\, ds\, du
$$

$$
= \int_\tau^\infty \int_0^u k(s)x(u - s)\, ds\, du, \tag{9.1.13}
$$

which is impossible because k is not identically zero on $[0, \tau]$, $x(t) > 0$ on $[0, \tau]$ and for $u = \tau$ the inner integral in (9.1.13) yields

$$
\int_0^\tau k(\tau - s)x(s)\, ds \geqslant \int_0^\tau k(s)x(\tau - s)\, ds > 0.
$$

The proof is complete.

In the next corollary we give an explicit condition in terms of k so that eqn (9.1.1) cannot have a solution x which is positive on $[0, \infty)$.

Corollary 9.1.1. *Suppose that* $k \in C[\mathbb{R}^+, \mathbb{R}^+]$ *and let* $T \geqslant 0$. *If*

$$\int_0^\infty sk(s)\,ds > 1/e \qquad (9.1.14)$$

then eqn (9.1.1) *cannot have a solution* x *which is positive (or negative) on* $[0, \infty)$.

Proof. By virtue of Theorem 9.1.2 it is enough to show that condition (9.1.2) is satisfied under condition (9.1.14). Indeed,

$$\min_{\lambda > 0} \frac{e^{\lambda s}}{\lambda} = es \qquad \text{for } s > 0$$

and so for all $\lambda > 0$,

$$-\lambda + \int_0^\infty e^{\lambda s} k(s)\,ds > \lambda \left[-1 + \int_0^\infty esk(s)\,ds \right] > 0.$$

The proof is complete.

Remark 9.1.1. *The function* $k(s) = \beta(s + 1)^{-3}$, $0 < \beta < 2e^{-1}$ *is an example where* (9.1.2) *is satisfied but not* (9.1.14).

9.2 Comparison results for integrodifferential equations of Volterra type

Our goal in this section is to compare the positive solutions $x(t)$ and $y(t)$ of the integrodifferential inequalities

$$\dot{x}(t) + p(t)x(t) + \int_0^t k(t, s)x(s)\,ds < 0, \qquad 0 \leqslant t < T \qquad (9.2.1)$$

$$\dot{y}(t) + p(t)y(t) + \int_0^t k(t, s)y(s)\,ds \geqslant 0, \qquad 0 \leqslant t < T \qquad (9.2.2)$$

where

$$0 < T \leqslant \infty, \qquad p \in C[[0, T), \mathbb{R}] \qquad \text{and} \qquad k \in C[[0, T) \times [0, T), \mathbb{R}^+].$$
$$(9.2.3)$$

The main result in this section is the following comparison theorem.

Theorem 9.2.1. *Assume that* (9.2.3) *holds and suppose that* $x, y \in C^1[[0, T), \mathbb{R}]$ *are solutions of* (9.2.1) *and* (9.2.2) *such that*

$$x(t) > 0 \qquad \text{for } 0 \leqslant t < T \qquad \text{and} \qquad y(0) > 0.$$

Then $y(t) > 0$ for $0 \leqslant t < T$ and $x(t)/y(t)$ is monotonically decreasing on $[0, T)$. In particular, if $x(0) \leqslant y(0)$ then $x(t) \leqslant y(t)$ for $0 \leqslant t < T$.

Proof. As $y(0) > 0$, there exists $T_1 \in (0, T]$ such that $y(t) > 0$ for $0 \leqslant t < T_1$. Set

$$z(t) = x(t)/y(t)$$

and observe that

$$\dot{z}(t) = \frac{\dot{x}(t)}{y(t)} - \frac{x(t)\dot{y}(t)}{y^2(t)} = \frac{1}{y(t)}\left[\dot{x}(t) - \frac{\dot{y}(t)}{y(t)}x(t)\right]$$

and that from (9.2.2),

$$x(t)\dot{y}(t) \geqslant -p(t)x(t)y(t) - x(t)\int_0^t k(t, s)y(s)\,ds.$$

Hence

$$\dot{z}(t) < \frac{1}{y(t)}\left[-\int_0^t k(t, s)\left[\frac{x(t)}{y(t)} - \frac{x(s)}{y(s)}\right]y(s)\,ds\right], \qquad 0 \leqslant t < T_1. \quad (9.2.4)$$

We now claim that $\dot{z}(t) < 0$ for every $t \in [0, T_1)$. Otherwise the set

$$S = \{t \in [0, T_1): \dot{z}(t) \geqslant 0\}$$

is non-empty. Let $t_0 = \inf S$. From (9.2.3), it is clear that $\dot{z}(0) < 0$ and thus $t_0 > 0$. Moreover,

$$\dot{z}(t) < 0 \qquad \text{for } 0 \leqslant t < t_0 \qquad \text{and} \qquad \dot{z}(t_0) = 0. \quad (9.2.5)$$

Hence

$$z(t_0) = x(t_0)/y(t_0) \leqslant z(s) = x(s)/y(s) \qquad \text{for } 0 \leqslant s \leqslant t_0.$$

From (9.2.4) we deduce that $\dot{z}(t_0) < 0$, which is incompatible with (9.2.5). Thus the set S is empty, which proves our claim that $z(t)$ is a decreasing function on $[0, T_1)$. From this it follows that $y(t)$ is positive on $[0, T)$. Otherwise there exists a $T_1 \in (0, T)$ such that $y(t) > 0$ for $0 \leqslant t < T_1$ and $y(t_1) = 0$. As $x(T_1) > 0$, it follows that

$$z(t) = x(t)/y(t) \to +\infty \qquad \text{as } t \to T_1. \quad (9.2.6)$$

On the other hand, $z(t)$ is a decreasing function on $[0, T_1)$, which contradicts (9.2.6). Thus $y(t) > 0$ and $z(t)$ is monotone decreasing on $[0, T)$. The proof of the theorem is complete.

In the next corollary the inequality (9.2.1) is replaced by

$$\dot{x}(t) + p(t)x(t) + \int_0^t k(t, s)x(s)\,ds \leqslant 0, \qquad 0 \leqslant t < T. \qquad (9.2.1')$$

Corollary 9.2.1. *Assume that (9.2.3) is satisfied and that*

$$p(t) + \int_0^t k(t, s)\,ds > 0 \qquad for \ 0 \leqslant t < T. \qquad (9.2.7)$$

Suppose that $x, y \in C^1[[0, T), \mathbb{R}]$ *are solutions of (9.2.1') and (9.2.2) such that*

$$x(t) > 0 \qquad for \ 0 \leqslant t < T \qquad and \qquad y(0) > 0.$$

Then the conclusion of Theorem 9.2.1 remains true.

Proof. Let $T_1 \in (0, T)$ be an arbitrary but fixed point and define $m = \min\{x(t): 0 \leqslant t \leqslant T_1\} > 0$. Let $\varepsilon \in (0, m)$ and set $x_\varepsilon(t) = x(t) - \varepsilon$ for $0 \leqslant t \leqslant T_1$. Then (9.2.1') and (9.2.7) imply that

$$\dot{x}_\varepsilon(t) < -\int_0^t k(t, s)x_\varepsilon(s)\,ds - p(t)x_\varepsilon(t), \qquad 0 \leqslant t \leqslant T_1,$$

since $x_\varepsilon(t) > 0$ for $0 \leqslant t \leqslant T_1$. Then by Theorem 9.2.1 it follows that $y(t) > 0$ and $x_\varepsilon(t)/y(t)$ is a monotone decreasing function on $[0, T_1]$. Thus, for $0 \leqslant t_1 < t_2 \leqslant T$,

$$\frac{x_\varepsilon(t_1)}{y(t_1)} = \frac{x(t_1) - \varepsilon}{y(t_1)} > \frac{x_\varepsilon(t_2)}{y(t_2)} = \frac{x(t_2) - \varepsilon}{y(t_2)},$$

that is,

$$\frac{x(t_1)}{y(t_1)} = \lim_{\varepsilon \to 0} \frac{x_\varepsilon(t_1)}{y(t_1)} \geqslant \lim_{\varepsilon \to 0} \frac{x_\varepsilon(t_2)}{y(t_2)} = \frac{x(t_2)}{y(t_2)}.$$

Therefore $y(t) > 0$ and $x(t)/y(t)$ is decreasing on $[0, T_1]$, from which the result follows.

By applying Theorem 9.2.1 we can now establish a sufficient condition for the existence of zeros of the solutions of the integrodifferential equation

$$\dot{x}(t) + p(t)x(t) + \int_0^t k(t, s)x(s)\,ds = 0, \qquad t \geqslant 0. \qquad (9.2.8)$$

Theorem 9.2.2. *Assume that* (9.2.3) *is satisfied and define*

$$K_0 = \inf_{0 \leqslant s \leqslant t < \infty} k(t, s) \quad and \quad M_0 = \sup_{0 \leqslant t < \infty} p^-(t),$$

where $p^-(t) = \max\{0, -p(t)\}$ *for* $t \geqslant 0$. *If* $M_0 < \infty$ *and*

$$M_0^2 - 4K_0 < 0 \tag{9.2.9}$$

then every solution $x \colon [0, \infty) \to \mathbb{R}$ *of eqn* (9.2.8) *has at least one zero on* $(0, \infty)$.

Proof. For the sake of contradiction we assume that eqn (9.2.8) has a solution $x \colon [0, \infty) \to \mathbb{R}$ such that $x(t) \neq 0$ for all $t > 0$. Without loss of generality we can assume that $x(t) > 0$ for all $t \geqslant 0$. Then it follows from (9.2.8) that for $t \geqslant 0$,

$$0 = \dot{x}(t) + p(t)x(t) + \int_0^t k(t, s)x(s)\, ds \geqslant \dot{x}(t) - M_0 x(t) + K_0 \int_0^t x(s)\, ds$$

and so for every $\varepsilon > 0$,

$$\dot{x}(t) < (M_0 + \varepsilon)x(t) - K_0 \int_0^t x(s)\, ds, \qquad t \geqslant 0.$$

Therefore by Theorem 9.2.1 we know that the solution $y_\varepsilon \colon [0, \infty) \to \mathbb{R}$ of

$$\dot{y}_\varepsilon(t) = (M_0 + \varepsilon)y_\varepsilon(t) - K_0 \int_0^t y_\varepsilon(s)\, ds, \qquad t \geqslant 0 \tag{9.2.10}$$

with initial condition $y_\varepsilon(0) = x(0)$ satisfies $y_\varepsilon(t) \geqslant x(t) > 0$ for all $t \geqslant 0$. From (9.2.10) we see that $y_\varepsilon(t)$ is a positive solution of the second-order differential equation

$$\ddot{y}(t) - (M_0 + \varepsilon)\dot{y}(t) + K_0 y(t) = 0.$$

Thus its characteristic equation

$$\lambda^2 - (M_0 + \varepsilon)\lambda + K_0 = 0$$

has a real root or, equivalently,

$$(M_0 + \varepsilon)^2 - 4K_0 > 0.$$

This contradicts (9.2.9), as $\varepsilon \to 0$, and the proof is complete.

Another corollary of Theorem 9.2.1 is the following.

Corollary 9.2.2. *Assume that (9.2.3) is satisfied and that there exists a $\lambda > 0$ such that for all $t \geq 0$,*

$$p(t) + \int_0^t k(t, s)\, e^{\lambda(t-s)}\, ds \leq \lambda. \qquad (9.2.11)$$

Then eqn (9.2.8) has a positive solution on $[0, \infty)$.

Proof. Set

$$y(t) = e^{-\lambda t}, \qquad t \geq 0.$$

Then $y(t) > 0$ and by (9.2.11) we obtain

$$\dot{y}(t) + p(t)\, y(t) + \int_0^t k(t, s)\, y(s)\, ds$$

$$= -\lambda\, e^{-\lambda t} + p(t)\, e^{-\lambda t} + \int_0^t k(t, s)\, e^{-\lambda s}\, ds \leq 0 \qquad \text{for all } t \geq 0. \quad (9.2.12)$$

Therefore, by applying Corollary 9.2.1 to eqn (9.2.8) and to inequality (9.2.12), we see that eqn (9.2.8) has a positive solution. The proof is complete.

Remark 9.2.1. *When $k(t, s) = k(t - s)$ for $0 \leq s \leq t < \infty$, then condition (9.2.11) reduces to*

$$p(t) + \int_0^t k(u)\, e^{\lambda u}\, du \leq \lambda \qquad \text{for all } t \geq 0. \qquad (9.2.13)$$

In particular when $p(t) \equiv 0$, then (9.2.13) reduces to

$$\int_0^\infty k(u)\, e^{\lambda u}\, du \leq \lambda$$

which was used in Theorem 9.1.3 to prove that eqn (9.1.1) has a positive solution on $[0, \infty)$. Thus Corollary 9.2.2 is a generalization of Theorem 9.1.3.

9.3 Positive solutions of integrodifferential equations with unbounded delay

Consider the linear integrodifferential equation with unbounded delay

$$\dot{x}(t) + bx(t) + \int_{-\infty}^t k(t - s)\, x(s)\, ds = 0, \qquad t \geq 0 \qquad (9.3.1)$$

where

$$b \in \mathbb{R}, \qquad k \in C[[0, \infty), \mathbb{R}^+] \qquad \text{and} \qquad 0 < \int_0^\infty k(s) \, e^{-\gamma_0 s} \, ds < \infty \qquad (9.3.2)$$

where γ_0 is some real number.

Let B^+ denote the space of initial functions

$$B^+ = \left\{ \phi \in C[(-\infty, 0], \mathbb{R}^+] : \int_{-\infty}^0 k(t - s) \phi(s) \, ds \right.$$

$$\left. \text{is a continuous function on } [0, \infty) \right\}.$$

Note that the set B^+ contains the function

$$\phi(t) = M \, e^{\gamma_0 t} \qquad \text{for } -\infty < t \leqslant 0 \text{ with } M \in (0, \infty).$$

With eqn (9.3.1) we associate an initial function of the form

$$x(t) = \phi(t) \qquad \text{for } -\infty < t \leqslant 0 \text{ with } \phi \in B^+. \qquad (9.3.3)$$

When (9.3.2) holds, then the initial value problem (9.3.1) and (9.3.3) has a unique solution on $(-\infty, \infty)$ (see Burton 1983). If we look for a positive solution of eqn (9.3.1) of the form $x(t) = e^{\lambda t}$, we see that λ is a root of the *characteristic equation*

$$\lambda + b + \int_0^\infty k(s) \, e^{-\lambda s} \, ds = 0. \qquad (9.3.4)$$

The main result in this section is the following necessary and sufficient condition for the existence of a solution of eqn (9.3.1) which is positive for $t > 0$.

Theorem 9.3.1. *Assume that* (9.3.2) *holds. Then the following statements are equivalent.*

(a) *There is no $\phi \in B^+$ such that the initial value problem* (9.3.1) *and* (9.3.3) *has a solution which is positive for $t > 0$.*

(b) *The characteristic equation* (9.3.4) *has no real roots.*

Proof.
   (a) $\Rightarrow$ (b). Otherwise (9.3.4) has a real root λ_0. Then $x(t) = e^{\lambda_0 t}$ is a positive solution of eqn (9.3.1) for $-\infty < t < \infty$. Moreover, the initial function ϕ for this solution is $\phi(t) = e^{\lambda_0 t}$ for $-\infty < t \leqslant 0$ and clearly $\phi \in B^+$.

(b) $\Rightarrow$ (a). Assume, for the sake of contradiction, that for some $\phi \in B^+$ the solution $x(t)$ of (9.3.1) and (9.3.3) is positive for $t > 0$. Then from eqn (9.3.1) we see that

$$\dot{x}(t) + bx(t) \leqslant 0, \qquad t \geqslant 0$$

and so

$$x(t) \leqslant x(0) \, e^{-bt}, \qquad t \geqslant 0. \tag{9.3.5}$$

Therefore the Laplace transform of $x(t)$,

$$X(s) = \int_0^\infty e^{-st} x(t) \, dt,$$

exists for all $\mathrm{Re}\, s > -b$. From (9.3.2) and (9.3.5) it follows that the Laplace transform of the integral term in eqn (9.3.1) exists for all $\mathrm{Re}\, s > b + \gamma_0$. Moreover,

$$\int_0^\infty e^{-st} \left[\int_{-\infty}^t k(t-u) x(u) \, du \right] dt = G(s) + K(s) X(s)$$

for all $\mathrm{Re}\, s > -b + \gamma_0$, where

$$G(s) = \int_0^\infty e^{-st} \left[\int_{-\infty}^0 k(t-u) \phi(u) \, du \right] dt,$$

$$K(s) = \int_0^\infty e^{-st} k(t) \, dt.$$

Hence, by taking Laplace transforms on both sides of eqn (9.3.1), we obtain

$$[s + b + K(s)] X(s) = x(0) - G(s) \qquad \text{for } \mathrm{Re}\, s > b + \gamma_0. \tag{9.3.6}$$

Let us denote by σ_x, σ_k, and σ_g the abscissae of convergence of the Laplace transforms $X(s)$, $K(s)$, and $G(s)$ of the functions $x(t)$, $k(t)$, and

$$g(t) = \int_{-\infty}^0 k(t-u) x(u) \, du,$$

respectively. Then $X(s)$, $K(s)$, and $G(s)$ are analytic functions for

$$\mathrm{Re}\, s > \sigma_x, \qquad \mathrm{Re}\, s > \sigma_k, \qquad \text{and} \qquad \mathrm{Re}\, s > \sigma_g,$$

respectively. From the hypothesis that the characteristic equation (9.3.4) has no real roots it follows that

$$s + b + K(s) > 0 \qquad \text{for } s \in \mathbb{R}$$

and therefore the function

$$\frac{x(0) - G(s)}{s - b + K(s)}$$

is analytic for all Re $s > \max\{\sigma_k, \sigma_g\}$. Hence we can extend (9.3.6) to hold for all Re $s > \max\{\sigma_x, \sigma_k, \sigma_g\}$. Then

$$X(s) = \frac{x(0) - G(s)}{s + b + K(s)} \qquad (9.3.7)$$

for all Re $s > \max\{\sigma_x, \sigma_k, \sigma_g\}$. Our strategy is to show that (9.3.7) is valid for all Re $s > -\infty$ and then to prove that this leads to a contradiction. Set

$$\sigma_0 = \max\{\sigma_k, \sigma_g\}.$$

First, we claim that

$$\sigma_x \leqslant \sigma_0. \qquad (9.3.8)$$

Otherwise by Theorem 1.3.1, the point $s = \sigma_x$ is a singularity of $X(s)$. Then from (9.3.7) we see that

$$\infty = \lim_{s \to s_x^-} X(s) = \frac{x(0) - G(s_x)}{s_x + b + K(s_x)} < \infty,$$

which is a contradiction. Thus (9.3.8) holds and so (9.3.7) holds for all Re $s > \sigma_0$. Now we claim that

$$\sigma_k = \sigma_g = -\infty. \qquad (9.3.9)$$

Otherwise, one of the following three cases holds:

(i) $-\infty \leqslant \sigma_g < \sigma_k < \infty$;

(ii) $-\infty \leqslant \sigma_k < \sigma_g < \infty$;

(iii) $-\infty < \sigma_k = \sigma_g < \infty$.

We prove that (i) leads to a contradiction. A similar argument may be used to show that (ii) and (iii) also lead to contradictions. It follows from (9.3.8) and (i) that

$$\sigma_x \leqslant \sigma_0 = \sigma_k.$$

Then by Theorem 1.3.1, $K(\sigma_k-) = \infty$ and (9.3.7) yields the contradiction

$$0 < X(\sigma_k-) = \lim_{s \to \sigma_k^-} \frac{x(0) - G(s)}{s + b + K(s)} = 0.$$

From (9.3.8) and (9.3.9) we see that (9.3.7) is valid for all Re $s > -\infty$. As $X(s) > 0$ for all $s \in (-\infty, \infty)$, (9.3.7) yields that

$$x(0) \geq x(0) - G(s) > s + b + K(s) \qquad \text{for } s \in (-\infty, \infty). \quad (9.3.10)$$

Now for $s \leq 0$,

$$e^{-st} \geq \tfrac{1}{2}s^2 t^2$$

and so

$$s + b + K(s) \geq s + b + \tfrac{1}{2}s^2 \int_0^\infty t^2 k(t)\, dt \to \infty \qquad \text{as } s \to -\infty.$$

This contradicts (9.3.10) and the proof of the theorem is complete.

Remark 9.3.1. *It is an elementary observation that Theorem 9.3.1 remains true if the initial condition (9.3.3) is replaced by the (possibly discontinuous) initial condition*

$$\left. \begin{array}{l} x(t) = \phi(t) \qquad \text{for } -\infty < t < 0 \text{ and } x(0) = x_0 \\ \text{where } \phi \in B^+ \text{ and } x_0 \in \mathbb{R} \end{array} \right\} \qquad (9.3.3')$$

The above remark enable us to obtain the following necessary condition for the existence of a positive solution for the integrodifferential equation

$$\dot{y}(t) + by(t) + \int_0^t k(t - s)y(s)\, ds = 0, \qquad t \geq 0. \quad (9.3.11)$$

See also Theorem 9.1.2.

Corollary 9.3.1. *Assume that (9.3.2) holds and that the equation (9.3.11) has a positive solution on* $[0, \infty)$. *Then eqn (9.3.4) has a real root.*

Proof. Let $y(t)$ be a positive solution of eqn (9.3.11) on $(0, \infty)$. Then the function

$$x(t) = \begin{cases} y(t), & t > 0 \\ 0, & t \leq 0 \end{cases}$$

is a solution of eqn (9.3.1) with initial function $\phi(t) = 0$ for $-\infty < t \leq 0$. On the other hand, $\phi \in B^+$ and $x(t) > 0$ for $0 < t < \infty$. Therefore by Theorem 9.3.1, eqn (9.3.4) has a real root. The proof is complete.

9.4 Oscillation in Volterra's integrodifferential equation

Consider the Volterra-type integrodifferential equation of population dynamics

$$\dot{N}(t) = N(t)\left[a - bN(t) - \int_{-\infty}^{t} k(t-s)N(s)\,ds \right], \qquad t \geqslant 0 \quad (9.4.1)$$

where

$$\left. \begin{array}{c} a \in (0, \infty), \quad b \in [0, \infty), \quad k \in C[[0, \infty), \mathbb{R}^+] \\[2mm] 0 < \displaystyle\int_0^\infty k(s)\,ds < \infty \end{array} \right\}. \qquad (9.4.2)$$

This equation arises in models for the variation of the population of a species where the death rate depends not only on the population at time t, but also on the population at all previous times $s \leqslant t$ in a manner distributed in the past by the delay kernel $k(s)$ (see Cushing 1977).

Let B^+ denote the space of initial functions

$$B^+ = \left\{ \phi \in C[(-\infty, 0], (0, \infty)] : \int_{-\infty}^{0} k(t-s)\,\phi(s)\,ds \right.$$

$$\left. \text{is a continuous function on } [0, \infty) \right\}.$$

By a solution of eqn (9.4.1) on $(-\infty, \infty)$ we mean a function

$$N \in C[(-\infty, \infty), \mathbb{R}] \cap C^1[[0, \infty), \mathbb{R}]$$

which satisfies (9.4.1) for $t \geqslant 0$ and such that the function $\phi(t) = N(t)$ for $t \leqslant 0$ is in B^+. Clearly every solution of (9.4.1) is positive for all t. With eqn (9.4.1) we associate an initial function of the form

$$N(t) = \phi(t) \qquad \text{for } t \leqslant 0, \qquad \text{where } \phi \in B^+. \qquad (9.4.3)$$

When (9.4.2) holds, the initial value problem (9.4.1) and (9.4.3) has a unique solution $N(t)$ on $(-\infty, \infty)$ (see Burton 1983).

Observe that eqn (9.4.1) has a unique positive equilibrium N^* and that

$$N^* = \frac{a}{b + \int_0^\infty k(s)\,ds}.$$

Let $N(t)$ be the unique positive solution of the initial value problem (9.4.1) and (9.4.3) and set

$$N(t) = N^* e^{x(t)} \qquad \text{for } -\infty < t < \infty.$$

Then $x(t)$ satisfies the initial value problem

$$\dot{x}(t) + bN^*[e^{x(t)} - 1] + N^* \int_{-\infty}^{t} k(t - s)[e^{x(s)} - 1]\, ds = 0, \qquad t \geq 0 \quad (9.4.4)$$

and

$$x(t) = \ln \frac{\phi(t)}{N^*}, \qquad -\infty < t \leq 0. \qquad (9.4.5)$$

The *linearized equation* associated with eqn (9.4.4) is

$$\dot{y}(t) + bN^* y(t) + N^* \int_{-\infty}^{t} k(t - s)\, y(s)\, ds = 0, \qquad t \geq 0. \quad (9.4.6)$$

If we look for a positive solution of eqn (9.4.6) of the form

$$N(t) = e^{\lambda t}, \qquad -\infty < t < \infty$$

we see that λ satisfies the *characteristic equation* of (9.4.6), namely,

$$\lambda + bN^* + N^* \int_{0}^{\infty} k(s)\, e^{-\lambda s}\, ds = 0. \qquad (9.4.7)$$

In Theorem 9.3.1 we proved that if (9.4.7) has no real roots then (9.4.6) has no positive solutions on $(-\infty, \infty)$. The next theorem shows that the same result is true for eqn (9.4.4). In this sense the following result may be thought of as being a linearized oscillation result for Volterra-type integro-differential equations.

Theorem 9.4.1. *Assume that (9.4.2) holds and that eqn (9.4.7) has no real roots. Let $N(t)$ be the unique solution of (9.4.1) and (9.4.3). Then $N(t) - N^*$ has at least one zero in the interval $(-\infty, \infty)$.*

Proof. Assume, for the sake of contradiction, that $N(t) - N^*$ has no zero in the interval $(-\infty, \infty)$. We assume that $N(t) > N^*$ for all t. The case where $N(t) < N^*$ for all t is similar and is omitted. Set

$$N(t) = N^* e^{x(t)} \qquad \text{for } -\infty < t < \infty.$$

Then $x(t) > 0$ for all t and $x(t)$ satisfies (9.4.4). Since $e^x - 1 \geq x$ for $x \geq 0$, it follows from (9.4.4) that

$$\dot{x}(t) + bN^* x(t) + N^* \int_{-\infty}^{t} k(t - s) x(s)\, ds \leq 0, \qquad t \geq 0$$

and

$$bN^* + N^* \int_{-\infty}^{t} k(t-s) \frac{x(s)}{x(\max\{s,0\})} \frac{x(\max\{s,0\})}{x(t)} \, ds \leq -\frac{\dot{x}(t)}{x(t)}$$

for $t \geq 0$. Set $\alpha(t) = -\dot{x}(t)/x(t)$ for $t \geq 0$. Then $\alpha(t) > 0$ for $t \geq 0$ and

$$\alpha(t) \geq bN^* + N^* \int_{-\infty}^{t} k(t-s) \frac{x(s)}{x(\max\{s,0\})} \exp\left(\int_{\max\{s,0\}}^{t} \alpha(u) \, du\right) ds,$$

$$t \geq 0. \quad (9.4.8)$$

Define the sequence of functions $\{\beta_n(t)\}$ for $n \geq 0$ as follows:

$$\beta_0(t) = 0 \quad \text{for } t \geq 0,$$

$$\beta_{n+1}(t) = bN^* + N^* \int_{-\infty}^{t} k(t-s) \frac{x(s)}{x(\max\{s,0\})} \exp\left(\int_{\max\{s,0\}}^{t} \beta_n(u) \, du\right)$$

$$\text{for } t \geq 0 \text{ and } n \geq 0. \quad (9.4.9)$$

Then it can be easily seen that the functions $\beta_n(t)$ are well-defined and continuous on $[0, \infty)$ for all $n \geq 0$. On the other hand,

$$0 \leq \beta_0(t) \leq \alpha(t) \quad \text{for } 0 \leq t < \infty$$

and clearly

$$0 \leq \beta_0(t) \leq \beta_1(t) \leq \cdots \leq \beta_n(t) \leq \cdots \leq \alpha(t), \quad 0 \leq t < \infty.$$

Thus the limit $\beta(t) = \lim_{n \to +\infty} \beta_n(t)$ exists and is an integrable function on any compact subinterval of $[0, \infty)$. Moreover, for $t \geq s \geq 0$

$$0 \leq \beta(t) \leq \alpha(t)$$

and

$$\exp\left(\int_{\max\{s,0\}}^{t} \beta(u) \, du\right) = \lim_{n \to +\infty} \exp\left(\int_{\max\{s,0\}}^{t} \beta_n(u) \, du\right).$$

Combining these facts, we see that $\beta(t)$ satisfies the equation

$$\beta(t) = bN^* + N^* \int_{-\infty}^{t} k(t-s) \frac{x(s)}{x(\max\{s,0\})} \exp\left(\int_{\max\{s,0\}}^{t} \beta(u) \, du\right), \quad t \geq 0.$$

Set

$$y(t) = \begin{cases} x(0) \exp\left(\int_0^t \beta(u) \, du\right), & 0 \leq t < \infty, \\ x(t), & -\infty < t < 0. \end{cases}$$

Then $y(t)$ is a positive and continuous function on $(-\infty, \infty)$ and is continuously differentiable on $[0, \infty)$. Moreover,

$$\frac{y(s)}{y(\max\{s, 0\})} = \frac{x(s)}{x(\max\{s, 0\})}, \qquad s \geq 0$$

and

$$\beta(t) = \frac{-\dot{y}(t)}{y(t)} \quad \text{and} \quad \exp\left(\int_{\max\{s, 0\}}^{t} \beta(u) \, du\right) = \frac{y(\max\{s, 0\})}{y(t)}, \qquad t \geq s \geq 0.$$

Thus $y(t)$ satisfies

$$\frac{-\dot{y}(t)}{y(t)} = bN^* + \int_{-\infty}^{t} k(t - s) \frac{x(s)}{x(\max\{0, s\})} \frac{y(\max\{s, 0\})}{y(t)} \, ds$$

or equivalently,

$$\dot{y}(t) = -bN^* y(t) - \int_{-\infty}^{t} k(t - s) y(s) \, ds \qquad \text{for } t \geq 0, \qquad (9.4.10)$$

where we used the fact that $x(s) = y(s)$ for all $s \leq 0$ and $x(s) = x(\max\{0, s\})$ for all $s \geq 0$. Since eqn (9.4.10) has a solution $y(t)$ that is positive on $(-\infty, \infty)$, it follows from Theorem 9.3.1 that its characteristic equation (9.4.11) has a real root. This is a contradiction and the proof of the theorem is complete.

The next result is a partial converse of Theorem 9.4.1.

Theorem 9.4.2. *Assume that* (9.4.2) *holds and that there exists $\delta_0 > 0$ such that the equation*

$$\lambda + (1 + \delta_0)bN^* + (1 + \delta_0)N^* \int_{0}^{\infty} k(t) e^{\lambda t} \, dt = 0 \qquad (9.4.11)$$

has a real root. Then eqn (9.4.1) *has a positive solution $N(t)$ such that*

$$N(t) > N^* \qquad for \ -\infty < t < \infty. \qquad (9.4.12)$$

Proof. Since (9.4.11) has a real root and (9.4.2) is a satisfied, it follows that (9.4.7) has a negative root. Moreover, there exists $\delta \in (0, \delta_0]$ such that the equation

$$\lambda + (1 + \delta)bN^* + (1 + \delta)N^* \int_{0}^{\infty} k(t) e^{\lambda t} \, dt = 0 \qquad (9.4.13)$$

has exactly two negative real roots $-\alpha_1$ and $-\alpha_2$ such that $0 < \alpha_1 < \alpha_2$. By virtue of (9.4.13) it can be easily seen that

$$\alpha_i = (1 + \delta)bN^* + (1 + \delta)N^* \int_{-\infty}^{t} k(t - s) \frac{e^{-\alpha_i s}}{e^{-\alpha_i \max\{s,0\}}} \exp\left(\int_{\max\{s,0\}}^{t} \alpha_i \, du\right) ds$$

(9.4.14)

for all $t \geqslant 0$ and $i = 1, 2$. Define two sequences $\{\beta_n\}_{n=0}^{\infty}$ and $\{x_n\}_{n=0}^{\infty}$ as follows:

$$\beta_0(t) = \alpha_1 \qquad \text{for } t \geqslant 0,$$

$$x_0(t) = \begin{cases} \varepsilon \exp\left(-\int_0^t \beta_0(u) \, du\right) & \text{for } t \geqslant 0 \\ \varepsilon & \text{for } t < 0, \end{cases}$$

$$\beta_{n+1}(t) = \begin{cases} bN^* \dfrac{e^{x_n(t)} - 1}{x_n(t)} + N^* \displaystyle\int_{-\infty}^{t} k(t - s) \dfrac{e^{x_n(s)} - 1}{x_n(\max\{s, 0\})} \\ \qquad\qquad\qquad \times \exp\left(\displaystyle\int_{\max\{s,0\}}^{t} \beta_n(u) \, du\right) ds & \text{for } t \geqslant 0 \\ \\ \alpha_1 & \text{for } t < 0, \end{cases}$$

$$x_{n+1}(t) = \begin{cases} \varepsilon \exp\left(-\int_0^t \beta_{n+1}(u) \, du\right) & \text{for } t \geqslant 0 \\ \varepsilon & \text{for } t < 0, \end{cases}$$

for all $n \geqslant 0$, where $\varepsilon \in (0, 1)$ is such that

$$\frac{e^\varepsilon - 1}{\varepsilon} \leqslant 1 + \delta.$$

(9.4.15)

Note that $\beta_n(t)$ is a well-defined and locally integrable function on $(-\infty, \infty)$ for all $n \geqslant 0$. We claim that for all $n \geqslant 0$,

$$0 \leqslant \beta_n(t) \leqslant \alpha_2 \qquad \text{for } t \geqslant 0.$$

(9.4.16)

The proof of the claim is by induction. First, (9.4.16) is satisfied for $n = 0$. Assume that (9.4.16) is satisfied for an index $n \geqslant 1$. Then by definition,

$$0 < \varepsilon e^{-\alpha_2 t} \leqslant x_n(t) \leqslant \varepsilon \qquad \text{for } t \geqslant 0,$$

(9.4.17)

$$x_n(t) = \varepsilon \qquad \text{for } t < 0.$$

Thus (9.4.15) yields

$$\frac{e^{x_n(u)} - 1}{x_n(u)} \leq \frac{e^\varepsilon - 1}{\varepsilon} \leq \delta \qquad \text{for } u \geq 0.$$

Hence

$$\beta_{n+1}(t) \leq \begin{cases} bN^*(1 + \delta) + N^*(1 + \delta) \displaystyle\int_{-\infty}^t k(t - s) \dfrac{x_n(s)}{x_n(\max\{s, 0\})} \\ \qquad\qquad \times \exp\left(\displaystyle\int_{\max\{s,0\}}^t \alpha_2 \, du\right) ds, & t \geq 0, \\[4mm] \alpha_2, & t < 0. \end{cases}$$

Since $x_n(s)/x_n(\max\{s, 0\}) = 1$ for all $s \geq 0$, the last inequality and (9.4.5) yield (9.4.16) and hence the claim is proved.

We now show that the limit $\beta(t) = \lim_{n \to +\infty} \beta_n(t)$ exists for all $t \in (-\infty, \infty)$. By the definition of $\{\beta_n(t)\}$ we have

$$\beta_{n+1}(t) = bN^* \frac{e^{x_n(t)} - 1}{x_n(t)} + N^* \int_0^t k(t - s) \frac{e^{x_n(s)} - 1}{x_n(s)} \exp\left(\int_s^t \beta_n(u) \, du\right) ds$$

$$+ N^* \frac{e^\varepsilon - 1}{\varepsilon} \int_{-\infty}^0 k(t - s) \exp\left(\int_0^t \beta_n(u) \, du\right) ds$$

$$= bN^* \frac{e^{x_n(t)} - 1}{x_n(t)} + N^* \int_0^t k(t - s) \frac{e^{x_n(s)} - 1}{x_n(s)} \exp\left(-\int_s^t \beta_n(u) \, du\right) ds$$

$$+ N^* \frac{e^\varepsilon - 1}{\varepsilon} \int_0^\infty k(u) \, du \exp\left(-\int_0^t \beta_n(u) \, du\right) \qquad (9.4.18)$$

for all $t \geq 0$ and $n \geq 0$. Let $T > 0$ be an arbitrary but fixed number. By virtue of (9.4.17) we see that for all $n \geq 1$ and $t \in [0, T]$,

$$\left| \frac{e^{x_n(t)} - 1}{x_n(t)} - \frac{e^{x_{n-1}(t)} - 1}{x_{n-1}(t)} \right| \leq a|x_n(t) - x_{n-1}(t)|$$

$$= a\varepsilon \left| \exp\left(-\int_0^t \beta_n(u) \, du\right) - \exp\left(-\int_0^t \beta_{n-1}(u) \, du\right) \right|$$

$$\leq \varepsilon ab \int_0^t |\beta_n(u) - \beta_{n-1}(u)| \, du,$$

where $a > 0$ and $b > 0$ are some constants. Moreover, for all $T \geq t \geq s \geq 0$

and $n \geqslant 1$,

$$\left| \frac{e^{x_n(s)} - 1}{x_n(s)} \exp\left(\int_s^t \beta_n(u) \, du \right) - \frac{e^{x_{n-1}(s)} - 1}{x_{n-1}(s)} \exp\left(\int_s^t \beta_{n-1}(u) \, du \right) \right|$$

$$\leqslant \left| \frac{e^{x_n(s)} - 1}{x_n(s)} - \frac{e^{x_{n-1}(s)} - 1}{x_{n-1}(s)} \right| \exp\left(\int_s^t \beta_n(u) \, du \right) + \frac{e^{x_{n-1}(s)} - 1}{x_{n-1}(s)} \right|$$

$$\times \exp\left(\int_s^t \beta_n(u) \, du \right) - \exp\left(\int_s^t \beta_{n-1}(u) \, du \right) \right|$$

$$\leqslant c_1 \, e^{\alpha_2(t-s)} \int_0^s |x_n(u) - x_{n-1}(u)| \, du$$

$$+ c_2 \, e^{\alpha_2(t-s)} \int_s^t |\beta_n(u) - \beta_{n-1}(u)| \, du$$

$$\leqslant c \, e^{\alpha_2(t-s)} \int_0^t |\beta_n(u) - \beta_{n-1}(u)| \, du$$

where $c = c_1 + c_2$ and $c_1, c_2 \in (0, \infty)$ are some constants. Combining these inequalities with (9.4.18), we find that for all $n \geqslant 1$ and $t \in [0, T]$,

$$|\beta_{n+1}(t) - \beta_n(t)| \leqslant c_1 \int_0^t |\beta_n(u) - \beta_{n-1}(u)| \, du$$

$$+ c_2 \int_0^t k(t-s) \, e^{\alpha_2(t-s)} \int_0^s |\beta_n(u) - \beta_{n-1}(u)| \, du.$$

Then for some $d > 0$ and for all $t \in [0, T]$ we have

$$|\beta_{n+1}(t) - \beta_n(t)| \leqslant d \int_0^t |\beta_n(u) - \beta_{n-1}(u)| \, du, \qquad n \geqslant 1.$$

By induction this yields that there exists a constant $m > 0$ such that

$$|\beta_{n+1}(t) - \beta_n(t)| \leqslant md \frac{t^n}{n!} \qquad \text{for all } t \in [0, T] \text{ and for all } n \geqslant 1.$$

This implies that $\{\beta_n(t)\}_{n=0}^{\infty}$ converges to a function $\beta(t)$ uniformly on $[0, T]$; hence by the definition of $\{x_n(t)\}_{n=0}^{\infty}$ we have

$$x(t) = \lim_{n \to +\infty} x_n(t) = \begin{cases} \varepsilon \exp\left(-\int_0^t \beta(u) \, du \right) & \text{for } t \geqslant 0 \\ \varepsilon & \text{for } t < 0 \end{cases}$$

and this convergence is uniform on $(-\infty, T]$. Thus $\beta(t)$ and $x(t)$ satisfy

$$
\beta(t) =
\begin{cases}
bN^* \dfrac{e^{x(t)} - 1}{x(t)} + N^* \displaystyle\int_{-\infty}^{t} c(t-s) \dfrac{e^{x(s)} - 1}{x(\max\{s, 0\})} \\[2mm]
\qquad\qquad\qquad\qquad \times \exp\left(\displaystyle\int_{\max\{s,0\}}^{t} \beta(u)\, du \right) ds & \text{for } t \geqslant 0 \\[4mm]
\alpha_1 & \text{for } t < 0.
\end{cases}
$$

Since $\beta(t) = -\dot{x}(t)/x(t)$, $t \geqslant 0$, we find that $x(t)$ satisfies eqn (9.4.1) on $[0, T]$ with initial condition $x(t) = \varepsilon$, $t \geqslant 0$. On the other hand, (9.4.17) yields

$$
0 < \varepsilon\, e^{-\alpha_2 t} \leqslant x(t) \leqslant \varepsilon, \qquad t \in [0, T].
$$

Thus eqn (9.4.1) has a solution $x(t)$ which is positive on $(-\infty, T]$. As T is an arbitrary positive number, eqn (9.4.1) has a solution which is positive on $(-\infty, \infty)$. The proof of the theorem is complete.

9.5 Notes

The results in Section 9.1 are from Ladas *et al.* (1990*a*). The results in Section 9.2 are extracted from Györi (1990*a*). The results in Sections 9.3 and 9.4 are from Györi and Ladas (1990). Some of the original work on oscillations of integrodifferential equations was done by Gopalsamy (1983, 1986*b*). See also Mingarelli (1983). For the basic theory of integrodifferential equations see Burton (1983) and Corduneanu (1971). For applications see Cushing (1977).

9.6 Open problems

9.6.1   Obtain a linearized oscillation result for the integrodifferential equation

$$
\dot{x}(t) + \int_0^t K(t-s)f(x(s))\, ds = 0, \qquad t \geqslant T
$$

under appropriate conditions on K and f.

9.6.2   Obtain sufficient conditions for the oscillation of all solutions of the integrodifferential equation (9.0.1).

9.6.3   Extend the results of Section 9.1 to higher-order integrodifferential equations.

9.6.4   Assume that $k \in C[\mathbb{R}^+, \mathbb{R}^+]$ and (9.1.14) holds. Does every solution of eqn (9.1.1) oscillate?

9.6.5   Generalize Theorem 9.3.1 by relaxing the hypothesis that $k(t)$ has a constant sign.

10

OSCILLATIONS OF SECOND- AND HIGHER-ORDER EQUATIONS

In Sections 10.1 and 10.2 we present a linearized oscillation theory for the second-order non-linear delay differential equation

$$\ddot{y}(t) + \alpha \dot{y}(t) + \beta f(y(t - r)) = 0 \qquad (10.0.1)$$

where

$$\alpha \in (-\infty, \infty), \quad \beta \in (0, \infty), \quad r \in [0, \infty), \quad f \in C[\mathbb{R}, \mathbb{R}], \quad uf(u) > 0$$

$$\text{for } u \neq 0$$

and

$$\left. \begin{array}{ll} \lim\limits_{u \to 0} \dfrac{f(u)}{u} = 1 & \text{if } \alpha > 0 \\[2mm] \lim\limits_{|u| \to \infty} \dfrac{f(u)}{u} = 1 & \text{if } \alpha \leqslant 0 \end{array} \right\}.$$

An important corollary of our results is that every solution of eqn (10.0.1) oscillates if and only if every solution of the linear equation

$$\ddot{z}(t) + \alpha \dot{z}(t) + \beta z(t - r) = 0 \qquad (10.0.2)$$

oscillates, that is, if and only if the characteristic equation of eqn (10.0.2),

$$\lambda^2 + \alpha \lambda + \beta \, e^{-\lambda r} = 0$$

has no real roots.

In section 10.3 we study some simple differential inequalities of the forms

$$y^{(n)}(t) + (-1)^{n+1} p^n y(t - n\tau) \lesseqgtr 0$$

$$y^{(n)}(t) - p^n y(t + n\tau) \lesseqgtr 0,$$

where $n \geqslant 1$ and $p, \tau \in (0, \infty)$, and obtain necessary and sufficient conditions for the existence of positive or negative solutions. These results are useful in the study of the oscillations of higher-order differential equations.

In Sections 10.4–10.8 we present some basic oscillation results for nth-order neutral delay differential equations, including sufficient conditions for oscillation, comparison results and linearized oscillations. Consider the neutral equation

$$\frac{d^n}{dt^n}[x(t) + P(t)x(t - \tau)] + Q(t)x(t - \sigma) = 0 \qquad (10.0.3)$$

where

$$n \geqslant 1, \qquad P, Q \in C[[t_0, \infty), \mathbb{R}] \qquad \text{and} \qquad \tau, \sigma \in [0, \infty).$$

Let $\rho = \max\{\tau, \sigma\}$. By a *solution* of eqn (10.0.3) we mean a function $x \in C[[t_1 - \rho, \infty), \mathbb{R}]$, for some $t_1 \geqslant t_0$, such that $[x(t) - P(t)x(t - \tau)]$ is n times continuously differentiable on $[t_1, \infty)$ and such that eqn (10.0.3) is satisfied for $t \geqslant t_1$. Let $t_1 \geqslant t_0$ be given initial point, let $\phi \in C[[t_1 - \rho, t_1], \mathbb{R}]$ be a given initial function and let z_k, $k = 0, 1, \ldots, n - 1$, be given initial constants. By using the method of steps we can see that eqn (10.0.3) has a unique solution $x \in C[[t_1 - \rho, \infty), \mathbb{R}]$ such that

$$x(t) = \phi(t) \qquad \text{for } t \in [t_1 - \rho, t_1]$$

and

$$\frac{d^k}{dt^k}[x(t) - P(t)\phi(t - \tau)]_{t=t_1} = z_k \qquad \text{for } k = 0, 1, \ldots, n - 1.$$

10.1 Linearized oscillations for second-order differential equations—oscillations of the sunflower equation

In this section we obtain linearized oscillation results for the second-order delay differential equation

$$\ddot{y}(t) + \alpha\dot{y}(t) + \beta f(y(t - r)) = 0, \qquad t \geqslant 0 \qquad (10.1.1)$$

where

$$\alpha, \beta, r \in (0, \infty), \qquad f \in C[\mathbb{R}, \mathbb{R}], \qquad (10.1.2)$$

$$\lim_{u \to 0} \frac{f(u)}{u} = 1 \qquad (10.1.3)$$

and, for some $A \in (0, \infty)$,

$$uf(u) > 0 \qquad \text{for } u \neq 0 \qquad \text{and} \qquad u \in [-A, A]. \qquad (10.1.4)$$

The case where $\alpha \leqslant 0$ is treated in the next section.

The first result in this section provides sufficient conditions for the oscillation of all solutions of eqn (10.1.1) which lie eventually in the strip $\mathbb{R}^+ \times [-A, A]$.

Theorem 10.1.1. *Assume that* (10.1.2)–(10.1.4) *are satisfied and that the characteristic equation*

$$\lambda^2 + \alpha\lambda + \beta\, e^{-\lambda r} = 0 \tag{10.1.5}$$

of the linearized equation

$$\ddot{z}(t) + \alpha\dot{z}(t) + \beta z(t - r) = 0 \tag{10.1.6}$$

has no negative roots. Then every solution of eqn (10.1.1) *whose graph lies eventually in the strip* $\mathbb{R}^+ \times [-A, A]$ *is oscillatory.*

We also establish the following partial converse of Theorem 10.1.1.

Theorem 10.1.2. *Assume* (10.1.2) *and* (10.1.4) *hold and that*

$$\left.\begin{array}{ll} f(u) \leqslant u & \text{for } u \in [0, A] \quad \text{and} \\ f \text{ is non-decreasing for } u \in [0, A] \end{array}\right\}. \tag{10.1.7}$$

Suppose also that the characteristic equation (10.1.5) *of the 'majorant' equation* (10.1.6) *has a negative root. Then eqn* (10.1.1) *has a non-oscillatory solution whose graph lies eventually in the strip* $\mathbb{R}^+ \times (0, A]$.

By combining Theorems 10.1.1 and 10.1.2 we obtain the following necessary and sufficient conditions for the oscillation of eqn (10.1.1).

Corollary 10.1.1. *Assume that conditions* (10.1.2). (10.1.3), (10.1.4) *and* (10.1.7) *are satisfied. Then every solution of eqn* (10.1.1) *whose graph lies eventually in the strip* $\mathbb{R}^+ \times [-A, A]$ *oscillates if and only if the characteristic equation* (10.1.5) *of the linearized equation* (10.1.6) *has no real roots.*

When $f(u) = \sin u$, eqn (10.1.1) is the so-called *sunflower equation*, which describes the motion of the top of the sunflower plant. The study of the movement of the tip of growing plants goes back to the early 1800s (see Israelsson and Johnsson 1967 for references). The sunflower equation

$$\ddot{y}(t) + \frac{a}{r}\,\dot{y}(t) + \frac{b}{r}\sin y(t - r) = 0, \qquad t \geqslant 0 \tag{10.1.8}$$

was introduced in 1967 by Israelsson and Johnsson (1967) as a model for the geotropic circumnutations of *Helianthus annus*. Somolinos (1978) showed that, under appropriate initial conditions and with the parameters a, b and r in a certain range, the solutions of eqn (10.1.8) remain in the strip $\mathbb{R}^+ \times [-A, A]$ where $A = \pi - \varepsilon$ and $\varepsilon > 0$ is sufficiently small. By applying Theorem 10.1.1 with $A \in (0, \pi)$ we therefore conclude that these solutions oscillate provided that the characteristic equation

$$\lambda^2 + \frac{a}{r}\lambda + \frac{b}{r}e^{-\lambda r} = 0 \qquad (10.1.9)$$

has no negative roots.

Proof of Theorem 10.1.1. Assume, for the sake of contradiction, that eqn (10.1.1) has an eventually positive solution $y(t) \in (0, A]$. Then in view of (10.1.4),

$$\ddot{y}(t) + \alpha\dot{y}(t) < 0,$$

which implies that the functions $u(t) \equiv \dot{y}(t) + \alpha y(t)$ and $v(t) \equiv \dot{y}(t)\,e^{\alpha t}$ are decreasing. Let

$$L_1 = \lim_{t\to\infty} u(t) \qquad \text{and} \qquad L_2 = \lim_{t\to\infty} v(t).$$

Clearly $L_1 \in \mathbb{R}$, for otherwise $\lim_{t\to\infty}\dot{y}(t) = -\infty$, which would imply that $\lim_{t\to\infty} y(t) = -\infty$.

First, we claim that $L_2 < 0$. Otherwise, $L_2 \geq 0$ and so eventually $\dot{y}(t) \geq 0$. Thus, $l_0 \equiv \lim_{t\to\infty} y(t)$ exists and is positive. Also $l_1 \equiv \lim_{t\to\infty} \dot{y}(t) = L_1 - \alpha l_0$ exists. Clearly, l_1 must be non-negative, for otherwise l_0 would be $-\infty$. But from eqn (10.1.1) we have

$$\lim_{t\to\infty} \ddot{y}(t) = -\alpha l_1 - \beta f(l_0) < 0, \qquad (10.1.10)$$

which implies that both l_1 and l_0 are $-\infty$. Hence our claim that $L_2 < 0$ has been established. This implies that $\dot{y}(t) < 0$, so $\lim_{t\to\infty} y(t) \equiv l_0 \geq 0$. Then $\lim_{t\to\infty} \dot{y}(t) \equiv l_1$ exists, and it must be zero. Otherwise $l_1 < 0$ and so $l_0 = -\infty$. From (10.1.10) we now have $\lim_{t\to\infty} \ddot{y}(t) = -\beta f(l_0)$, and l_0 must be zero; otherwise $\lim_{t\to\infty} \ddot{y}(t) < 0$, and hence $l_0 = -\infty$, which is a contradiction. Therefore we have established that

$$y(t) > 0, \qquad \dot{y}(t) < 0, \qquad \text{and} \qquad \lim_{t\to\infty} y(t) = \lim_{t\to\infty} \dot{y}(t) = 0. \qquad (10.1.11)$$

Next, we claim that there exist positive numbers m and $\varepsilon < \frac{1}{2}$ such that

$$\lambda^2 + \alpha\lambda + \beta(1-\varepsilon)e^{-\lambda r} \geq \tfrac{1}{2}m \qquad \text{for } \lambda < 0. \qquad (10.1.12)$$

Indeed, by hypothesis, (10.1.5) has no negative roots. Set

$$F(\lambda) \equiv \lambda^2 + \alpha\lambda + \beta\,e^{-\lambda r}.$$

Then $F(-\infty) = \infty$ and $F(0) = \beta > 0$. It follows that

$$\lambda^2 + \alpha\lambda + \beta\,e^{-\lambda r} \geq m,$$

where $m = \min_{\lambda<0} F(\lambda)$. Observe that

$$\lim_{\lambda \to -\infty} (\lambda^2 + \alpha\lambda + \tfrac{1}{2}\beta\,e^{-\lambda r}) = \infty,$$

and so there exists $\lambda_0 < 0$ such that

$$\lambda^2 + \alpha\lambda + \tfrac{1}{2}\beta\,e^{-\lambda r} \geqslant m/2 \qquad \text{for } \lambda \leqslant \lambda_0. \qquad (10.1.13)$$

Choose

$$\varepsilon = \min\left\{\frac{1}{2}, \frac{m\,e^{\lambda_0 r}}{2\beta}\right\}.$$

Thus, for $\lambda \leqslant \lambda_0$, using (10.1.13) we have

$$\lambda^2 + \alpha\lambda + \beta(1 - \varepsilon)\,e^{-\lambda r} \geqslant \lambda^2 + \alpha\lambda + \tfrac{1}{2}\beta\,e^{-\lambda r} \geqslant m/2,$$

while for $\lambda_0 < \lambda < 0$ we get

$$\lambda^2 + \alpha\lambda + \beta(1 - \varepsilon)\,e^{-\lambda r} \geqslant \lambda^2 + \alpha\lambda + \beta\,e^{-\lambda r} - \tfrac{1}{2}m\,e^{(\lambda_0 - \lambda)r} \geqslant m - m/2 = m/2,$$

and the proof of (10.1.12) is complete.

Now, by integrating eqn (10.1.1) from t to ∞ and by using (10.1.11), we find

$$\dot{y}(t) + \alpha y(t) = \beta \int_t^\infty f(y(s - r))\,\mathrm{d}s. \qquad (10.1.14)$$

Choose t_1 so large that

$$\frac{f(y(s - r))}{y(s - r)} \geqslant 1 - \varepsilon \qquad \text{for } s \geqslant t_1,$$

which is possible in view of (10.1.13) and (10.1.11). Then for $t \geqslant t_1$ eqn (10.1.14) yields the linear inequality

$$\dot{y}(t) + \alpha y(t) \geqslant \beta(1 - \varepsilon) \int_t^\infty y(s - r)\,\mathrm{d}s. \qquad (10.1.15)$$

Next, we establish the following claim.

Claim. Let $y(t)$ be an eventually positive solution of (10.1.15). Assume that $\alpha \in \mathbb{R}$ and $\beta(1 - \varepsilon) > 0$. Then

$$\dot{z}(t) + \alpha z(t) = \beta(1 - \varepsilon) \int_t^\infty z(s - r)\,\mathrm{d}s \qquad (10.1.16)$$

has a solution $z(t)$ such that, eventually, $0 < z(t) \leqslant y(t)$.

Proof. By setting $u(t) = y(t) \, e^{\alpha t}$ into (10.1.15), we find that $u(t) > 0$ and that

$$\dot{u}(t) \geq \beta(1 - \varepsilon) \, e^{\alpha t} \int_t^\infty e^{-\alpha(s-r)} u(s - r) \, ds.$$

By integrating from T to t, we get

$$u(t) \geq u(T) + \int_T^t \left[\beta(1 - \varepsilon) \, e^{\alpha s} \int_s^\infty e^{-\alpha(\xi - r)} u(\xi - r) \, d\xi \right] ds \quad (10.1.17)$$

for $t \geq T$. We shall employ the Knaster–Tarski fixed-point theorem (see Theorem 1.7.3). Let X be the set of all real-valued non-decreasing functions x defined on $[T, \infty)$ and such that

$$x(t) = u(t) \qquad \text{for } T \leq t \leq T + r \qquad (10.1.18)$$

$$x(t) \leq u(t) \qquad \text{for } t > T + r. \qquad (10.1.19)$$

If x_1 and x_2 belong to X, we shall say that $x_1 \leq x_2$ if and only if $x_1(t) \leq x_2(t)$ for $t \geq T$. Clearly, with this ordering, X is a partially ordered set. Define the mapping S on X as follows:

$$(Sx)(t) = \begin{cases} u(t), & T \leq t \leq T + r \\ \\ u(T + r) + \displaystyle\int_{T+r}^t \left[\beta(1 - \varepsilon) \, e^{\alpha s} \right. \\ \qquad\qquad\qquad \left. \times \displaystyle\int_s^\infty e^{-\alpha(\xi - r)} x(\xi - r) \, d\xi \right] ds, & t > T + r. \end{cases}$$

In view of (10.1.19) and (10.1.17),

$$(Sx)(t) \leq (Su)(t) \leq u(t) \qquad \text{for } t \geq T + r,$$

while, in view of the definition of S,

$$(Sx)(t) = u(t) \qquad \text{for } T \leq t \leq T + r.$$

Also, Sx is a non-decreasing function for $t \geq T$. Thus $S: X \rightarrow X$. Finally, $\inf X \in X$, and every non-empty subset of X has a supremum that belongs to X. Hence all the hypotheses of the Knaster–Tarski fixed-point theorem are satisfied, so S has a fixed point $x \in X$. That is,

$$(Sx)(t) = x(t) \qquad \text{for } t \geq T.$$

Then

$$x(t) = \begin{cases} u(t), & T \leqslant t \leqslant T+r \\ u(T+r) + \displaystyle\int_{T+r}^{t} \left[\beta(1-\varepsilon) \, e^{\alpha s} \right. \\ \qquad \left. \times \displaystyle\int_{s}^{\infty} e^{-\alpha(\xi-r)} x(\xi-r) \, d\xi \right] ds, & t > T+r. \end{cases}$$

Clearly, $x(t) > 0$ for $t \geqslant T$, and differentiating for $t \geqslant T+r$, we see that x satisfies the equation

$$\dot{x}(t) = \beta(1-\varepsilon) \, e^{\alpha t} \int_{t}^{\infty} e^{-\alpha(\xi-r)} x(\xi-r) \, d\xi, \qquad t \geqslant T+r.$$

By setting $x(t) = z(t) \, e^{\alpha t}$, $t \geqslant T+r$, we see that $z(t) > 0$ and that z satisfies (10.1.16). Also, $z(t) \, e^{\alpha t} = x(t) \leqslant y(t) \, e^{\alpha t}$, and the proof of the claim is complete.

Now, by differentiating both sides of (10.1.16), we see that $z(t)$ is a bounded positive solution of

$$\ddot{z}(t) + \alpha \dot{z}(t) + \beta(1-\varepsilon) z(t-r) = 0.$$

This contradicts (10.1.12) and the proof of the theorem is complete.

Proof of Theorem 10.1.2. Let μ be a negative root of (10.1.5). Then there exists a $T \geqslant t_0$ such that $e^{\mu T} \leqslant A$. Thus eqn (10.1.6) has the non-oscillatory solution

$$y(t) = e^{\mu t} \in (0, A] \qquad \text{for } t \geqslant T.$$

Hence, in the light of (10.1.7), the differential inequality

$$\ddot{y}(t) + \alpha \dot{y}(t) + \beta f(y(t-r)) \leqslant 0, \qquad t \geqslant T, \tag{10.1.20}$$

is also satisfied by $y(t) = e^{\mu t}$ and (10.1.11) holds. By integrating (10.1.20) from t to ∞ and by using (10.1.11), we get

$$\dot{y}(t) + \alpha y(t) \geqslant \beta \int_{t}^{\infty} f(y(s-r)) \, ds. \tag{10.1.21}$$

Now, by slight modification in the proof of the claim which we established in the proof of Theorem 10.1.1 we see that the equation

$$\dot{z}(t) + \alpha z(t) = \beta \int_{t}^{\infty} f(z(s-r)) \, ds$$

has a solution $z(t)$ such that

$$0 < z(t) \leqslant e^{\mu t}, \qquad t \geqslant T.$$

Clearly $z(t)$ is a positive solution of eqn (10.1.1) that lies in the strip $\mathbb{R}^+ \times (0, A]$. The proof is complete.

10.2 Linearized oscillations for second-order differential equations—continued

In this section we obtain linearized oscillation results which complement the results in the previous section. Consider the delay differential equation

$$\ddot{y}(t) + \alpha \dot{y}(t) + \beta f(y(t - r)) = 0, \qquad t \geqslant 0 \qquad (10.2.1)$$

where

$$\alpha \in (-\infty, 0], \qquad \beta \in (0, \infty), \qquad r \in [0, \infty), \qquad f \in C[\mathbb{R}, \mathbb{R}], \qquad (10.2.2)$$

$$\lim_{|u| \to \infty} \frac{f(u)}{u} = 1 \qquad (10.2.3)$$

$$uf(u) > 0 \qquad \text{for } u \neq 0. \qquad (10.2.4)$$

The first result in this section provides sufficient conditions for the oscillation of all solutions of eqn (10.2.1).

Theorem 10.2.1. *Assume that* (10.2.2)–(10.2.4) *are satisfied and that the characteristic equation*

$$\lambda^2 + \alpha\lambda + \beta\, e^{-\lambda r} = 0 \qquad (10.2.5)$$

of the linearized equation

$$\ddot{z}(t) + \alpha \dot{z}(t) + \beta z(t - r) = 0 \qquad (10.2.6)$$

has no positive roots. Then every solution of eqn (10.2.1) *oscillates.*

The next result is a partial converse of Theorem 10.2.1.

Theorem 10.2.2. *Assume that* (10.2.2), (10.2.4) *hold and that there exists* $M > 0$ *such that*

$$f(u) \leqslant u \quad \text{for } u > M \quad \text{and} \quad f \text{ is non-decreasing for } u > M. \qquad (10.2.7)$$

Suppose also that the characteristic equation (10.2.5) *of the 'majorant' equation* (10.2.6) *has a positive root. Then eqn* (10.2.1) *has an unbounded non-oscillatory solution.*

By combining Theorems 10.2.1 and 10.2.2 we obtain the following necessary and sufficient condition for the oscillation of all solutions of eqn (10.2.1).

Corollary 10.2.1. *Assume that conditions* (10.2.2), (10.2.3), (10.2.4) *and* (10.2.7) *are satisfied. Then every solution of eqn* (10.2.1) *oscillates if and only if the characteristic equation* (10.2.5) *of the linearized equation* (10.2.6) *has no real roots.*

By combining Corollary 10.2.1 with Corollary 10.1.2 of the previous section we obtain the following general linearized oscillation result for eqn (10.2.1).

Corollary 10.2.2. *Assume that* (10.2.4) *holds,*

$$\alpha \in (-\infty, \infty), \quad \beta \in (0, \infty), \quad r \in [0, \infty), \quad f \in C[\mathbb{R}, \mathbb{R}] \qquad (10.2.8)$$

and that

$$\left. \begin{array}{ll} \lim_{u \to 0} \dfrac{f(u)}{u} = 1 & if \ \alpha > 0 \\[3mm] \lim_{|u| \to \infty} \dfrac{f(u)}{u} = 1 & if \ \alpha \leqslant 0 \end{array} \right\}. \qquad (10.2.9)$$

Then every solution of eqn (10.2.1) *oscillates if and only if the characteristic equation* (10.2.5) *of the linearized equation* (10.2.6) *has no real roots.*

In the special case where the delay r is zero, the above corollary reduces to the following explicit necessary and sufficient conditions for the oscillation of all solutions of eqn (10.2.1).

Corollary 10.2.3. *Assume that* (10.2.4), (10.2.8) *with* $r = 0$ *and* (10.2.9) *hold. Then every solution of*

$$\ddot{y}(t) + \alpha \dot{y}(t) + \beta f(y(t)) = 0 \qquad (10.2.10)$$

oscillates if and only if

$$\beta > \tfrac{1}{4}\alpha^2.$$

For the proofs of Theorems 10.2.1 and 10.2.2 we need the following two lemmas.

Lemma 10.2.1. *Assume that conditions* (10.2.) *and* (10.2.4) *are satisfied. Let*

$y(t)$ be an eventually positive solution of eqn (10.2.1). Then eventually

$$\dot{y}(t) > 0 \tag{10.2.11}$$

and

$$\lim_{t \to \infty} y(t) = \infty. \tag{10.2.12}$$

Proof. In view of (10.2.2) and (10.2.4). $\ddot{y}(t) + \alpha\dot{y}(t) < 0$, which implies that the functions

$$u(t) \equiv \dot{y}(t) + \alpha y(t) \tag{10.2.13}$$

$$v(t) \equiv \dot{y}(t) \, e^{\alpha t} \tag{10.2.14}$$

are both strictly decreasing. In particular, it follows from (10.2.14) that $\dot{y}(t)$ is eventually of constant sign. We claim that (10.2.11) holds. Otherwise, $\dot{y}(t) < 0$ and $\ddot{y}(t) < -\alpha\dot{y}(t) \leqslant 0$, which leads to the contradiction that $y(t) < 0$. Next, we claim that (10.2.12) holds. Otherwise, $\lim_{t \to \infty} y(t)$ exists and is a positive number l. But then eqn (10.2.1) implies that $\lim_{t \to \infty} \dot{u}(t) = -\beta f(L) < 0$, so

$$\lim_{t \to \infty} u(t) = -\infty. \tag{10.2.15}$$

On the other hand, as we see from (10.2.13)

$$u(t) > \alpha y(t) \geqslant \alpha \frac{l}{2} \in \mathbb{R},$$

which contradicts (10.2.19) and completes the proof.

Lemma 10.2.2. *Assume that* (10.2.4) *holds and that the characteristic equation* (10.2.5) *has no positive roots. Then there exists positive numbers μ and ε such that*

$$\lambda^2 + \alpha\lambda + \beta(1 - \varepsilon) \, e^{-\lambda r} \geqslant \mu \qquad for \; \lambda \geqslant 0. \tag{10.2.16}$$

Proof. Set $F(\lambda) = \lambda^2 + \alpha\lambda + \beta \, e^{-\lambda r}$. Since $F(\infty) = \infty$, $F(0) = \beta > 0$ and $F(\lambda)$ has no positive roots, it follows that $m \equiv \min_{\lambda \geqslant 0} F(\lambda) > 0$ and so $F(\lambda) \geqslant m$ for $\lambda \geqslant 0$. Also, there exists $\lambda_0 > 0$ such that

$$\lambda^2 + \alpha\lambda + \tfrac{1}{2}\beta \, e^{-\lambda r} \geqslant m/2 \qquad for \; \lambda \geqslant \lambda_0. \tag{10.2.17}$$

Choose $\varepsilon = \min\{\tfrac{1}{2}, m/2\beta\}$. Then for $\lambda \geqslant \lambda_0$, by using (10.2.17), we have

$$\lambda^2 + \alpha\lambda + \beta(1 - \varepsilon) \, e^{-\lambda r} \geqslant \lambda^2 + \alpha\lambda + \tfrac{1}{2}\beta \, e^{-\lambda r} \geqslant m/2 \tag{10.2.18}$$

while for $0 \leqslant \lambda < \lambda_0$ we get

$$\lambda^2 + \alpha\lambda + \beta(1 - \varepsilon) \, e^{-\lambda r} \geqslant \lambda^2 + \alpha\lambda + \beta \, e^{-\lambda r} - \tfrac{1}{2}m \, e^{-\lambda r} \geqslant m - \tfrac{1}{2}m = m/2.$$

$$\tag{10.2.19}$$

From (10.2.18) and (10.2.19) we see that (10.2.16) holds with $\mu = \frac{1}{2}m$. The proof is complete.

Proof of Theorem 10.2.1. Assume, for the sake of contradiction, that eqn (10.2.1) has an eventually positive solution. Then, by Lemma 10.2.1, (10.2.11) and (10.2.12) are satisfied. By Lemma 10.2.2, there exist positive numbers μ and ε with $\varepsilon \leqslant \frac{1}{2}$ such that (10.2.16) holds. Choose $T \geqslant t_0$ so large that

$$y(t - r) > 0 \quad \text{and} \quad \frac{f(y(t - r))}{y(t - r)} > 1 - \varepsilon \quad \text{for } t \geqslant T, \qquad (10.2.20)$$

which is possible because of (10.2.3) and (10.2.12). Then, from eqn (10.2.1) we see that

$$\ddot{y}(t) + \alpha \dot{y}(t) + \beta(1 - \varepsilon) y(t - r) \leqslant 0, \qquad t \geqslant T, \qquad (10.2.21)$$

that is,

$$(e^{\alpha t} \dot{y}(t))\,{}^{\cdot} + \beta(1 - \varepsilon)\, e^{\alpha t} y(t - r) \leqslant 0, \qquad t \geqslant T. \qquad (10.2.22)$$

By integrating (10.2.22) from $t \geqslant T$ to ∞ and by taking into account the fact that

$$\lim_{t \to \infty} e^{\alpha t} \dot{y}(t) \leqslant c \in [0, \infty)$$

(indeed, $e^{\alpha t} \dot{y}(t) > 0$ and $(e^{\alpha t} \dot{y}(t))\,{}^{\cdot} < 0$), we conclude that

$$e^{\alpha t} \dot{y}(t) \geqslant \beta(1 - \varepsilon) \int_t^\infty e^{\alpha s} y(s - r)\, ds, \qquad t \geqslant T$$

and so

$$\dot{y}(t) \geqslant \beta(1 - \varepsilon)\, e^{-\alpha t} \int_t^\infty e^{\alpha s} y(s - r)\, ds, \qquad t \geqslant T.$$

By integrating this inequality from $T + r$ to $t > T + r$, we obtain

$$y(t) \geqslant y(T + r) + \int_{T+r}^t \left[\beta(1 - \varepsilon)\, e^{-\alpha s} \int_s^\infty e^{\alpha \xi} y(\xi - r)\, d\xi \right] ds, \qquad t \geqslant T + r.$$
$$(10.2.23)$$

Now we apply the Knaster–Tarski fixed-point theorem (see Theorem 1.7.3). Let X be the set of all non-negative and non-decreasing functions x defined on $[T, \infty)$ such that

$$x(t) = y(t) \qquad \text{for } T \leqslant t \leqslant T + r \qquad (10.2.24)$$

$$x(t) \leqslant y(t) \qquad \text{for } t > T + r. \qquad (10.2.25)$$

If x_1 and x_2 belong to X, we shall say that $x_1 \leqslant x_2$ if and only if $x_1(t) \leqslant x_2(t)$ for $t \geqslant T$. Clearly X is a partially ordered set. Define the mapping S on X as follows:

$$
(Sx)(t) = \begin{cases} y(t), & T \leqslant t \leqslant T+r \\ y(T+r) + \displaystyle\int_{T+r}^{t}\left[\beta(1-\varepsilon)\,\mathrm{e}^{-\alpha s} \right. \\ \qquad\qquad \left. \times \displaystyle\int_{s}^{\infty} \mathrm{e}^{\alpha\xi}x(\xi - r)\,\mathrm{d}\xi\right]\mathrm{d}s, & t > T+r. \end{cases}
$$

In view of (10.2.23) and (10.2.25),

$$(Sx)(t) \leqslant (Sy)(t) \leqslant y(t) \qquad \text{for } t > T+r,$$

while in view of the definition of S,

$$(Sx)(t) = y(t) \qquad \text{for } T \leqslant t \leqslant T+r.$$

Also, Sx is a non-negative and non-decreasing function for $t \geqslant T$. So $S\colon X \to X$. Clearly, all other conditions of the Knaster–Tarski fixed-point theorem are satisfied and consequently S has a fixed point $x \in X$. Thus

$$(Sx)(t) = x(t) \qquad \text{for } t \geqslant T$$

or

$$
x(t) = \begin{cases} y(t), & T \leqslant t \leqslant T+r \\ y(T+r) + \displaystyle\int_{T+r}^{t}\left[\beta(1-\varepsilon)\,\mathrm{e}^{-\alpha s} \right. \\ \qquad\qquad \left. \times \displaystyle\int_{s}^{\infty} \mathrm{e}^{\alpha\xi}x(\xi - r)\,\mathrm{d}\xi\right]\mathrm{d}s, & t > T+r. \end{cases}
$$

Obviously $x(t) > 0$ and $\dot{x}(t) > 0$ for $t \geqslant T$. Differentiating for $t \geqslant T+r$, we see that x satisfies the differential equation

$$(\mathrm{e}^{\alpha t}\dot{x}(t))^{\cdot} + \beta(1-\varepsilon)\,\mathrm{e}^{\alpha t}x(t-r) = 0, \qquad t \geqslant T+r$$

and so

$$\ddot{x}(t) + \alpha\dot{x}(t) + \beta(1-\varepsilon)x(t-r) = 0, \qquad t > T+r.$$

This in turn implies that the characteristic equation

$$\lambda^2 + \alpha\lambda + \beta(1-\varepsilon)\,\mathrm{e}^{-\lambda r} = 0$$

has a positive root, which contradicts (10.2.16). The proof is complete.

Proof of Theorem 10.2.2. Let μ be a positive root of eqn (10.2.5). Then eqn (10.1.10) has the unbounded non-oscillatory solution $y(t) = e^{\mu t}$. Hence, in view of (10.2.7), the linear differential inequality

$$\ddot{y}(t) + \alpha \dot{y}(t) + \beta f(y(t - r)) \leqslant 0$$

has $y(t)$ as a solution for large t. Now by an argument similar to that used in the proof of Theorem 10.2.1 we see that the corresponding linear equation (10.2.6) also has a non-oscillatory solution $z(t)$, which, by Lemma 10.1.1, is unbounded. The proof is complete.

10.3 Differential inequalities of higher order

Consider the following delay and advanced differential inequalities and equations

$$y^{(n)}(t) + (-1)^{n+1} p^n y(t - n\tau) \leqslant 0 \tag{10.3.1}$$

$$y^{(n)}(t) + (-1)^{n+1} p^n y(t - n\tau) \geqslant 0 \tag{10.3.2}$$

$$y^{(n)}(t) + (-1)^{n+1} p^n y(t - n\tau) = 0 \tag{10.3.3}$$

$$y^{(n)}(t) - p^n y(t + n\tau) \leqslant 0 \tag{10.3.4}$$

$$y^{(n)}(t) - p^n y(t + n\tau) \geqslant 0 \tag{10.3.5}$$

$$y^{(n)}(t) - p^n y(t + n\tau) = 0 \tag{10.3.6}$$

where

$$n \geqslant 1 \quad \text{and} \quad p, \tau \in (0, \infty). \tag{10.3.7}$$

The main results in this section are the following two theorems.

Theorem 10.3.1. *Assume that* (10.3.7) *holds. Then*

$$p\tau > 1/e \tag{10.3.8}$$

is a necessary and sufficient condition so that:

(a) *when n is odd,* (10.3.1) *has no eventually positive solutions,* (10.3.2) *has no eventually negative solutions, and every solution of* (10.3.3) *oscillates.*

(b) *when n is even,* (10.3.1) *has no eventually negative bounded solutions,* (10.3.2) *has no eventually positive bounded solutions, and every bounded solution of* (10.3.3) *oscillates.*

Theorem 10.3.2. *Assume that* (10.3.7) *holds. Then* (10.3.8) *is a necessary and sufficient condition so that:*

(a) *when n is odd, (10.3.4) has no eventually negative solutions, (10.3.5) has no eventually positive solutions, and every solution of (10.3.6) oscillates.*

(b) *when n is even, (10.3.4) has no eventually negative unbounded solutions, (10.3.5) has no eventually positive unbounded solutions, and every solution of (10.3.6) oscillates.*

Proof of Theorem 10.3.1. The 'necessary' part of the theorem follows from the observation that when

$$p\tau \leqslant 1/e$$

the corresponding inequality or equation has a solution (with the right properties) of the form $y(t) = e^{\lambda t}$, where λ is a real root of the characteristic equation

$$\lambda^n + (-1)^{n+1} p^n e^{-\lambda n\tau} = 0.$$

Next, we establish the 'sufficient' part of the theorem for the inequality (10.3.1). The proof for the inequality (10.3.2) follows in a similar way or by making the change of variable $y(t) = -x(t)$, and is omitted. Also the result about the equation (10.3.3) follows easily from the previous result and its proof is omitted.

First, assume that n is odd and, for the sake of contradiction, assume that (10.3.1) has an eventually positive solution $y(t)$. It follows from (10.3.1) that then $y^{(n)}(t)$ is eventually negative. As n is odd, this implies that there exists an even integer l, $0 \leqslant l < n$, such that eventually,

$$\left.\begin{array}{ll} y^{(k)}(t) > 0 & \text{for } k = 1, 2, \ldots, l \\ (-1)^k y^{(k)}(t) > 0 & \text{for } k = l+1, \ldots, n \end{array}\right\}. \tag{10.3.9}$$

We claim that $l = 0$, that is, eventually,

$$(-1)^k y^{(k)}(t) > 0 \qquad \text{for } k = 1, 2, \ldots, n. \tag{10.3.10}$$

Otherwise, $l > 0$ and by integrating (10.3.1) $n - l$ times from t_1 to t, for t_1 sufficiently large, we obtain

$$\begin{aligned}
y^{(l)}(t) &\leqslant \sum_{k=0}^{n-l-1} \frac{(t-t_1)^k}{k!} y^{(l+k)}(t_1) - \int_{t_1}^{t} \frac{(t-s)^{n-l-1}}{(n-l-1)!} p^n y(s - n\tau)\, ds \\
&\leqslant \sum_{k=0}^{n-l-1} \frac{(t-t_1)^k}{k!} y^{(l+k)}(t_1) - \frac{y(t_1 - n\tau)}{(n-l-1)!} p^n \int_{t_1}^{t} (t-s)^{n-l-1}\, ds \\
&= \sum_{k=0}^{n-l-1} \frac{(t-t_1)^k}{k!} y^{(l+k)}(t_1) - \frac{y(t_1 - n\tau)}{(n-l)!} p^n (t-t_1)^{n-l}.
\end{aligned}$$

This implies that

$$y^{(l)}(t) \to -\infty \quad \text{as } t \to \infty,$$

which contradicts (10.3.9) and proves (10.3.10). Set

$$x(t) = y^{(n-1)}(t) - py^{(n-2)}(t-\tau) + \cdots + p^{n-1}y(t-(n-1)\tau)). \quad (10.3.11)$$

Then eventually

$$x(t) > 0. \quad (10.3.12)$$

Also observe that in view of (10.3.1) and (10.3.11)

$$\dot{x}(t) + px(t-\tau) \leqslant 0. \quad (10.3.13)$$

But because of (10.3.8) and Theorem 2.3.3, inequality (10.3.13) cannot have an eventually positive solution. This contradicts (10.3.12) and completes the proof in this case.

Finally, assume that n is even and, for the sake of contradiction, assume that eqn (10.3.1) has an eventually negative, bounded solution $y(t)$. Then eventually $y^{(n)}(t) < 0$ and because $y(t)$ is bounded, it follows that (10.3.10) holds. Define $x(t)$ by (10.3.11) and observe that (10.3.12) and (10.3.13) are true here too. In view of Theorem 2.3.3 this is a contradiction and the proof is complete.

Proof of Theorem 10.3.2. We present the details of the 'sufficient' part of the proof for the inequality (10.3.5) when n is odd. The remaining part of the proof follows in a way similar to that in the proof of Theorem 10.3.1 and is omitted. Hence assume, for the sake of contradiction, that eqn (10.3.5) has an eventually positive solution $y(t)$. Then $y^{(n)}(t)$ is eventually positive. It follows that there exists an odd integer m, $0 < m \leqslant n$, such that eventually,

$$\left. \begin{array}{ll} y^{(j)}(t) > 0 & \text{for } j = 0, 1, 2, \ldots, m \\ (-1)^j y^{(j)}(t) > 0 & \text{for } j = m+1, m+2, \ldots, n \end{array} \right\}. \quad (10.3.14)$$

We claim that $m = n$, that is, eventually,

$$y^{(j)}(t) > 0 \quad \text{for } j = 0, 1, 2, \ldots, n. \quad (10.3.15)$$

Otherwise $m < n$ and by integrating (10.3.5) $n - m$ times from t_0 to t, for t_0 sufficiently large, we obtain

$$y^{(m)}(t) \geqslant \sum_{i=0}^{n-m-1} \frac{(t-t_0)^i}{i!} y^{(m+i)}(t_0) + \int_{t_0}^{t} \frac{(t-s)^{n-m-1}}{(n-m-1)!} p^n y(s+n\tau) \, ds$$

$$\geqslant \sum_{i=0}^{n-m-1} \frac{(t-t_0)^i}{i!} y^{(m+i)}(t_0) + \frac{y(t_0+n\tau)}{(n-m-1)!} \int_{t_0}^{t} (t-s)^{n-m-1} \, ds$$

$$= \sum_{i=0}^{n-m-1} \frac{(t-t_0)^i}{i!} y^{(m+i)}(t_0) + p^n \frac{y(t_0+n\tau)}{(n-m)!} (t-t_0)^{n-m}.$$

This implies that

$$y^{(m)}(t) \to \infty \qquad \text{as } t \to \infty,$$

which is impossible because by (10.3.14), $y^{(m)}(t)$ is a decreasing function. Set

$$x(t) = y^{(n-1)}(t) + py^{(n-2)}(t+\tau) + \cdots + p^{n-1}y(t+(n-1)\tau),$$

which by (10.3.15), for sufficiently large t, is positive. Observe that

$$\dot{x}(t) - px(t+\tau) = y^{(n)}(t) - p^n y(t+n\tau) \geq 0.$$

But $p\tau > 1/e$, which by Theorem 2.3.4 implies that the above inequality cannot have an eventually positive solution. This contradicts the fact that $x(t)$ is eventually positive and the proof is complete.

10.4 Sufficient conditions for the oscillation of higher-order neutral equations

In this section we obtain sufficient conditions for the oscillation of all solutions of the higher-order neutral differential equation

$$\frac{d^n}{dt^n}[y(t) + P(t)y(t-\tau)] + Q(t)y(t-\sigma) = 0, \qquad (10.4.1)$$

where

$$n \geq 1, \quad P \in C[[t_0, \infty), \mathbb{R}], \quad Q \in C[[t_0, \infty), \mathbb{R}^+] \quad \text{and} \quad \tau, \sigma \in [0, \infty).$$
$$(10.4.2)$$

Lemma 10.4.1. *Assume that* (10.4.2) *is satisfied and suppose that*

$$\int_{t_0}^{\infty} Q(s)\, ds = \infty. \qquad (10.4.3)$$

Let $y(t)$ *be an eventually positive solution of eqn* (10.4.1) *and set*

$$z(t) = y(t) + P(t)y(t-\tau). \qquad (10.4.4)$$

Then the following statements are true.

(a) $z^{(n-1)}$ *is decreasing and* $z^{(i)}(t)$ *for* $i = 0, 1, \ldots, n-2$ *is strictly monotonic.*

(b) *Assume that there exist numbers* P_1 *and* $P_2 \in \mathbb{R}$ *such that* $P(t)$ *is in one of the following ranges:*

(i) $P_1 \leq P(t) \leq 0$;

(ii) $0 \leq P(t) \leq P_2 < 1$;

Then either

$$\lim_{t \to \infty} z^{(i)}(t) = -\infty \qquad \textit{for } i = 0, 1, \dots, n - 1 \qquad (10.4.5)$$

or

$$\left.\begin{aligned}
&\lim_{t \to \infty} z^{(i)}(t) = 0 \quad \textit{for } i = 1, 2, \dots, n - 1, \\
&z^{(i)}(t) z^{(i+1)}(t) < 0 \quad \textit{for } i = 1, 2, \dots, n - 2, \\
&z^{(n-1)}(t) > 0, z^{(n)}(t) \leq 0 \quad \textit{and } z^{(n)}(t) \textit{ is not identically zero.}
\end{aligned}\right\} \qquad (10.4.6)$$

Proof.

(a) From eqn (10.4.1) we find that

$$z^{(n)}(t) = -Q(t) y(t - \sigma) \leq 0. \qquad (10.4.7)$$

Therefore, $z^{(n-1)}(t)$ is decreasing. In view of (10.4.3) it follows that $z^{(n)}(t) \not\equiv 0$, so $z^{(n-1)}(t)$ is eventually positive or eventually negative. Hence $z^{(i)}(t)$ for $i = 0, 1, \dots, n - 2$, is strictly monotonic.

(b) As $z^{(n-1)}(t)$ decreases, it follows that either

$$\lim_{t \to \infty} z^{(n-1)}(t) = -\infty \qquad (10.4.8)$$

or

$$\lim_{t \to \infty} z^{(n-1)}(t) \equiv L \in \mathbb{R}. \qquad (10.4.9)$$

If (10.4.8) holds, then it is obvious that (10.4.5) is satisfied. Next, assume that (10.4.9) holds. By integrating eqn (10.4.1), we find

$$L - z^{(n-1)}(t_0) + \int_{t_0}^{\infty} Q(t) y(t - \sigma) \, dt = 0,$$

which, in view of (10.4.2) and (10.4.3), implies that

$$\lim_{t \to \infty} \inf y(t) = 0.$$

Let $\{t_n\}$ be a sequence of points such that

$$\lim_{n \to \infty} t_n = \infty \qquad \text{and} \qquad \lim_{n \to \infty} y(t_n) = 0.$$

First, assume that $p_1 \leq p(t) \leq 0$. Then from the monotonic nature of $z(t)$ and from the fact that

$$z(t_n) \leq y(t_n) \to 0 \qquad \text{as } n \to \infty$$

and

$$z(t_n + \tau) \geqslant p_1 y(t_n) \to 0 \qquad \text{as } n \to \infty$$

we conclude that

$$\lim_{t \to \infty} z(t) = 0.$$

$$(10.4.10)$$

Next, assume that $0 \leqslant p(t) \leqslant p_2 < 1$. Then $z(t) > 0$. We also claim that $z(t)$ is decreasing. Otherwise $z(t)$ is increasing. Then

$$z(t_n) - z(t_n - \tau) = y(t_n) + [p(t_n) - 1]y(t_n - \tau) - p(t_n - \tau)y(t_n - 2\tau) \geqslant 0$$

and so

$$y(t_n) + (p_2 - 1)y(t_n - \tau) \geqslant 0,$$

which implies that

$$\lim_{n \to \infty} y(t_n - \tau) = 0.$$

Then from (10.4.4) we see that $\lim_{n \to \infty} z(t_n) = 0$. This is a contradiction and so our claim that $z(t)$ is decreasing is true. Hence $\lim_{t \to \infty} z(t)$ exists and is finite. By applying Lemma 1.5.2 we see that (10.4.10) holds. Finally, from (10.4.10) we can easily conclude that consecutive derivatives of $z(t)$ alternate in sign, that is, for each $i = 0, 1, \ldots, n - 2$, $z^{(i)}(t)z^{(i+1)}(t) < 0$. The proof is complete.

Lemma 10.4.2. *Assume that (10.4.2) and (10.4.3) hold. Let $y(t)$ be an eventually positive solution of eqn (10.4.1) and set*

$$z(t) = y(t) + P(t)y(t - \tau).$$

Then the following statements are true.

(a) *If $P(t)$ satisfies the hypotheses of Lemma 10.4.1(b) and n is even, then $z(t)$ is eventually negative.*

(b) *If $-1 \leqslant P(t) \leqslant 0$ and n is odd, then $z(t)$ is eventually positive.*

(c) *If $P(t) \leqslant -1$, then $z(t)$ is eventually negative.*

Proof.

   (a) By Lemma 10.4.1, for n even, either (10.4.5) or (10.4.6) implies that $z(t) < 0$.

   (b) Assume, for the sake of contradiction, that $z(t) < 0$. Then

$$y(t) < -P(t)y(t - \tau) \leqslant y(t - \tau),$$

which implies that $y(t)$ is bounded. Hence $z(t)$ is also bounded. Therefore by Lemma 10.4.1, (10.4.6) holds. But since n is odd, $z^{(n)}(t) \leqslant 0$, and consecutive derivatives of $z(t)$ alternate in sign, it follows that $z(t) > 0$. This contradicts our assumption and the proof is complete in this case.

(c) From Lemma 10.4.1, we know $z(t)$ is eventually positive or eventually negative. Assume, for the sake of contradiction, that eventually

$$z(t) > 0. \tag{10.4.11}$$

Hence

$$y(t) > -P(t)y(t - \tau) \geqslant y(t - \tau)$$

and so $y(t)$ is bounded from below by a positive constant, say m. Then from eqn (10.4.1) we find

$$z^{(n)}(t) + mQ(t) \leqslant 0.$$

By integrating from t_1 to t, for t_1 sufficiently large, we find

$$z^{(n-1)}(t) - z^{(n-1)}(t_1) + m \int_{t_1}^{t} Q(s) \, ds \leqslant 0,$$

which in view of (10.4.3), implies

$$\lim_{t \to \infty} z^{(n-1)}(t) = -\infty.$$

Therefore

$$\lim_{t \to \infty} z(t) = -\infty,$$

which contradicts (10.4.11) and the proof is complete.

The following two theorems give a detailed description of the non-oscillatory solutions of eqn (10.4.1) for n odd and even, respectively.

Theorem 10.4.1. *Assume that n is odd and (10.4.2) and (10.4.3) are satisfied. Suppose also that there exists real numbers P_1, P_2 and P_3 such that $P(t)$ is in one of the following ranges:*

$$P_1 \leqslant P(t) \leqslant -1 \tag{10.4.12}$$

$$-1 < P_2 \leqslant P(t) \leqslant 0 \tag{10.4.13}$$

$$0 \leqslant P(t) \leqslant P_3 < 1. \tag{10.4.14}$$

Let $y(t)$ be an eventually positive solution of eqn (10.4.1) and set

$$z(t) = y(t) + P(t)y(t - \tau). \tag{10.4.15}$$

Then

(a) *The following statements are equivalent*:

(i) (10.4.5) *holds*;

(ii) $P(t)$ *satisfies* (10.4.12) *and* $P(t) \not\equiv -1$;

(iii) $\lim_{t \to \infty} y(t) = \infty$. $\qquad\qquad\qquad\qquad\qquad$ (10.4.16)

(b) *The following statements are equivalent*:

(i) (10.4.6) *holds*;

(ii) $P(t)$ *satisfies* (10.4.13) *or* (10.4.14);

(iii) $\lim_{t \to \infty} y(t) = 0$. $\qquad\qquad\qquad\qquad\qquad$ (10.4.17)

Proof. From Lemma 10.4.1, we see that either (10.4.5) or (10.4.6) holds. First, assume that (10.4.5) holds. Then $P(t)$ must be negative and $y(t)$ unbounded. Therefore, there exists a t^* such that

$$z(t^*) < 0 \qquad \text{and} \qquad y(t^*) = \max_{s \leqslant t^*} y(s).$$

Then

$$0 > z(t^*) = y(t^*) + P(t^*)y(t^* - \tau) \geqslant y(t^*)[1 + P(t^*)],$$

which implies that $P(t^*) < -1$, so (10.4.12) holds and $P(t) \not\equiv -1$. Also observe that

$$z(t) = y(t) + P(t)y(t - \tau) > P(t)y(t - \tau) \geqslant P_1 y(t - \tau)$$

and so (10.4.5) implies that (10.4.16) holds. Next, assume that (10.4.6) holds. If (10.4.14) holds, then from (10.4.15) we see that (10.4.17) holds. Also if (10.4.13) holds, it follows from Lemma 1.5.5 that (10.4.18) is satisfied. Finally, we claim that (10.4.12) cannot hold. Otherwise, by Lemma 10.4.2, $z(t) < 0$. But, by Lemma 10.4.1 and the fact that n is odd, it follows that $z(t) > 0$, which is a contradiction. On the basis of the foregoing discussion, the proof of (a) and (b) follow immediately.

Theorem 10.4.2. *Assume n is even and* (10.4.2) *and* (10.4.3) *are satisfied. Suppose also that there exists a real number P such that*

$$-1 < P \leqslant P(t) \leqslant 0. \qquad\qquad\qquad\qquad (10.4.18)$$

Then every non-oscillatory solution of eqn (10.4.1) *tends to zero as $t \to \infty$.*

Proof. As the negative of a solution of eqn (10.4.1) is also a solution, it suffices to prove the theorem for an eventually positive solution $y(t)$ of eqn

(10.4.1). Set $z(t) = y(t) + P(t)y(t - \tau)$. By Lemma 10.4.2(a), we have $z(t) < 0$ and hence $y(t) < -P(t)y(t - \tau) < y(t - \tau)$. Therefore $y(t)$ is a bounded function. Assume, for the sake of contradiction, that $\lim \sup_{t \to \infty} y(t) \equiv s > 0$. Let $\{t_k\}$ be a sequence of points such that

$$\lim_{k \to \infty} t_k = \infty \quad \text{and} \quad \lim_{k \to \infty} y(t_k) = s.$$

Then, for sufficiently large k,

$$z(t_k) = y(t_k) + P(t_k)y(t_k - \tau) \geqslant y(t_k) + P_1 y(t_k - \tau)$$

and hence

$$\lim_{k \to \infty} \sup y(t_k - \tau) \geqslant s/(-P_1) > s,$$

which is a contradiction. So $s = 0$ and the proof is complete.

We are now ready to establish sufficient conditions for the oscillations of all solutions of eqn (10.4.1).

Theorem 10.4.3. *Assume that n is odd, (10.4.2) and (10.4.3) hold and*

$$P(t) \leqslant -1.$$

Suppose also that there exists a positive constant r such that

$$\frac{Q(t)}{P(t + \tau - \sigma)} \leqslant -r \tag{10.4.19}$$

$$r^{1/n} \frac{\tau - \sigma}{n} > \frac{1}{e} \tag{10.4.20}$$

Then every solution of eqn (10.4.1) oscillates.

Proof. Otherwise eqn (10.4.1) has an eventually positive solution $y(t)$. Set $z(t) = y(t) + P(t)y(t - \tau)$. Then $z^{(n)}(t) = -Q(t)y(t - \sigma) \leqslant 0$ and by Lemma 10.4.2(c), $z(t) < 0$. As $z(t) > P(t)y(t - \tau)$, we find

$$\frac{1}{P(t + \tau - \sigma)} Q(t)z(t + \tau - \sigma) \leqslant Q(t)y(t - \sigma) = -z^{(n)}(t)$$

and so

$$z^{(n)}(t) + \frac{Q(t)}{P(t + \tau - \sigma)} z[t + (\tau - \sigma)] \leqslant 0.$$

But by (10.4.19), (10.4.20) and Theorem 10.3.2 the above inequality has no eventually negative solution. This is a contradiction and the proof is complete.

Theorem 10.4.4. *Assume that n is odd, (10.4.2) and (10.4.3) hold and*

$$P(t) \equiv -1.$$

Then every solution of eqn (10.4.1) oscillates.

Proof. Otherwise eqn (10.4.1) has an eventually positive solution $y(t)$. Set $z(t) = y(t) - y(t - \tau)$. Then by Lemma 10.4.2(b), $z(t)$ is positive, while by Lemma 10.4.2(c), $z(t)$ is negative. This is a contradiction and the proof is complete.

Theorem 10.4.5. *Assume that n is even, $-1 \leqslant P(t) < 0$ and suppose that there exists a positive constant r such that*

$$\frac{Q(t)}{P(t + \tau - \sigma)} \leqslant -r$$

$$r^{1/n} \frac{\sigma - \tau}{n} > \frac{1}{e} \tag{10.4.22}$$

Then every solution of eqn (10.4.1) oscillates.

Proof. Otherwise eqn (10.4.1) has an eventually positive solution $y(t)$. Set $z(t) = y(t) + P(t)y(t - \tau)$. Then by Lemma 10.4.2(a) we have

$$z(t) < 0. \tag{10.4.23}$$

From (10.4.23),

$$y(t) < -P(t)y(t - \tau) \leqslant y(t - \tau),$$

which implies that $y(t)$ is bounded. Hence $z(t)$ is also bounded. Since $z(t) > P(t)y(t - \tau)$, we find

$$\frac{1}{P(t + \tau - \sigma)} Q(t)z(t + \tau - \sigma) \leqslant Q(t)y(t - \sigma) = -z^{(n)}(t),$$

that is,

$$z^{(n)}(t) + \frac{Q(t)}{P(t + \tau - \sigma)} z[t + (\tau - \sigma)] \leqslant 0.$$

But by Theorem 10.3.1, this inequality cannot have an eventually negative bounded solution. This contradicts (10.4.23) and completes the proof of the theorem.

Theorem 10.4.6. *Assume n even and suppose that there exists a real number P_1 such that*

$$0 \leqslant P(t) \leqslant P_1 < 1.$$

Then every solution of eqn (10.4.1) oscillates.

Proof. Otherwise eqn (10.4.1) has an eventually positive solution $y(t)$. Set $z(t) = y(t) + P(t)y(t - \tau)$. Then $z(t) \geqslant y(t) > 0$. But by Lemma 10.4.2(a), $z(t) < 0$. This is a contradiction and the proof is complete.

Corollary 10.4.1. *Consider the n-th-order delay differential equation*

$$y^{(n)}(t) + Q(t)y(t - \sigma) = 0 \tag{10.4.24}$$

and assume that n is even, $Q \in C[[t_0, \infty), \mathbb{R}]$, and

$$\int_{t_0}^{\infty} Q(t) \, dt = \infty.$$

Then every solution of eqn (10.4.24) oscillates.

10.5 A comparison theorem for neutral equations of odd order

Consider the neutral delay differential equations

$$\frac{d^n}{dt^n} [x(t) - P_1(t)x(t - \tau_1)] + Q_1(t)x(t - \sigma_1) = 0 \tag{10.5.1}$$

$$\frac{d^n}{dt^n} [y(t) - P_2(t)y(t - \tau_2)] + Q_2(t)y(t - \sigma_2) = 0 \tag{10.5.2}$$

where $n \geqslant 1$ is an odd integer,

$$P_1, P_2, Q_1, Q_2 \in C[[t_0, \infty), \mathbb{R}^+], \quad \tau_1, \tau \in (0, \infty) \quad \text{and} \quad \sigma_1, \sigma_2 \in [0, \infty).$$

$$\tag{10.5.3}$$

Our goal in this section is to establish a comparison result according to which, if every solution of eqn (10.5.1) oscillates, then every solution of eqn (10.5.2) also oscillates.

The following two lemmas, which are interesting in their own right, will be used in the proof of the comparison theorem. The first lemma is an extension of Lemma 6.4.1(c) to nth-order equations and its proof is omitted. The second lemma states that, under appropriate hypotheses, if an integral inequality has a positive solution, then the corresponding equation also has a positive solution. It should be noted that both lemmas are true for n odd or even.

Lemma 10.5.1. *Assume that n is a positive integer*

$$P, Q \in C[[t_0, \infty), \mathbb{R}^+], \qquad \tau, \sigma \in \mathbb{R}^+, \tag{10.5.4}$$

$$0 \leqslant P(t) \leqslant 1 \tag{10.5.5}$$

and

$$\int_{t_0}^{\infty} Q(t) \, dt = \infty. \tag{10.5.6}$$

Let x(t) be an eventually positive solution of the equation

$$\frac{d^n}{dt^n} [x(t) - P(t)x(t - \tau)] + Q(t)x(t - \sigma) = 0 \tag{10.5.7}$$

and set $z(t) = x(t) - P(t)x(t - \tau)$. Then eventually,

$$z^{(n)}(t) \leqslant 0, \qquad (-1)^i z^{(n-i)}(t) < 0 \qquad \text{for } i = 1, 2, \ldots, n, \tag{10.5.8}$$

$$\lim_{t \to \infty} z^{(i)}(t) = 0 \qquad \text{for } i = 0, 1, \ldots, n - 1. \tag{10.5.9}$$

Lemma 10.5.2. *Assume that $n \geqslant 1$ is a positive integer,*

$$P, Q \in C[[T, \infty), \mathbb{R}^+], \qquad \tau \in (0, \infty), \qquad \sigma \in [0, \infty) \tag{10.5.10}$$

and either

$$P(t) > 0 \qquad \text{for } t \geqslant T \tag{10.5.11}$$

or

$$\sigma > 0 \qquad \text{and} \qquad Q(t) \not\equiv 0 \quad \text{on any interval of length } \sigma. \tag{10.5.12}$$

Let $\rho = \max\{\tau, \sigma\}$ and assume that the integral inequality

$$P(t)z(t - \tau) + \int_t^{\infty} \int_{s_{n-1}}^{\infty} \cdots \int_{s_1}^{\infty} Q(s)z(s - \sigma) \, ds \, ds_1 \ldots ds_{n-1} \leqslant z(t),$$
$$t \geqslant T \tag{10.5.13}$$

has a continuous positive solution $z: [T - \rho, \infty) \to (0, \infty)$ such that

$$z(T) < z(t) \qquad \text{for } T - \rho \leqslant t < T. \tag{10.5.14}$$

Then there exists a continuous positive solution $x: [T - m, \infty) \to (0, \infty)$ of the corresponding integral equation

$$P(t)x(t - \tau) + \int_t^{\infty} \int_{s_{n-1}}^{\infty} \cdots \int_{s_1}^{\infty} Q(s)x(s - \sigma) \, ds \, ds_1 \ldots ds_{n-1} = x(t),$$
$$t \geqslant T. \tag{10.5.15}$$

Proof. Define the set of functions

$$W = \{w \in C[[T - \rho, \infty), \mathbb{R}^+] : 0 \leqslant w(t) \leqslant z(t) \quad for\ t \geqslant T - \rho\}$$

and define the mapping S on W as follows:

$$(Sw)(t) = \begin{cases} P(t)w(t - \tau) + \displaystyle\int_t^\infty \int_{s_{n-1}}^\infty \cdots \\ \qquad\qquad \times \displaystyle\int_{s_1}^\infty Q(s)w(s - \sigma)\,ds\,ds_1 \ldots ds_{n-1}, \quad t \geqslant T \\ (Sw)(T) + z(t) - z(T), \qquad\qquad\qquad T - \rho \leqslant t < T. \end{cases}$$

Clearly, the function $Sw: [T - \rho, \infty) \to \mathbb{R}^+$ is continuous. Also, $w_1, w_2 \in W$ with $w_1 \leqslant w_2$ implies $Sw_1 \leqslant Sw_2$. Note that $Sz \leqslant z$ and so $w \in W$ implies $Sw \leqslant Sz \leqslant z$. Thus

$$S: W \to W.$$

We now define the following sequence in W:

$$z_0 = z \quad \text{and} \quad z_k = Sz_{k-1} \quad \text{for } k = 1, 2, \ldots$$

It is clear by induction that for $k = 1, 2, \ldots$

$$0 \leqslant z_k(t) \leqslant z_k(t) \leqslant z(t) \quad \text{for } t \geqslant T - \rho.$$

Set

$$x(t) = \lim_{k \to \infty} z_k(t) \quad \text{for } t \geqslant T - \rho.$$

Then it follows by Lebesgue's dominated convergence theorem that $x(t)$ satisfies (10.5.15).

Next, we claim that $x(t)$ is a continuous function. Indeed, for $T - \rho \leqslant t \leqslant T$,

$$z_k(t) - z_m(t) = [z_k(T) + z(t) - z(T)] - [z_m(T) + z(t) - z(T)]$$
$$= z_k(t) - z_m(T)$$

and so the sequence of continuous functions $\{z_k(t)\}$ converges uniformly on $[T - \rho, T]$. Hence x is continuous on $[T - \rho, T]$. In view of (10.5.15) it is now clear that $x(t)$ is continuous on $[T - \rho, \infty)$.

Finally, it remains to show that $x(t) > 0$ for $t \geqslant T - \rho$. Indeed, if $T - \rho \leqslant t < T$, it remains to show that $x(t) > 0$ for $t \geqslant T - \rho$. Indeed, if $T - \rho \leqslant t < T$, then from (10.5.14) and the definition of $z_k(t)$ we see that

$$z_k(t) \geqslant z(t) - z(T) > 0$$

and so

$$x(t) \geqslant z(t) - z(T) > 0 \quad \text{for } T - \rho \leqslant t < T.$$

We now claim that $x(t) > 0$ for $t \geq T$. Otherwise

$$t^* = \inf\{t \geq T : x(t) = 0\} \in [T, \infty).$$

Then

$$x(t) > 0 \qquad \text{for } T - \rho \leq t < t^* \qquad (10.5.16)$$

and $x(t^*) = 0$. But either (10.5.11) or (10.5.12) holds, and because of (10.5.16), (10.5.15) yields

$$0 = x(t^*) = P(t^*)x(t^* - \tau)$$

$$+ \int_{t^*}^{\infty} \int_{s_{n-1}}^{\infty} \cdots \int_{s_1}^{\infty} Q(s)x(s - \sigma)\, ds\, ds_1 \dots ds_{n-1} > 0.$$

This is a contradiction and the proof is complete.

The main result in this section is the following comparison theorem.

Theorem 10.5.1. *Assume $n \geq 1$ is an odd integer and (10.5.3) holds. Suppose that $P_1 \in C^1[[t_0, \infty), \mathbb{R}^+]$,*

$$\tau_1 \leq \tau_2, \qquad \sigma_1 \leq \sigma_2 \qquad (10.5.17)$$

and for t sufficiently large,

$$P_1(t) \leq P_2(t - \sigma_2)\frac{Q_2(t)}{Q_2(t - \tau_2)}, \qquad Q_1(t) \leq Q_2(t), \qquad (10.5.18)$$

$$P_1(t) \text{ is bounded}, \qquad \dot{P}_1(t) \geq 0, \qquad 0 \leq P_2(t) \leq 1, \qquad (10.5.19)$$

$$Q_2(t) > 0, \qquad \int_{t_0}^{\infty} Q_2(t)\, dt = \infty$$

and either $P_1(t) > 0$ or $\sigma_1 > 0$ and $Q_1(t) \not\equiv 0$ on any interval of length σ_1. Finally, assume that every solution of eqn (10.5.1) oscillates. Then every solution of eqn (10.5.2) also oscillates.

Proof. Assume, for the sake of contradiction, that eqn (10.5.2) has an eventually positive solution $y(t)$. Set $z(t) = y(t) - P_2(t)y(t - \tau_2)$. Then by Lemma 10.5.1, (10.5.8) and (10.5.9) hold. It also follows by direct substitution that $z(t)$ satisfies the equation

$$z^{(n)}(t) - P_2(t - \sigma_2)\frac{Q_2(t)}{Q_2(t - \tau_2)} z^{(n)}(t - \tau_2) + Q_2(t)z(t - \sigma_2) = 0. \qquad (10.5.20)$$

By using (10.5.18) and (10.5.8), eqn (10.5.20) yields the inequality

$$z^{(n)}(t) - P_1(t)z^{(n)}(t - \tau_2) + Q_1(t)z(t - \sigma_2) \leq 0 \qquad (10.5.21)$$

for t sufficiently large, say $t \geqslant T$. We also choose T so large that (10.5.8) holds for $t \geqslant T - \max\{\tau_2, \sigma_2\}$. By integrating (10.5.21) from t to t_1 and then by letting $t_1 \to \infty$ we find, for $t \geqslant T$,

$$-z^{(n-1)}(t) - \int_t^\infty P_1(s)z^{(n)}(s - \tau_2)\, \mathrm{d}s + \int_t^\infty Q_1(s)z(s - \sigma_2)\, \mathrm{d}s \leqslant 0.$$

By integrating by parts the first integral and then by using (10.5.18), (10.5.19) and the monotonic character of $z^{(i)}(t)$, we obtain the inequality

$$-z^{(n)}(t) + P_1(t)z^{(n-1)}(t - \tau_1) + \int_t^\infty Q_1(s)z(s - \sigma_1)\, \mathrm{d}s \leqslant 0, \qquad t \geqslant T.$$

By integrating again and again we are led to the inequality

$$P_1(t)z(t - \tau_1) + \int_t^\infty \int_{s_{n-1}}^\infty \cdots \int_{s_1}^\infty Q_1(s)z(s - \sigma_1)\, \mathrm{d}s\, \mathrm{d}s_1 \ldots \mathrm{d}s_{n-1} \leqslant z(t),$$

$$t \geqslant T. \quad (10.5.22)$$

From (10.5.8) and (10.5.9), $z(t)$ is a positive and strictly decreasing function, so (10.5.14) holds. It follows by Lemma 10.5.2 that the equation

$$P_1(t)x(t - \tau_1) + \int_t^\infty \int_{s_{n-1}}^\infty \cdots \int_{s_1}^\infty Q_1(s)\, \mathrm{d}s\, \mathrm{d}s_1 \ldots \mathrm{d}s_{n-1} = x(t), \qquad t \geqslant T$$

has a positive solution, eqn (10.5.1) also has a positive solution. This contradicts the hypothesis that every solution of eqn (10.5.1) oscillates and the proof is complete.

10.6 A linearized oscillation theorem for neutral equations of odd order

Consider the non-linear neutral delay differential equation of odd order $n \geqslant 1$,

$$\frac{\mathrm{d}^n}{\mathrm{d}t^n}\left[x(t) - p(t)g(x(t - \tau))\right] + q(t)h(x(t - \sigma)) = 0 \qquad (10.6.1)$$

under the following hypotheses:

$$p, q \in C[[t_0, \infty), \mathbb{R}^+], \quad g, h \in C[\mathbb{R}, \mathbb{R}], \quad \tau \in (0, \infty), \quad \sigma \in [0, \infty), (10.6.2)$$

$$\left.\begin{array}{c} \displaystyle\lim_{t \to \infty} \inf p(t) \equiv p_0 \in (0, 1), \quad \lim_{t \to \infty} \sup p(t) \equiv P_0 \in (0, 1), \\[2mm] \displaystyle\lim_{t \to \infty} q(t) = q_0 \in (0, \infty), \end{array}\right\} \quad (10.6.3)$$

$$0 \leqslant \frac{g(u)}{u} \leqslant 1 \quad \text{for } u \neq 0, \quad \lim_{u \to \infty} \frac{g(u)}{u} = 1, \qquad (10.6.4)$$

$uh(u) > 0$ for $u \neq 0$, $|h(u)| \geq h_0 > 0$ for $|u|$ sufficiently large,   (10.6.5)

and

$$\lim_{u \to \infty} \frac{h(u)}{u} = 1.$$ (10.6.6)

With eqn (10.6.1) we associate the linear equation

$$\frac{d^n}{dt^n}[y(t) - p_0 y(t - \tau)] + q_0 y(t - \sigma) = 0.$$ (10.6.7)

Our aim in this section is to establish conditions for the oscillation of all solutions of eqn (10.6.1) in terms of the oscillation of all solutions of the linear equation (10.6.7), and vice versa.

Theorem 10.6.1. *Assume that (10.6.2)–(10.6.6) are satisfied and that every solution of eqn (10.6.7) oscillates. Then every solution of eqn (10.6.1) also oscillates.*

Proof. Assume, for the sake of contradiction, that eqn (10.6.1) has a non-oscillatory solution $x(t)$. We assume that $x(t)$ is eventually positive. The case where $x(t)$ is eventually negative is similar and will be omitted. Set

$$z(t) = x(t) - p(t)g(x(t - \tau)).$$ (10.6.8)

Then

$$z^{(n)}(t) = -q(t)h(x(t - \sigma)) \leq 0.$$ (10.6.9)

Therefore $z^{(n-1)}(t)$ is decreasing, so either

$$\lim_{t \to \infty} z^{(n-1)}(t) = -\infty$$ (10.6.10)

or

$$\lim_{t \to \infty} z^{(n-1)}(t) \equiv l \in \mathbb{R}.$$ (10.6.11)

We claim that (10.6.11) holds. Otherwise (10.6.10) holds, which implies that

$$\lim_{t \to \infty} z^{(i)}(t) = -\infty \qquad \text{for } i = 0, 1, \dots, n - 1.$$ (10.6.12)

Then $x(t)$ is an unbounded function, so there exists a sequence of points $\{t_k\}$ such that

$$\lim_{k \to \infty} t_k = \infty, \quad \lim_{k \to \infty} x(t_k) = \infty \quad \text{and} \quad x(t_k) = \max_{s \leq t_k} x(s) \quad \text{for } k = 1, 2, \dots$$

Now by using (10.6.3) and (10.6.4) we see that

$$z(t_k) = x(t_k) - p(t_k)g(x(t_k - \tau)) = x(t_k) - p(t_k)\frac{g(x(t_k - \tau))}{x(t_k - \tau)}x(t_k - \tau)$$

$$\geqslant x(t_k)[1 - p(t_k)] \to \infty \qquad \text{as } k \to \infty,$$

which contradicts (10.6.12), so (10.6.11) holds. From (10.6.9) and (10.6.11) it follows that for $i = 0, 1, \ldots, n - 1$ the function $z^{(i)}(t)$ is monotonic. By integrating both sides of (10.6.9) from t_1 to ∞, for t_1 sufficiently large, we obtain

$$l - z^{(n-1)}(t_1) = -\int_{t_1}^{\infty} q(s)h(x(s - \sigma))\, ds. \qquad (10.6.13)$$

We claim that

$$\liminf_{t \to \infty} x(t) = 0. \qquad (10.6.14)$$

Otherwise, there exists a positive constant c and a $t_2 \geqslant t_1$ such that $x(t) \geqslant c$ for $t \geqslant t_2$. Then from (10.6.3) and (10.6.5) and for t sufficiently large, $q(t)h(x(t - \sigma))$ is bounded from below by a positive constant. This contradicts (10.6.13), so (10.6.14) holds. Let $\{t_k\}$ be a sequence of points such that

$$\lim_{k \to \infty} t_k = \infty \qquad \text{and} \qquad \lim_{k \to \infty} x(t_k) = 0.$$

From (10.6.8) we see that

$$z(t_k) \leqslant x(t_k) \to 0 \qquad \text{as } k \to \infty$$

$$z(t_k + \tau) \geqslant -p(t_k + \tau)\frac{g(x(t_k))}{x(t_k)}x(t_k) \to 0 \qquad \text{as } k \to \infty.$$

As $z(t)$ is monotonic, it follows that

$$\lim_{t \to \infty} z(t) = 0.$$

From this and (10.6.1) it is clear that

$$\lim_{t \to \infty} z^{(i)}(t) = 0 \qquad \text{for } i = 0, 1, \ldots, n - 1 \qquad (10.6.15)$$

and

$$z^{(n)}(t) \leqslant 0, \qquad z^{(n-1)}(t) > 0, \ldots, z'(t) < 0, \qquad z(t) > 0. \qquad (10.6.16)$$

We now claim that

$$\lim_{t \to \infty} x(t) = 0. \qquad (10.6.17)$$

To this end, observe that for t sufficiently large,

$$z(t) = x(t) - P(t)x(t - \tau) \qquad (10.6.18)$$

where

$$P(t) = p(t)\frac{g(x(t - \tau))}{x(t - \tau)} \leqslant p(t).$$

Let $\tilde{p} \in (P_0, 1)$. Then from (10.6.3) and for t sufficiently large,

$$0 \leqslant P(t) \leqslant p(t) \leqslant \tilde{p} < 1$$

and (10.6.17) follows by applying Lemma 1.5.2 to (10.6.18).

Next, we rewrite eqn (10.6.1) in the form

$$\frac{d^n}{dt^n}[x(t) - P(t)x(t - \tau)] + Q(t)x(t - \sigma) = 0,$$

where for t sufficiently large,

$$P(t) = p(t)\frac{g(x(t - \tau))}{x(t - \tau)} \quad \text{and} \quad Q(t) = q(t)\frac{h(x(t - \sigma))}{x(t - \sigma)}.$$

Note that because of (10.6.17), (10.6.3), (10.6.4) and (10.6.6),

$$\liminf_{t \to \infty} P(t) = p_0 \quad \text{and} \quad \lim_{t \to \infty} Q(t) = q_0.$$

Now one can see by direct substitution that for t sufficiently large, $z(t)$, which can be written in the form (10.6.18), is a solution of the neutral equation

$$z^{(n)}(t) - P(t - \sigma)\frac{Q(t)}{Q(t - \tau)}z^{(n)}(t - \tau) + Q(t)z(t - \sigma) = 0 \quad (10.6.19)$$

and

$$\liminf_{t \to \infty}\left[P(t - \sigma)\frac{Q(t)}{Q(t - \tau)}\right] = p_0.$$

Then for any positive number ε in the interval $0 < \varepsilon < \frac{1}{2}\min\{p_0, q_0\}$, eqn (10.6.19) yields the inequality

$$z^{(n)}(t) - p_0(1 - \varepsilon)z^{(n)}(t - \tau) + q_0(1 - \varepsilon)z(t - \sigma) \leqslant 0.$$

By integrating this inequality from t to t_1 and then by letting $t_1 \to \infty$, we find

$$-z^{(n-1)}(t) + p_0(1 - \varepsilon)z^{(n-1)}(t - \tau) + q_0(1 - \varepsilon)\int_t^\infty z(s - \sigma)\,ds \leqslant 0.$$

By repeating the same procedure n times and by noting the fact that n is odd, we are led to the inequality

$$p_0(1 - \varepsilon)z(t - \tau) + q_0(1 - \varepsilon) \int_t^\infty \int_{s_{n-1}}^\infty \cdots$$

$$\times \int_{s_1}^\infty z(s - \sigma) \, ds \, ds_1 \ldots ds_{n-1} \leqslant z(t), \qquad t \geqslant T,$$

where T is sufficiently large and $z \colon [T - \rho, \infty) \to (0, \infty)$, with $\rho = \max\{\tau, \sigma\}$, is a continuous and strictly decreasing function. It follows by Lemma 10.5.2 that the equation

$$p_0(1 - \varepsilon)v(t - \tau) + q_0(1 - \varepsilon) \int_t^\infty \int_{s_{n-1}}^\infty \cdots$$

$$\times \int_{s_1}^\infty v(s - \sigma) \, ds \, ds_1 \ldots ds_{n-1} = v(t)$$

has a continuous positive solutions $v \colon [T - \rho, \infty) \to (0, \infty)$. Then $v(t)$ is also a positive solution of the neutral equation

$$\frac{d^n}{dt^n} [v(t) - p_0(1 - \varepsilon)v(t - \tau)] + q_0(1 - \varepsilon)v(t - \sigma) = 0. \quad (10.6.20)$$

Hence, by Theorem 6.3.1, the characteristic equation of eqn (10.6.20),

$$\lambda^n - p_0(1 - \varepsilon)\lambda^n + q_0(1 - \varepsilon) \, \mathrm{e}^{-\lambda\sigma} = 0,$$

has a real root. By an argument similar to that in the proof of Theorem 6.5.2 and because of the fact that the ε is arbitrarily small, it follows that eqn (10.6.7) has a positive solution. This contradicts the hypothesis and completes the proof of the theorem.

Consider the neutral delay differential equation

$$\frac{d^n}{dt^n} [x(t) - p(t)x(t - \tau)] + q(t)h(x(t - \sigma)) = 0, \qquad (10.6.21)$$

where $n \geqslant 1$ is an odd integer,

$$p, q \in C[[t_0, \infty), \mathbb{R}^+], \quad \tau \in (0, \infty), \quad \sigma \in [0, \infty) \text{ and } h \in C[\mathbb{R}, \mathbb{R}]. \quad (10.6.22)$$

The next theorem is a partial converse of Theorem 10.6.1 and shows that, under appropriate hypotheses, eqn (10.6.21) has a positive solution provided that an associated linear equation has a positive solution.

Theorem 10.6.2. *Assume that (10.6.22) holds and that there exists positive constants p_0, q_0 and δ such that*

$$0 < p(t) \leqslant p_0 < 1, \qquad 0 \leqslant q(t) \leqslant q_0 \qquad (10.6.23)$$

and

$$\left. \begin{array}{llll} either & 0 \leqslant h(u) \leqslant u & for & 0 \leqslant u \leqslant \delta \\ or & 0 \geqslant h(u) \geqslant u & for & -\delta \leqslant u \leqslant 0 \end{array} \right\}. \qquad (10.6.24)$$

Suppose also that $h(u)$ is non-decreasing in u and that the characteristic equation of eqn (10.6.7),

$$\lambda^n - p_0 \lambda^n e^{-\lambda \tau} + q_0 e^{-\lambda \sigma} = 0 \qquad (10.6.25)$$

has a real root. Then eqn (10.6.21) has a non-oscillatory solution.

Proof. Assume that $0 \leqslant h(u) \leqslant u$ for $0 \leqslant u \leqslant \delta$. The case where $0 \geqslant h(u) \geqslant u$ for $-\delta \leqslant u \leqslant 0$ is similar and is omitted. Let λ_0 be a root of eqn (10.6.25). As $p_0 \in (0, 1)$, it follows that $\lambda_0 < 0$. Set $y(t) = \exp(\lambda_0 t)$. Then there exists a $T \geqslant t_0$ such that

$$y(t) > 0, \ \dot{y}(t) < 0, \ \ddot{y}(t) > 0, \ldots, y^{(n-1)}(t) > 0, \ y^{(n)}(t) < 0,$$

$$\lim_{t \to \infty} y^{(i)}(t) = 0 \qquad \text{for } i = 0, 1, 2, \ldots, n,$$

and $y(t)$ satisfies the neutral equation

$$\frac{d^n}{dt^n} [y(t) - p_0 y(t - \tau)] + q_0 y(t - \sigma) = 0. \qquad (10.6.26)$$

By integrating n times both sides of (10.6.26) from $t \geqslant T$ to ∞ we obtain

$$p_0 y(t - \tau) + \int_t^\infty \int_{s_{n-1}}^\infty \cdots \int_{s_1}^\infty q_0 y(s - \sigma) \, ds \, ds_1 \ldots ds_{n-1} = y(t), \qquad t \geqslant T.$$

In view of (10.6.23) and (10.6.24), this inequality implies that

$$p(t) y(t - \tau) + \int_t^\infty \int_{s_{n-1}}^\infty \cdots \int_{s_1}^\infty q_0(s) h(y(s - \sigma)) \, ds \, ds_1, \ldots ds_{n-1} \leqslant y(t),$$

$$t \geqslant T.$$

By a slight modification in the proof of Lemma 10.5.2 and because of the increasing nature of h, it follows that the corresponding equation

$$p(t) x(t - \tau) + \int_t^\infty \int_{s_{n-1}}^\infty \cdots \int_{s_1}^\infty q(s) h(x(s - \sigma)) \, ds \, ds_1, \ldots ds_{n-1} = x(t),$$

$$t \geqslant T$$

has a continuous positive solution. Hence eqn (10.6.21) also has a positive solution. This contradicts the hypothesis and completes the proof of the theorem.

By combining Theorems 10.6.1 and 10.6.2 we obtain the following necessary and sufficient condition for the oscillation of every solution of eqn (10.6.21).

Corollary 10.6.1. *Assume that there exist positive constants* h_0, p_0, q_0 *and* δ *such that* (10.6.22), (10.6.5), (10.6.6) *and* (10.6.24) *are satisfied,*

$$0 < p(t) \leqslant p_0 \equiv \lim_{t \to \infty} p(t) < 1, \qquad 0 \leqslant q(t) \leqslant q_0 \equiv \lim_{t \to \infty} q(t),$$

and $h(u)$ *is non-decreasing in* u. *Then every solution of eqn* (10.6.21) *oscillates if and only if every solution of the linear equation* (10.6.7) *oscillates if and only if eqn* (10.6.25) *has no negative real roots.*

10.7 A comparison theorem for neutral equations of even order

The main result in this section is the following comparison theorem for the neutral delay differential equations

$$\frac{d^n}{dt^n}[y(t) - P_1(t)y(t - \tau_1)] - Q_1(t)y(t - \sigma_1) = 0 \qquad (10.7.1)$$

$$\frac{d^n}{dt^n}[y(t) - P_2(t)y(t - \tau_2)] - Q_2(t)y(t - \sigma_2) = 0 \qquad (10.7.2)$$

where n is an even integer and

$$P_1 \in C^1[[t_0, \infty), \mathbb{R}^+], \quad P_2, Q_1, Q_2 \in C[[t_0, \infty), \mathbb{R}^+] \quad \text{and} \quad \tau_1, \tau_2, \sigma_1, \sigma_2 \in \mathbb{R}^+.$$

$$(10.7.3)$$

Theorem 10.7.1. *Assume that* (10.7.3) *holds,*

$$P_1, P_2 \text{ are bounded}, \quad \dot{P}_1(t) \geqslant 0 \qquad \text{and} \qquad Q_2(t) > 0 \quad \text{for } t \geqslant t_0, \qquad (10.7.4)$$

$$0 < \tau_1 \leqslant \tau_2 \quad \text{and} \quad \sigma_1 \leqslant \sigma_2,$$

$$\left. \begin{array}{l} P_1(t) \leqslant P_2(t - \sigma_2)\dfrac{Q_2(t)}{Q_2(t - \tau_2)} \quad \text{and} \quad Q_1(t) \leqslant Q_2(t) \quad \text{for } t \geqslant t_0, \end{array} \right\} \qquad (10.7.5)$$

$$\int_{t_0}^{\infty} Q_2(t)\,dt = \infty \qquad (10.7.6)$$

and either $P_1(t) > 0$ for $t \geq t_0$ or

$$\sigma_1 > 0 \quad \text{and} \quad Q_1(t) \not\equiv 0 \quad \text{on an interval of length } \sigma_1. \quad (10.7.7)$$

Suppose also that every bounded solution of eqn (10.7.1) oscillates. Then every bounded solution of eqn (10.7.2) also oscillates.

Proof. Assume, for the sake of contradiction, that eqn (10.7.2) has a bounded and eventually positive solution $y(t)$. Set $z(t) = y(t) - P_2(t)y(t - \tau_2)$. Then one can show that

$$z^{(n)}(t) \geq 0, \quad (-1)^{(n-i)} z^{(i)}(t) > 0 \quad \text{for } i = 1, \ldots, n \quad (10.7.8)$$

$$\lim_{t \to \infty} z^{(i)}(t) = 0 \quad \text{for } i = 0, 1, \ldots, n - 1. \quad (10.7.9)$$

It also follows by direct substitution that $z(t)$ satisfies the equation

$$z^{(n)}(t) - P_2(t - \sigma_2) \frac{Q_2(t)}{Q_2(t - \tau_2)} z^{(n)}(t - \tau_2) - Q_2(t) z(t - \sigma_2) = 0. \quad (10.7.10)$$

By using (10.7.5) and the fact that eventually $z^{(n)}(t) \geq 0$ and $z(t) > 0$, eqn (10.7.10) yields the inequality

$$z^{(n)} - P_1(t) z^{(n)}(t - \tau_2) - Q_2(t) z(t - \sigma_2) \geq 0 \quad (10.7.11)$$

for t sufficiently large, say for $t \geq T$. We also choose T so large that (10.7.8) holds for $t \geq T - \max\{\tau_2, \sigma_2\}$. By integrating (10.7.11) from t to ∞ we find

$$-z^{(n-1)}(t) - \int_t^\infty P_1(s) z^{(n)}(s - \sigma_2)\, ds - \int_t^\infty Q_1(s) z(s - \sigma_2)\, ds \geq 0, \quad t \geq T.$$

Then by integrating by parts the first integral and by using (10.7.4), (10.7.5) and the monotonic character of $z^{(i)}(t)$, we obtain the inequality

$$-z^{(n-1)}(t) + P_1(t) z^{(n-1)}(t - \tau_1) - \int_t^\infty Q_1(s) z(s - \sigma_1)\, ds \geq 0, \quad t \geq T.$$

By repeating the same procedure n times and by noting the fact that n is even, we are led to the inequality

$$P_1(t) z(t - \tau_1) + \int_t^\infty \int_{s_{n-1}}^\infty \cdots \int_{s_1}^\infty Q_1(s) z(s - \sigma_1)\, ds\, ds_1 \ldots ds_{n-1} \leq z(t),$$
$$t \geq T.$$

As all the hypotheses of Lemma 10.5.2 are satisfied, it follows that the equation

$$P(t) x(t - \tau_1) + \int_t^\infty \int_{s_{n-1}}^\infty \cdots \int_{s_1}^\infty Q_1(s) x(s - \sigma_1)\, ds\, ds_1 \ldots ds_{n-1} = x(t),$$
$$t \geq T$$

and so also eqn (10.7.1) has a bounded positive solution. This contradicts the hypothesis that every bounded solution of eqn (10.7.1) oscillates and the proof is complete.

10.8 A linearized oscillation theorem for neutral equations of even order

In this section we established a linearized oscillation result for neutral delay differential equations of even order. Consider the NDDE

$$\frac{d^n}{dt^n}[x(t) - p(t)g(x(t - \tau))] - q(t)h(x(t - \sigma)) = 0, \qquad (10.8.1)$$

where n is an even integer,

$$p, q \in C[[t_0, \infty), \mathbb{R}^+], \quad g, h \in C[\mathbb{R}, \mathbb{R}], \quad \tau \in (0, \infty) \quad \text{and} \quad \sigma \in [0, \infty).$$
$$(10.8.2)$$

Theorem 10.8.1. *Assume that* (10.8.2) *holds,*

$$\left. \begin{array}{c} \lim\sup_{t \to \infty} p(t) = P_0 \in (0, 1), \qquad \lim\inf_{t \to \infty} p(t) = p_0 \in (0, 1), \\[2mm] \lim_{t \to \infty} q(t) = q_0 \in (0, \infty), \end{array} \right\} \qquad (10.8.3)$$

$$0 \leqslant \frac{g(u)}{u} \leqslant 1 \quad \text{for } u \neq 0, \qquad \lim_{u \to \infty} \frac{g(u)}{u} = 1, \qquad (10.8.4)$$

$$uh(u) > 0 \quad \text{for } u \neq 0 \quad \text{and} \quad \lim_{u \to \infty} \frac{h(u)}{u} = 1. \qquad (10.8.5)$$

Suppose that every bounded solution of the linearized equation

$$\frac{d^n}{dt^n}[y(t) - p_0 y(t - \tau)] - q_0 y(t - \sigma) = 0 \qquad (10.8.6)$$

oscillates. Then every bounded solution of eqn (10.8.1) *also oscillates.*

Proof. Assume, for the sake of contradiction, that eqn (10.8.1) has a bounded non-oscillatory solution $x(t)$. We assume that $x(t)$ is eventually positive. The case where $x(t)$ is eventually negative is similar and is omitted. Set $z(t) = x(t) - p(t)g(x(t - \tau))$. Then

$$z^{(n)}(t) = q(t)h(x(t - \sigma)). \qquad (10.8.7)$$

As $x(t)$ is bounded, $z(t)$ is also bounded. Then by (10.8.7) it follows that

$$\lim_{t \to \infty} z^{(n-1)}(t) = l \in \mathbb{R} \quad \text{exists}$$

By integrating both sides of (10.8.7) from t_1 to ∞, for t_1 sufficiently large, we obtain

$$l - z^{(n-1)}(t_1) = \int_{t_1}^{\infty} q(s)h(x(s - \sigma))\, ds,$$

which in view of (10.8.3), (10.8.5) and the boundedness of $x(t)$ implies that

$$\liminf_{t \to \infty} x(t) = 0.$$

Then, by an argument similar to that in the proof of Theorem 10.6.1, it follows that

$$\lim_{t \to \infty} z^{(i)}(t) = 0 \qquad \text{for } i = 0, 1, 2, \ldots, n-1, \tag{10.8.8}$$

$$z^{(n)}(t) \geqslant 0, \qquad z^{(n-1)}(t) < 0, \ldots, z(t) > 0 \tag{10.8.9}$$

$$\lim_{t \to \infty} x(t) = 0. \tag{10.8.10}$$

Next, we rewrite eqn (10.8.1) in the form

$$\frac{d^n}{dt^n}[x(t) - P(t)x(t - \tau)] - Q(t)x(t - \sigma) = 0,$$

where for t sufficiently large,

$$P(t) = \frac{g(x(t - \tau))}{x(t - \tau)} \qquad \text{and} \qquad Q(t) = \frac{h(x(t - \sigma))}{x(t - \sigma)}.$$

Note also that

$$\liminf_{t \to \infty} P(t) = p_0 \qquad \text{and} \qquad \lim_{t \to \infty} Q(t) = q_0.$$

It is easy to see by direct substitution that, for t sufficiently large, $z(t)$ is a solution of the neutral equation

$$z^{(n)}(t) - P(t - \sigma)\frac{Q(t)}{Q(t - \tau)}z^{(n)}(t - \tau) - Q(t)z(t - \sigma) = 0. \tag{10.8.11}$$

Also

$$\liminf_{t \to \infty}\left[P(t - \sigma)\frac{Q(t)}{Q(t - \tau)}\right] = p_0.$$

Then for any positive number ε in the interval $0 < \varepsilon < \frac{1}{2}\min\{p_0, q_0\}$, eqn (10.8.11) yields the inequality

$$z^{(n)}(t) - (p_0 - \varepsilon)z^{(n)}(t - \tau) - (q_0 - \varepsilon)z(t - \sigma) \geqslant 0.$$

By integrating this inequality from t to ∞ and by using (10.8.8), we find

$$-z^{(n-1)}(t) + (p_0 - \varepsilon)z^{(n-1)}(t - \tau) - (q_0 - \varepsilon)\int_t^\infty z(s - \sigma)\,ds \geqslant 0.$$

By repeating the same procedure n times and by noting the fact n is even, we are led to the inequality

$$(p_0 - \varepsilon)z(t - \tau) + (q_0 - \varepsilon)\int_t^\infty \int_{s_{n-1}}^\infty \cdots \int_{s_1}^\infty z(s - \sigma)\,ds\,ds_1 \ldots ds_{n-1} \leqslant z(t),$$

$$t \geqslant T,$$

where T is sufficiently large and $z\colon [T - \rho, \infty) \to (0, \infty)$, with $\rho = \max\{\tau, \sigma\}$, is a continuous and strictly decreasing function with limit zero. It follows by Lemma 10.5.2 that the equation

$$(p_0 - \varepsilon)v(t - \tau) + (q_0 - \varepsilon)\int_t^\infty \int_{s_{n-1}}^\infty \cdots \int_{s_1}^\infty v(s - \sigma)\,ds\,ds_1 \ldots ds_{n-1} = v(t),$$

$$t \geqslant T$$

has a continuous bounded positive solution $v\colon [T - \rho, \infty) \to (0, \infty)$. Clearly $v(t)$ is also a bounded positive solution of the neutral equation

$$\frac{d^n}{dt^n}\left[v(t) - (p_0 - \varepsilon)v(t - \tau)\right] - (q_0 - \varepsilon)v(t - \sigma) = 0.$$

As ε is arbitrarily small, it follows that eqn (10.8.6) has a bounded non-oscillatory solution. This contradicts the hypothesis and completes the proof of the theorem.

Consider the neutral delay differential equation

$$\frac{d^n}{dt^n}\left[y(t) - p(t)y(t - \tau)\right] - q(t)h(y(t - \sigma)) = 0, \qquad (10.8.12)$$

where n is an even integer,

$$p, q \in C[[t_0, \infty), \mathbb{R}^+], \quad \tau \in (0, \infty), \quad \sigma \in [0, \infty) \quad \text{and} \quad h \in C[\mathbb{R}, \mathbb{R}]. \qquad (10.8.13)$$

The next theorem is a partial converse of Theorem 10.8.1 and shows that, under appropriate hypotheses, eqn (10.8.12)˙ has a bounded positive solution, provided that an associated linear equation with constant coefficients has a bounded positive solution.

Theorem 10.8.2. *Assume that* (10.8.13) *holds and that there exists positive constants* p_0, q_0 *and* δ *such that*

$$0 \leqslant p(t) \leqslant p_0 \qquad \text{and} \qquad 0 \leqslant q(t) \leqslant q_0 \qquad \text{for } t \geqslant t_0 \qquad (10.8.14)$$

and

$$\text{either} \quad 0 \leqslant h(u) \leqslant 0 \quad \text{for} \quad 0 \leqslant u \leqslant \delta$$
$$\left. \text{or} \quad \quad 0 \geqslant h(u) \geqslant u \quad \text{for} \quad -\delta \leqslant u \leqslant 0 \right\} . \tag{10.8.15}$$

Suppose also that $h(u)$ is non-decreasing in a neighbourhood of the origin and that the characteristic equation of eqn (10.8.6),

$$\lambda^n - p_0 \lambda^n e^{-\lambda \tau} - q_0 e^{-\lambda \sigma} = 0, \tag{10.8.16}$$

has a root in the interval $(-\infty, 0]$. Then eqn (10.8.12) has a bounded non-oscillatory solution.

Proof. Assume that $0 \leqslant h(u) \leqslant u$ for $0 \leqslant u \leqslant \delta$. The case where $0 \geqslant h(u) \geqslant u$ for $-\delta \leqslant u \leqslant 0$ is similar and is omitted. Let λ_0 be a root of eqn (10.8.16). As $q_0 > 0$, it follows that $\lambda_0 < 0$. Set $y(t) = \exp(\lambda_0 t)$. Then for T sufficiently large, $0 < y(t) \leqslant \delta$ for $t \geqslant T - \delta$ and h is non-decreasing in $[0, y(T - \delta)]$. Clearly,

$$y(t) > 0, \, y'(t) < 0, \dots, y^{(n-1)}(t) < 0, \, y^{(n)} > 0, \tag{10.8.17}$$

$$\lim_{t \to \infty} y^{(i)}(t) = 0 \quad \text{for } i = 0, 1, 2, \dots, n, \tag{10.8.18}$$

and $y(t)$ satisfies the neutral equation

$$\frac{d^n}{dt^n} [y(t) - p_0 y(t - \tau)] - q_0 y(t - \sigma) = 0. \tag{10.8.19}$$

By integrating n times both sides of (10.8.19) from $t \geqslant T$ to ∞ and by using (10.8.18), we obtain

$$y(t) - p_0 y(t - \tau) - \int_t^\infty \int_{s_{n-1}}^\infty \cdots \int_{s_1}^\infty q_0 y(s - \sigma) \, ds \, ds_1 \dots ds_{n-1} = 0,$$
$$t \geqslant T.$$

In view of (10.8.14) and (10.8.15), this equation implies that

$$p(t) y(t - \tau) + \int_t^\infty \int_{s_{n-1}}^\infty \cdots \int_{s_1}^\infty q(s) h(y(s - \sigma)) \, ds \, ds_1 \dots ds_{n-1} \leqslant y(t),$$
$$t \geqslant T.$$

Then by Lemma 10.5.2 it follows that the corresponding equation

$$p(t) x(t - \tau) + \int_t^\infty \int_{s_{n-1}}^\infty \cdots \int_{s_1}^\infty q(s) h(x(s - \sigma)) \, ds \, ds_1 \dots ds_{n-1} = x(t),$$
$$t \geqslant T$$

has a continuous bounded positive solution $x(t)$. Hence $x(t)$ is also a solution of eqn (10.8.12). The proof of the theorem is complete.

Finally, by combining Theorems 10.8.1 and 10.8.2 we obtain the following necessary and sufficient condition for the oscillation of every bounded solution of eqn (10.8.12).

Corollary 10.8.1. *Assume that* (10.8.13) *holds and that there exist* p_0, $q_0 \in (0, \infty)$ *such that*

$$0 < p(t) \leqslant p_0 = \lim_{t \to \infty} p(t) < 1 \qquad for\ t \geqslant t_0$$

$$0 \leqslant q(t) \leqslant q_0 = \lim_{t \to \infty} q(t) \qquad for\ t \geqslant t_0.$$

Suppose that

$$\lim_{u \to \infty} \frac{h(u)}{u} = 1, \qquad uh(u) > 0 \qquad for\ u \neq 0$$

and $h(u)$ *is non-decreasing in a neighbourhood of the origin. Then every bounded solution of eqn* (10.8.12) *oscillates if and only if every bounded solution of the linear equation* (10.8.6) *oscillates.*

10.9 Notes

The results in Sections 10.1 and 10.2 are from Kulenovic and Ladas (1988 and 1987a respectively). Theorems 10.3.1 and 10.3.2 are extracted from Ladas and Stavroulakis (1982b, 1984a).

The results in Sections 10.4–10.8 are from Chuanxi and Ladas (1990a–d).

For other results on higher-order equations see Domshlak (1982), Jaros and Kusano (1988), Kartsatos (1969), Kosmalla (1986), Kusano (1984), Ladde et al. (1987), Mishev and Bainov (1983), and Waltman (1968).

10.10 Open problems

10.101 Obtain linearized oscillation theorems for second-order delay equations of the form

$$\ddot{y}(t) \pm \alpha(t)\dot{y}(t) + \beta(t)f(y(t - r)) = 0$$

where the coefficients $\alpha(t)$ and $\beta(t)$ have positive limits.

10.10.2   Extend the results of Section 10.4 to equations where the coefficient $P(t)$ lies in different ranges.

10.10.3   Obtain sufficient conditions for the existence of positive solutions for higher-order neutral delay differential equations.

10.10.4   Obtain linearized oscillation theorems for neutral delay differential equations of the form (10.6.1) or (10.8.1) when the coefficient $p(t)$ takes values outside the interval $(0, 1)$.

10.10.5   Extend the results of Section 6.5 to higher-order neutral or non-neutral differential equations.

11

APPLICATIONS OF OSCILLATION THEORY TO THE ASYMPTOTIC BEHAVIOUR OF SOLUTIONS

Our aim in this chapter is to show that sometimes it is possible to investigate the global asymptotic behaviour of all solutions of a differential or a difference equation by investigating separately the oscillatory and the non-oscillatory solutions. We shall demonstrate this technique by applying it in Section 11.1–11.3 to some equations in mathematical biology and in Section 11.4 to a non-autonomous linear difference equation.

11.1 Global attractivity in population dynamics

In this section we establish sufficient conditions for the global attractivity of the positive equilibrium of the delay differential equations

$$\dot{N}(t) = -\mu N(t) + p\, e^{-\gamma N(t-\tau)}, \qquad t \geqslant 0 \tag{11.1.1}$$

$$\dot{N}(t) = -\delta N(t) + PN(t-\tau)\, e^{-aN(t-\tau)}, \qquad t \geqslant 0. \tag{11.1.2}$$

Equation (11.1.1) was used by Wazewska-Czyzewska and Lasota (1988) as a model for the survival of red blood cells in an animal; see also Arino and Kimmel (1986). Here $N(t)$ denotes the number of red blood cells at time t, μ is the probability of death of a red blood cell, p and γ are positive constants related to the production of red blood cells per unit time and τ is the time required to produce a red blood cell.

Equation (11.1.2) was used by Gurney et al. (1980) in describing the dynamics of Nicholson's blowflies; see also Wazewska-Czyzewska and Lasota (1988). Here $N(t)$ is the size of the population at time t, P is the maximum per capita daily egg production rate, $1/a$ is the size at which the population reproduces at its maximum rate, δ is the per capita daily adult death rate and τ is the generation time.

The equilibrium N^* of eqn (11.1.1) is positive and satisfies the equation

$$N^* = \frac{p}{\mu} e^{-\gamma N^*}. \tag{11.1.3}$$

The following theorem provides a sufficient condition for the equilibrium N^* of eqn (11.1.1) to be a global attractor.

Theorem 11.1.1. *Set*

$$M = \gamma N^*(1 - e^{-\mu\tau})$$

and assume that

$$e^M < 2. \tag{11.1.4}$$

Then the equilibrium solution N^* of eqn (11.1.1) is a global attractor.

Proof. The change of variables

$$N(t) = N^* + \frac{1}{\gamma}x(t)$$

reduces eqn (11.1.1) to the delay differential equation

$$\dot{x}(t) + \mu x(t) + \mu \gamma N^*[1 - e^{-x(t-\tau)}] = 0. \tag{11.1.5}$$

It suffices to show that every solution $x(t)$ of (11.1.5) tends to zero as $t \to \infty$. First, assume that $x(t)$ is an eventually non-negative solution of eqn (11.1.5). From eqn (11.1.5) we then see that $\dot{x}(t) \leq 0$, so $\lim_{t\to+\infty} x(t)$ exists and is a non-negative number l. We must prove that $l = 0$. Otherwise, $l > 0$ and from eqn (11.1.5) we see that

$$\lim_{t\to\infty} [\dot{x}(t)] = -\mu l - \mu \gamma N^*[1 - e^{-l}] < 0,$$

which implies that eventually $x(t)$ is negative. This is a contradiction and so the proof is complete for non-negative solutions. A similar argument shows that every eventually non-positive solution of eqn (11.1.5) tends to zero as $t \to \infty$.

Finally, assume that there exists a sequence of points $\{\xi_n\}$ such that

$$\tau < \xi_1 < \xi_2 < \cdots < \xi_n < \xi_{n+1} < \cdots,$$

$$\lim_{n\to\infty} \xi_n = \infty, \quad \text{and} \quad x(\xi_n) = 0 \quad \text{for } n = 1, 2, \ldots,$$

and such that in each interval (ξ_n, ξ_{n+1}) the function $x(t)$ assumes both positive and negative values. Let t_n and s_n be points in (ξ_n, ξ_{n+1}) for $n = 1, 2, \ldots$ such that $x(t_n) = \max x(t)$ for $\xi_n \leq t \leq \xi_{n+1}$ and $x(s_n) = \min x(t)$ for $\xi_n \leq t \leq \xi_{n+1}$. Then for $n = 1, 2, \ldots, x(t_n) > 0$ and $\dot{x}(t_n) = 0$, while $x(s_n) < 0$ and $\dot{x}(s_n) = 0$. Next, we claim that for each $n = 1, 2, \ldots,$

$$x(t) \text{ has a zero } T_n \in [\xi_n, t_n) \cap [t_n - \tau, t_n) \tag{11.1.6}$$

and

$$x(t) \text{ has a zero } S_n \in [\xi_n, s_n) \cap [s_n - \tau, s_n). \tag{11.1.7}$$

We prove (11.1.6). The proof of (11.1.7) is similar. If (11.1.6) were false, then $\xi_n < t_n - \tau < \xi_{n+1}$ and $x(t_n - \tau) > 0$. As $\dot{x}(t_n) = 0$, eqn (11.1.5) yields

$$\mu x(t_n) + \mu\gamma N^*[1 - e^{-x(t_n - \tau)}] = 0,$$

which implies that $x(t_n - \tau) < 0$. This is a contradiction and the proof of (11.1.6) and (11.1.7) is complete.

From eqn (11.1.5) we obtain

$$\frac{d}{dt}[x(t)\, e^{\mu t}] = -\mu\gamma N^*\, e^{\mu t} + \mu\gamma N^*\, e^{\mu t}\, e^{-x(t-\tau)}. \qquad (11.1.8)$$

By integrating (11.1.8) from S_n to s_n and using the fact that $0 < s_n - S_n \leqslant \tau$, we find

$$x(s_n) = -\gamma N^*[1 - e^{-\mu(s_n - S_n)}] + \mu\gamma N^*\, e^{\mu s_n} \int_{S_n}^{s_n} e^{\mu s}\, e^{-x(s-\tau)}\, ds$$

$$> -\gamma N^*(1 - e^{-\mu\tau}) = -M.$$

As this is true for every $n = 1, 2, \ldots$, it follows that

$$x(t) > -M, \qquad t \geqslant \xi_1. \qquad (11.1.9)$$

Next, by integrating (11.1.5) from T_n to t_n and using (11.1.9) and the fact that $0 < t_n - T_n \leqslant \tau$, we find

$$x(t_n) = -\gamma N^*[1 - e^{-\mu(t_n - T_n)}] + \mu\gamma N^*\, e^{-\mu t_n} \int_{T_n}^{t_n} e^{\mu s}\, e^{-x(s-\tau)}\, ds$$

$$< -\gamma N^*[1 - e^{-\mu(t_n - T_n)}] + \gamma N^*\, e^M[1 - e^{\mu(t_n - T_n)}]$$

$$= \gamma N^*[1 - e^{-\mu(t_n - T_n)}](e^M - 1) \leqslant \gamma N^*(1 - e^{-\mu\tau})(e^M - 1)$$

$$< \gamma N^*(1 - e^{-\mu\tau}) = M.$$

Thus $x(t) < M$, $t \geqslant \xi_1$. So far we have established that

$$-M < x(t) < M, \qquad t \geqslant \xi_1. \qquad (11.1.10)$$

By using (11.1.10) and an argument similar to that given above, we find

$$-M(-e^{-M} + 1) < x(t) < M(e^M - 1), \qquad t \geqslant \xi_1.$$

One can show by induction that

$$-L_n \leqslant x(t) \leqslant R_n, \qquad (11.1.11)$$

where $L_0 = R_0 = M$, and for $n = 0, 1, 2, \ldots$,

$$-L_{n+1} = M(e^{-L_n} - 1), \qquad R_{n+1} = M(e^{R_n} - 1) \qquad (11.1.12)$$

and

$$-M \leqslant L_n \leqslant -L_{n+1} < 0 < R_{n+1} \leqslant R_n \leqslant M. \qquad (11.1.13)$$

Set

$$L = \lim_{n \to \infty} L_n \quad \text{and} \quad R = \lim_{n \to \infty} R_n.$$

In view of (11.1.11), the proof that $\lim_{t \to \infty} x(t) = 0$ is complete if we show that

$$L = R = 0. \qquad (11.1.14)$$

To this end, from (11.1.12) and (11.1.13) we have $-L = M(e^{-L} - 1)$, $R = M(e^R - 1)$ and

$$-M \leqslant -L \leqslant 0 \leqslant R \leqslant M. \qquad (11.1.15)$$

Hence $-L$ and R are zeros of the function $\phi(\lambda) = M(e^\lambda - 1) - \lambda$ in the interval $-M \leqslant \lambda \leqslant M$. We have $\phi(-\infty) = \phi(\infty) = \infty$, $\phi(0) = 0$, ϕ is decreasing in $(-\infty, -\ln M)$ and ϕ is increasing in $(-\ln M, \infty)$. Note also that in view of the hypothesis (11.1.14), $M \in (0, 1)$ and $\phi(M) = M(e^M - 1) - M < M(2 - 1) - M = 0$. Therefore $\phi(\lambda)$ has exactly one zero in $(-\infty, M]$, namely $\lambda = 0$. Thus $-L$ and R, which are zeros of $\phi(\lambda)$ in $[-M, M]$, are both zero. This proves (11.1.14) and completes the proof of the theorem.

Next, we turn our attention to the problem of the global attractivity of the positive equilibrium of eqn (11.1.2). For $P > \delta$, the positive equilibrium N^* of eqn (11.1.2) is given by

$$N^* = \frac{1}{a} \ln \left[\frac{P}{\delta} \right]. \qquad (11.1.16)$$

The following theorem provides a sufficient condition for the equilibrium N^* to be a global attractor.

Theorem 11.1.2. *Assume that*

$$P > \delta \qquad (11.1.17)$$

and

$$(e^{\delta\tau} - 1)\left(\frac{P}{\delta} - 1 \right) < 1. \qquad (11.1.18)$$

Let $\phi \in C[[-\tau, 0], \mathbb{R}^+]$ *with* $\phi(0) > 0$ *and let* $N(t)$ *be the unique solution of eqn (11.1.2) with*

$$N(t) = \phi(t), \qquad -\tau \leqslant t \leqslant 0. \qquad (11.1.19)$$

Then

$$\lim_{t \to \infty} N(t) = N^*. \tag{11.1.20}$$

Proof. First we show, by the method of steps, that the unique solution of (11.1.2) and (11.1.19) is positive for $t \geqslant 0$. Indeed, on $0 \leqslant t \leqslant \tau$, the solution of (11.1.2) and (11.1.19) is given explicitly by

$$N(t) = \phi(0)\, e^{-\delta t} + e^{-\delta t} \int_0^t P\phi(s - \tau)\, e^{-a\phi(s-\tau)}\, e^{\delta s}\, ds > 0.$$

In a similar way we show that $N(t) > 0$ on $\tau \leqslant t \leqslant 2\tau$, etc. Thus

$$N(t) > 0 \qquad \text{for } t \geqslant 0. \tag{11.1.21}$$

The change of variables

$$N(t) = N^* + \frac{1}{a}\, x(t) \tag{11.1.22}$$

reduces eqn (11.1.2) to the delay differential equation

$$\dot{x}(t) + \delta x(t) + a\delta N^*[1 - e^{-x(t-\tau)}] - \delta x(t - \tau)\, e^{-x(t-\tau)} = 0. \tag{11.1.23}$$

From (11.1.21) and (11.1.22) it follows that

$$x(t) > -aN^* \qquad \text{for } t \geqslant 0. \tag{11.1.24}$$

To complete the proof it suffices to show that for every solution $x(t)$ of eqn (11.1.23) which satisfies (11.1.24),

$$\lim_{t \to \infty} x(t) = 0. \tag{11.1.25}$$

First we prove that $x(t)$ is bounded from above. Otherwise there exists a sequence of points $\{t_n\}$ such that $\lim_{n \to \infty} t_n = \infty$, $\lim_{n \to \infty} x(t_n) = \infty$ and $\dot{x}(t_n) \geqslant 0$ for $n = 1, 2, \ldots$ It follows from (11.1.23) that

$$x(t_n) + aN^* \leqslant e^{-x(t_n - \tau)}[x(t_n - \tau) + aN^*]$$

and so

$$\lim_{n \to \infty} e^{-x(t_n - \tau)}[x(t_n - \tau) + aN^*] = \infty. \tag{11.1.26}$$

Clearly, either $\lim_{n \to \infty} x(t_n - \tau) = \infty$ or there is a sub-sequence $\{t_{n_k}\}$ such that $\lim_{k \to \infty} x(t_{n_k} - \tau)$ exists and is a real number. In either case, (11.1.26) leads to a contradiction and the proof that $x(t)$ is bounded from above is complete. Now our strategy is to establish (11.1.25) in each of the following three cases.

Case 1: $x(t)$ is eventually non-negative. That is, there exists $t_0 \geq 0$ such that $x(t) \geq 0$ for $t \geq t_0$. Assume, for the sake of contradiction, that (11.1.25) is false. Then $\mu \equiv \lim \sup_{t \to \infty} x(t) \in (0, \infty)$ and there is a sequence of points $t_n \geq t_0$ such that $\lim_{n \to \infty} t_n = \infty$, $\mu = \lim_{n \to \infty} x(t_n)$ and

$$\dot{x}(t_n) \geq 0 \qquad \text{for } n = 1, 2, \ldots \tag{11.1.27}$$

Also clearly,

$$\lim_{n \to \infty} \sup x(t_n - \tau) \leq \mu. \tag{11.1.28}$$

It follows from (11.1.23) and (11.1.27) that

$$x(t_n) + aN^* \leq [x(t_n - \tau) + aN^*] e^{-x(t_n - \tau)} \leq x(t_n - \tau) + aN^*, \tag{11.1.29}$$

which together with (11.1.28) implies $\lim_{n \to \infty} x(t_n - \tau) = \mu$. Taking limits in (11.1.29), as $n \to \infty$, we obtain $\mu + aN^* \leq [\mu + aN^*] e^{-\mu} < \mu + aN^*$. This is a contradiction and the proof in this case is complete.

Case 2: $x(t)$ is eventually non-positive. More precisely, there exists $t_0 \geq 0$ such that $-aN^* < x(t) \leq 0$ for $t \geq t_0$. Assume, for the sake of contradiction, that (11.1.25) is false. Then $\lambda \equiv \lim \inf_{t \to \infty} x(t) < 0$ and there is a sequence of points $t_n \geq t_0$ such that

$$\lim_{n \to \infty} t_n = \infty, \qquad \lambda = \lim_{n \to \infty} x(t_n) \qquad \text{and} \qquad \dot{x}(t_n) \leq 0 \qquad \text{for } n = 1, 2, \ldots$$

Let

$$\varepsilon = \min_{t_0 \leq t \leq t_0 + \tau} [x(t) + aN^*].$$

Then in view of (11.1.24), $\varepsilon > 0$. Now we claim that

$$x(t) > -aN^* + \varepsilon/2 \qquad \text{for } t \geq t_0 + \tau. \tag{11.1.30}$$

Otherwise there exists a point $t_2 \geq t_0 + \tau$ such that $x(t_2) = -aN^* + \varepsilon/2$, $x(t) > -aN^* + \varepsilon/2$ for $t_0 + \tau \leq t < t_2$ and $\dot{x}(t_2) \leq 0$. Then (11.1.23) yields

$$\dot{x}(t_2) + \delta\varepsilon/2 = \delta e^{-x(t_2 - \tau)}[aN^* + x(t_2 - \tau)] > \delta e^{-x(t_2 - \tau)}\varepsilon/2 > \delta\varepsilon/2.$$

That is, $\dot{x}(t_2) > 0$ and this contradiction completes the proof of (11.1.30). In particular, (11.1.30) implies that

$$\lambda + aN^* > 0. \tag{11.1.31}$$

Now, from eqn (11.1.23) we find

$$x(t_n) + aN^* \geq e^{-x(t_n - \tau)}[x(t_n - \tau) + aN^*] \geq x(t_n - \tau) + aN^*. \tag{11.1.32}$$

Thus $x(t_n) \geqslant x(t_n - \tau)$, which implies that

$$\overline{\lim_{n \to \infty}} \, x(t_n - \tau) \leqslant \lambda \leqslant \varliminf_{n \to \infty} x(t_n - \tau),$$

so $\lim_{n \to \infty} x(t_n - \tau) = \lambda$. In view of (11.1.31), it follows from (11.1.32) that $\lambda + aN^* \geqslant e^{-\lambda}(\lambda + aN^*) > \lambda + aN^*$ and this contradiction completes the proof in Case 2.

Case 3: $x(t)$ oscillates about zero. Hence there exists a sequence of points $\{\xi_n\}$ such that $\tau < \xi_1 < \xi_2 < \cdots < \xi_n < \xi_{n+1} < \cdots$, $\lim_{n \to \infty} \xi_n = \infty$, $x(\xi_n) = 0$ for $n = 1, 2, \ldots$ and such that in each interval (ξ_n, ξ_{n+1}) the function $x(t)$ assumes both positive and negative values. Let t_n and s_n be points in (ξ_n, ξ_{n+1}) for $n = 1, 2, \ldots$ such that $x(t_n) = \max x(t)$ for $\xi_n \leqslant t \leqslant \xi_{n+1}$ and $x(s_n) = \min x(t)$ for $\xi_n \leqslant t \leqslant \xi_{n+1}$. Then for $n = 1, 2, \ldots, x(t_n) > 0$ and $\dot{x}(t_n) = 0$, while $x(s_n) < 0$ and $\dot{x}(s_n) = 0$. Also, $\mu \equiv \lim \sup_{t \to \infty} x(t) = \lim \sup_{n \to \infty} x(t_n)$ and $\lambda \equiv \lim \inf_{t \to \infty} x(t) = \lim \inf_{n \to \infty} x(s_n) \geqslant -aN^*$. Next, we claim that for each $n = 1, 2, \ldots,$

$$x(t) \text{ has a zero } T_n \in [\xi_n, t_n) \cap [t_n - \tau, t_n) \tag{11.1.33}$$

and

$$x(t) \text{ has a zero } S_n \in [\xi_n, s_n) \cap [s_n - \tau, s_n). \tag{11.1.34}$$

We prove (11.1.33). The proof of (11.1.34) is similar and is omitted. If (11.1.33) were false, then $\xi_n < t_n - \tau < \xi_{n+1}$ and $x(t_n - \tau) > 0$. As $\dot{x}(t_n) = 0$, eqn (11.1.23) yields

$$x(t_n) + aN^* = e^{-x(t_n - \tau)}[x(t_n - \tau) + aN^*] < x(t_n - \tau) + aN^*,$$

that is, $x(t_n) < x(t_n - \tau)$, which contradicts the definition of t_n. Multiplying both sides of (11.1.23) by $e^{\delta t}$ and then by integrating from T_n to t_n, we find

$$x(t_n) \, e^{\delta t} = -aN^*(e^{\delta t} - e^{\delta T_n}) + a\delta N^* \int_{T_n}^{t_n} e^{\delta s} \, e^{-x(s - \tau)} \, ds$$

$$+ \delta \int_{T_n}^{t_n} e^{\delta s} x(s - \tau) \, e^{-x(s - \tau)} \, ds. \tag{11.1.35}$$

Clearly, for any $\varepsilon > 0$ there exists n_0 such that

$$x(t) < \mu + \varepsilon \qquad \text{for } t \geqslant t_{n_0}. \tag{11.1.36}$$

Also,

$$x(s - \tau) \, e^{-x(s - \tau)} < \mu + \varepsilon \qquad \text{for } s \geqslant t_{n_0} \tag{11.1.37}$$

is obvious if $x(s - \tau)$ is negative, while if $x(s - \tau) \geqslant 0$ it follows by (11.1.36).

Now, in view of (11.1.37) and (11.1.24), eqn (11.1.35) yields

$$x(t_n)\, e^{\delta t_n} \leqslant -aN^*(e^{\delta t_n} - e^{\delta T_n}) + aN^*\, e^{aN^*}(e^{\delta t_n} - e^{\delta T_n}) + (\mu + \varepsilon)(e^{\delta t_n} - e^{\delta T_n})$$

or

$$x(t_n) \leqslant [1 - e^{-\delta(t_n - T_n)}][-aN^* + aN^*P/\delta + \mu + \varepsilon].$$

As ε is arbitrary and $0 < t_n - T_n \leqslant \tau$, we conclude that for $n \geqslant n_0$, $x(t_n) \leqslant (1 - e^{-\delta\tau})[aN^*[P/\delta - 1] + \mu]$, so $\mu \leqslant (1 - e^{-\delta\tau})[aN^*[P/\delta - 1] + \mu]$ or, after some simplifications, $\mu \leqslant aN^*(e^{\delta\tau} - 1)[P/\delta - 1]$. Set $K_1 = aN^*(e^{\delta\tau} - 1) \times [P/\delta - 1]$ and observe that in view of (11.1.18), $K_1 < aN^*$, so

$$e^{K_1} < e^{aN^*} = P/\delta. \tag{11.1.38}$$

Next, we prove that

$$\lambda = \liminf_{t \to \infty} x(t) \geqslant -K_1. \tag{11.1.39}$$

Indeed, by multiplying both sides of (11.1.23) by $e^{\delta t}$ and then by integrating from S_n to s_n, we find

$$x(s_n)\, e^{\delta s_n} = -aN^*(e^{\delta s_n} - e^{\delta S_n}) + \delta \int_{S_n}^{s_n} e^{\delta s}\, e^{-x(s-\tau)}[aN^* + x(s - \tau)]\, ds.$$

As $aN^* + x(s - \tau) > 0$, it follows that for $\varepsilon > 0$ and n sufficiently large, we have $e^{-x(s-\tau)}[aN^* + x(s - \tau)] \geqslant e^{-(K_1 + \varepsilon)}(aN^* + \lambda - \varepsilon)$, $S_n \leqslant s \leqslant s_n$, so

$$x(s_n)\, e^{\delta s_n} \geqslant -aN^*(e^{\delta s_n} - e^{\delta S_n}) + e^{-(K_1 + \varepsilon)}(aN^* + \lambda - \varepsilon)(e^{\delta s_n} - e^{\delta S_n})$$

or

$$0 < -x(s_n) \leqslant [aN^* - e^{-(K_1 + \varepsilon)}(aN^* + \lambda - \varepsilon)][1 - e^{-\delta(s_n - S_n)}]$$

and so

$$\lambda \geqslant -[aN^* - e^{-K_1}(aN^* + \lambda)](1 - e^{-\delta\tau})$$

or, after some simplifications,

$$-\lambda \leqslant \frac{aN^*(e^{K_1} - 1)(e^{\delta\tau} - 1)\, e^{-K_1}\, e^{-\delta\tau}}{1 - e^{-K_1}(1 - e^{-\delta\tau})}.$$

But

$$e^{-K_1}\, e^{-\delta\tau} \leqslant 1 - e^{-K_1}(1 - e^{-\delta\tau}),$$

hence

$$-\lambda \leqslant aN^*(e^{K_1} - 1)(e^{\delta\tau} - 1) \leqslant aN^*(e^{\delta\tau} - 1)[P/\delta - 1] = K_1$$

and the proof of (11.1.39) is complete.

So far we have established that

$$-K_1 \leqslant \lambda \leqslant \mu \leqslant K_1. \tag{11.1.40}$$

It follows by induction that there is a sequence of positive constants $\{K_n\}$ such that

$$-K_n \leqslant \lambda \leqslant \mu \leqslant K_n, \qquad n = 1, 2, \ldots, \tag{11.1.41}$$

$$0 < K_n \leqslant aN^* \tag{11.1.42}$$

and

$$K_{n+1} \equiv aN^*(e^{\delta\tau} - 1)(e^{K_n} - 1). \tag{11.1.43}$$

Our goal is to show that

$$\lambda = \mu = 0 \tag{11.1.44}$$

which, in view of (11.1.41), will be true if we show that some sub-sequence of K_n converges to zero. Assume, for the sake of contradiction, that no sub-sequence of $\{K_n\}$ converges to zero. Then

$$m = \liminf_{n \to \infty} K_n > 0 \tag{11.1.45}$$

and so also $M \equiv \limsup_{n \to \infty} K_n > 0$. From (11.1.42) and (11.1.45) we have $0 < m \leqslant M \leqslant aN^*$. Let $\{\rho_n\}$ and $\{\sigma_n\}$ be such that $m = \lim_{n \to \infty} K_{\rho_n}$ and $M = \lim_{n \to \infty} K_{\sigma_n}$. Then from (11.1.43) and for any $\varepsilon > 0$ we find, for n sufficiently large, that

$$m - \varepsilon \leqslant K_{\rho_n + 1} = aN^*(e^{\delta\tau} - 1)(e^{K_{\rho_n}} - 1)$$

and

$$M + \varepsilon \geqslant K_{\sigma_n + 1} = aN^*(e^{\delta\tau} - 1)(e^{K_{\sigma_n}} - 1).$$

By taking limits, as $n \to \infty$, and using the fact that ε is arbitrary, we obtain

$$m \leqslant aN^*(e^{\delta\tau} - 1)(e^m - 1) \tag{11.1.46}$$

and

$$M \geqslant aN^*(e^{\delta\tau} - 1)(e^M - 1). \tag{11.1.47}$$

Set $F(\lambda) = aN^*(e^{\delta\tau} - 1)(e^\lambda - 1) - \lambda$ and observe that $F(0) = 0$, $\lim_{\lambda \to \infty} F(\lambda) = \infty$ and, by (11.1.18),

$$F(aN^*) = aN^*(e^{\delta\tau} - 1)(e^{aN^*} - 1) - aN^* = aN^*[(e^{\delta\tau} - 1)[P/\delta - 1] - 1] < 0.$$

Thus $F(\lambda) = 0$ has a solution in the interval (aN^*, ∞). On the other hand,

$$F'(\lambda) = aN^*(e^{\delta\tau} - 1)\, e^{\lambda} - 1$$

has exactly one zero in $(-\infty, \infty)$ and so, by Rolle's theorem, $F(\lambda)$ cannot have more than two zeros in $(-\infty, \infty)$. In particular, $\lambda = 0$ is the only zero in $F(\lambda)$ in $[0, aN^*]$. But (11.1.46) and (11.1.47) imply that $F(m) \geqslant 0$ and $F(M) \leqslant 0$, that is, $F(\lambda)$ has a zero in $[m, M] \subseteq (0, aN^*]$. This is a contradiction and the proof of (11.1.44) is complete. The proof of the theorem is also complete.

11.2 Global attractivity in models of haematopoiesis

Mackey and Glass (1977) have proposed the equations

$$\dot{P}(t) = \frac{\beta_0 \Theta^n}{\Theta^n + [P(t-\tau)]^n} - \gamma P(t), \qquad (11.2.1)$$

$$\dot{P}(t) = \frac{\beta_0 \Theta^n P(t-\tau)}{\Theta^n + [P(t-\tau)]^n} - \gamma P(t) \qquad (11.2.2)$$

as models of haematopoiesis (blood cell production). In these equations, $P(t)$ denotes the density of mature cells in blood circulation, τ is the time delay between the production of immature cells in the bone marrow and their maturation for release in the circulating blood stream, and β_0, Θ, n, and γ are positive constants. In eqn (11.2.1) the production is a monotonic decreasing function of $P(t - \tau)$, while in eqn (11.2.2) the production is a single-humped function of $P(t - \tau)$. Details of the derivation of (11.2.1) and (11.2.2) can be found in Mackey and Glass (1977) and Mackey and an der Heiden (1982), where also the local asymptotic stability of the positive equilibrium of eqns (11.2.1) and (11.2.2) has been studied by the well-known technique of linearization. For more details of eqns (11.2.1) and (11.2.2) we refer to Glass and Mackey (1979).

Our aim in this section is to obtain sufficient conditions for the positive equilibrium of eqns (11.2.1) and (11.2.2) to be a global attractor. Such stability results are particularly important for eqn (11.2.2), which for some values of its parameters has been numerically observed to exhibit chaotic behaviour (see Mallet-Paret and Nussbaum 1986).

The change of variables

$$P(t) = \Theta x(t) \qquad (11.2.3)$$

transforms eqn (11.2.1) to the delay differential equation

$$\dot{x}(t) = \frac{\beta}{1 + [x(t-\tau)]^n} - \gamma x(t), \qquad (11.2.4)$$

where $\beta = (\beta_0/\Theta)$. Throughout this section we assume that β, n, and γ are positive constants while τ is a non-negative constant. Equation (11.2.4) has a unique positive equilibrium K, and K satisfies the equation

$$\frac{\beta}{1 + K^n} = \gamma K.$$

With equation (11.2.4) one associates an initial function of the form

$$x(t) = \phi(t) \quad \text{for } -\tau \leqslant t \leqslant 0, \quad \text{where } \phi \in C[[-\tau, 0], \mathbb{R}^+] \text{ and } \phi(0) > 0.$$

$$(11.2.5)$$

We first establish the following result.

Lemma 11.2.1.

(a) *The IVP (11.2.4) and (11.2.5) has a unique solution $x(t)$ which exists and is positive for all $t \geqslant 0$.*

(b) *If $x(t)$ does not oscillate about K then it converges to K, as $t \to +\infty$.*

(c) *If $x(t)$ oscillates about K then there exists $t_1 \geqslant 0$ such that*

$$K - \gamma(K + \beta\tau)\tau \leqslant x(t) \leqslant K + \beta\tau \quad \text{for } t \geqslant t_1. \quad (11.2.6)$$

Proof.

(a) The existence and the positivity of the solution follows by an argument similar to that in Theorem 1.1.5.

(b) Assume that $x(t)$ does not oscillate about K. Then there exists a $T \geqslant 0$ such that for all $t \geqslant T - \tau$ either $x(t) > K$ or $x(t) < K$. But eqn (11.2.4) can be written in the form

$$\dot{x}(t) = \gamma \left[K \frac{1 + K^n}{1 + (x(t - \tau))^n} - x(t) \right],$$

and hence $x(t)$ is strictly increasing if $x(t) < K$, and $x(t)$ is strictly decreasing if $x(t) > K$. Thus $L = \lim_{t \to +\infty} x(t)$ exists and $K = L$, for otherwise, $\lim_{t \to +\infty} \dot{x}(t) \neq 0$, which would imply that $x(t)$ is unbounded. The proof of statement (b) is complete.

(c) Assume that $x(t)$ is oscillatory about K. Then without loss of generality we can assume that $\tau > 0$. Let $\tau \leqslant t_1 < t_2 < \cdots < t_m < \cdots$ be a sequence of zeros of $x(t) - K$ with $t_m \to \infty$, as $m \to \infty$. Our strategy is to show that (11.2.6) holds in each interval (t_m, t_{m+1}) for $m = 1, 2, \ldots$ Let $s_m \in (t_m, t_{m+1})$ be a point where $x(t)$ obtains its maximum or its minimum in (t_m, t_{m+1}). It suffices to show that (11.2.6) holds for $t = s_m$. As $\dot{x}(s_m) = 0$, it follows from eqn (11.2.4) that

$$\frac{1}{1 + (x(s_m - \tau))^n} = \frac{1}{1 + K^n} \frac{x(s_m)}{K},$$

which implies that $(x(s_m) - K)(x(s_m - \tau) - K) < 0$. Therefore there exists a point $\xi_m \in (s_m - \tau, s_m)$ such that $x(\xi_m) = K$. By integrating both sides of (11.2.4) from ξ_m to s_m we find

$$x(s_m) - x(\xi_m) = \int_{\xi_m}^{s_m} \left(\frac{\beta}{1 + (x(s - \tau))^n} - \gamma x(s) \right) ds \leqslant \beta(s_m - \xi_m) \leqslant \beta\tau,$$

$$(11.2.7)$$

which shows that

$$x(s_m) \leqslant K + \beta\tau \qquad \text{for } t \geqslant t_1. \tag{11.2.8}$$

Now by using (11.2.8) in (11.2.7) we find

$$x(s_m) - K > \int_{\xi_m}^{s_m} \left[\frac{\beta}{1 + (K + \beta\tau)^n} - \gamma(K + \beta\tau) \right] ds \geqslant \int_{\xi_m}^{s_m} -\gamma(K + \beta\tau) \, ds$$

$$\geqslant -\gamma(K + \beta\tau)\tau,$$

which shows that

$$K - \gamma(K + \beta\tau)\tau < x(\xi_m).$$

The proof of Lemma 11.2.1 is complete.

In the next theorem we make use of the following lemma.

Lemma 11.2.2. *Consider the delay differential equation*

$$\dot{y}(t) + f(t, y(t)) + g(t, y(t), y(t - \tau)) = 0, \qquad t \geqslant t_0 \tag{11.2.9}$$

where $\tau > 0$ and for some $D \subseteq \mathbb{R}$,

$$f \in C[[t_0, \infty) \times D, \mathbb{R}] \qquad and \qquad g \in C[[t_0, \infty) \times D \times D, \mathbb{R}].$$

Assume that there exists an open set D_0 in D and constants

$$p_1, p_2, q_1, q_2 \in \mathbb{R}^+$$

such that for all $y, z \in D_0 - \{0\}$,

$$p_1 \leqslant \liminf_{t \to +\infty} \frac{f(t, y)}{y} \leqslant \limsup_{t \to +\infty} \frac{f(t, y)}{y} \leqslant p_2 \tag{11.2.10}$$

$$q_1 \leqslant \liminf_{t \to +\infty} \frac{g(t, y, z)}{z} \leqslant \limsup_{t \to +\infty} \frac{g(t, y, z)}{z} \leqslant q_2 \tag{11.2.11}$$

$$(p_2 + q_2)q_2\tau < p_1 + q_1. \tag{11.2.12}$$

Then every solution $y: [t_0 - \tau, \infty) \to \mathbb{R}$ of eqn (11.2.9) which satisfies the *relation*

$$\liminf_{t \to +\infty} y(t) \quad \text{and} \quad \limsup_{t \to +\infty} y(t) \in D_0 \qquad (11.2.13)$$

tends to zero as $t \to +\infty$.

Proof. Assume that $y: [t_0 - \tau, \infty) \to \mathbb{R}$ is a solution of eqn (11.2.9) such that (11.2.13) is satisfied. We define the following two functions for $t \geq t_0$:

$$p(t) = \begin{cases} \dfrac{f(t, y(t))}{y(t)} & \text{if } y(t) \neq 0 \\ p_1 & \text{if } y(t) = 0 \end{cases}$$

$$q(t) = \begin{cases} \dfrac{g(t, y(t), y(t - \tau))}{y(t - \tau)} & \text{if } y(t - \tau) \neq 0 \\ q_1 & \text{if } y(t - \tau) = 0. \end{cases}$$

Then clearly $y(t)$ satisfies the linear equation

$$\dot{y}(t) + p(t)y(t) + q(t)y(t - \tau) = 0, \qquad t \geq t_0. \qquad (11.2.14)$$

Since the functions $f(t, y(t))$, $g(t, y(t - \tau))$ and $y(t)$ are continuous on $[t_0, \infty)$ and (11.2.13) is satisfied, it can be easily seen that there exists a $T \geq t_0$ such that $p(t)$ and $q(t)$ are continuous for almost every $t \geq T$. Moreover, (11.2.10), (11.2.11), and (11.2.12) yield that there exist $\tilde{p}_1, \tilde{p}_2, \tilde{q}_1, \tilde{q}_2 \in \mathbb{R}_+$ such that

$$\tilde{p}_1 \leq p(t) \leq \tilde{p}_2 \quad \text{and} \quad \tilde{q}_1 \leq q(t) \leq \tilde{q}_2 \quad \text{for all } t \geq T, \quad (11.2.15)$$

$$(\tilde{p}_2 + \tilde{q}_2)\tilde{q}_2\tau < \tilde{p}_1 + \tilde{q}_1. \qquad (11.2.16)$$

Thus $p(t)$ and $q(t)$ are locally Lebesgue integrable functions. We define a Lyapunov-like non-negative functional $V(t)$ by the formula

$$V(t) = \left[y(t) - \int_{t-\tau}^{t} q(s + \tau)y(s) \, ds \right]^2$$

$$+ \int_{t-\tau}^{t} \left\{ [p(s + \tau) + q(s + 2\tau)] \int_{s}^{t} q(u + \tau)y^2(u) \, du \right\} ds.$$

Then $V(t)$ is an absolutely continuous function and for a.e. $t \geq T$,

$$\frac{dV(t)}{dt} = -2 \left[y(t) - \int_{t-\tau}^{t} q(s + \tau)y(s) \, ds \right] [p(t) + q(t + \tau)]y(t)$$

$$- [p(t) + q(t + \tau)] \int_{t-\tau}^{t} q(u + \tau)y^2(u) \, du$$

$$+ q(t + \tau)y^2(t) \int_{t-\tau}^{t} [p(s + \tau) + q(s + 2\tau)] \, ds,$$

where we used eqn (11.2.14).

By using the inequality

$$2y(t)\,y(s) \leqslant y^2(t) + y^2(s)$$

we obtain

$$\frac{dV(t)}{dt} \leqslant y^2(t)\bigg\{-2[p(t) + q(t + \tau)] + [p(t) + q(t + \tau)] \int_{t-\tau}^t q(s + \tau)\,ds$$

$$+ q(s + \tau) \int_{t-\tau}^t [p(s + \tau) + q(s + 2\tau)]\,ds\bigg\} \qquad \text{a.e. } t \geqslant T.$$

Now by virtue of (11.2.15) we find

$$\frac{dV(t)}{dt} \leqslant y^2(t)[-2(\tilde{p}_1 + \tilde{q}_1) + (\tilde{p}_2 + \tilde{q}_2)\tilde{q}_2\tau + \tilde{q}_2(\tilde{p}_2 + \tilde{q}_2)\tau]$$

$$= -2[(\tilde{p}_1 + \tilde{p}_2) - (\tilde{p}_2 + \tilde{q}_2)\tilde{q}_2\tau]y^2(t) \qquad \text{a.e. } t \geqslant T.$$

By integrating both sides from T to t we see that

$$V(t) + 2[(\tilde{p}_1 + \tilde{p}_2) - (\tilde{p}_1 + \tilde{q}_2)\tilde{q}_2\tau] \int_T^t y^2(s)\,ds \leqslant V(T)$$

and so (11.2.16) yields that $V(t)$ is bounded on $[T, \infty)$ and $y^2(t) \in L^1[T, \infty)$. Since $V(t)$ is bounded on $[T, \infty)$, one can easily see that there exists a constant $c_1 \geqslant 0$ such that

$$|y(t)| \leqslant c_1 + \int_{t-\tau}^t q(s + \tau)|y(s)|\,ds \leqslant c_1 + \tilde{q}_2\tau \max_{t-\tau \leqslant s \leqslant t} |y(s)|$$

for all $t \geqslant T$. But (11.2.16) yields that $\tilde{q}_2\tau < 1$ and hence by Lemma 1.5.3 we find that $y(t)$ is a bounded function on $[t_0, \infty)$. Therefore by virtue of (11.2.14) and (11.2.15) we find that $\dot{x}(t)$ is a bounded function on $[T, \infty)$. Thus the function $z(t) = y^2(t)$ is a non-negative and integrable function on $[T, \infty)$. Moreover, there exists an $M > 0$ such that

$$|z(t_2) - z(t_1)| \leqslant M|t_2 - t_1| \qquad \text{for all } t_1, t_2 \geqslant T.$$

Now for any $t \geqslant T$ and $\delta > 0$, we have

$$z(t) = \frac{1}{\delta} \int_t^{t+\delta} z(s)\,ds - \frac{1}{\delta} \int_t^{t+\delta} [z(s) - z(t)]\,ds$$

and clearly

$$\limsup_{t \to +\infty} z(t) \leqslant \frac{1}{\delta} \limsup_{t \to +\infty} \bigg(\int_t^{t+\delta} z(s)\,ds + \frac{M}{\delta} \int_t^{t+\delta} (s - t)\,ds \bigg) = \frac{M}{2}\,\delta.$$

Since $\delta > 0$ is an arbitrary number, this yields that $z(t) \to 0$, as $t \to +\infty$. The proof of the lemma is complete.

Next, we turn our attention to the problem of the global attractivity of the positive equilibrium K of eqn (11.2.1). As we proved in Lemma 11.2.1(b), every positive solution of eqn (11.2.4) that is not oscillatory about K converges to the equilibrium K. Thus when $\tau = 0$, every positive solution of eqn (11.2.4) does not oscillate about K and tends to the equilibrium K. Naturally, we expect the same behaviour for small delays τ.

Theorem 11.2.1. *Set*

$$L = K - \gamma(K + \beta\tau)\tau, \qquad U = K + \beta\tau,$$

$$q_1 = \frac{\beta n L^{n-1}}{(1 + U^n)^2} \qquad and \qquad q_2 = \frac{\beta n U^{n-1}}{(1 + L^n)^2}. \qquad (11.2.7)$$

Assume that

$$L > 0 \qquad and \qquad (\gamma + q_2)q_2\tau < \gamma + q_1. \qquad (11.2.18)$$

Then every positive solution $x(t)$ of eqn (11.2.4) satisfies

$$\lim_{t \to +\infty} x(t) = K, \qquad (11.2.19)$$

that is K is a global attractor of the positive solutions of (11.2.4).

Proof. We have already established (11.2.19) for the solutions of (11.2.4) that are not oscillatory about K. Therefore it remains to establish (11.2.19) for the solutions $x(t)$ of eqn (11.2.4) that oscillate about K. In view of Lemma 11.2.1(c),

$$L \leqslant x(t) \leqslant U \qquad \text{for } t \geqslant t_1.$$

Set

$$y(t) = x(t) - K,$$

$$f(t, y) = -\mu y \qquad and \qquad g(t, y, z) = \frac{\beta}{1 + (z + K)^n} - \gamma K$$

for all $t \geqslant t_0$ and $y, z \in D$, where $D = \{u: u \geqslant -K\}$. Then it is clear that $y(t)$ satisfies equation (11.2.9) and $f(t, y)$ and $g(t, y, z)$ satisfy (11.2.10) and (11.2.11) with $p_1 = p_2 = \gamma$ and q_1 and q_2 defined in (11.2.17). Thus the conclusion follows by applying Lemma 11.2.2.

Next, we turn our attention to the problem of the global attractivity of the positive equilibrium of eqn (11.2.2). The change of variables

$$P(t) = \Theta x(t) \tag{11.2.20}$$

transforms eqn (11.2.2) to the delay differential equation

$$\dot{x}(t) = \beta_0 \frac{x(t-\tau)}{1 + (x(t-\tau))^n} - \gamma x(t). \tag{11.2.21}$$

Throughout this section we assume that

$$\beta_0 > \gamma > 0 \quad \text{and} \quad n, \tau \in \mathbb{R}^+. \tag{11.2.22}$$

Then eqn (11.2.21) has a unique positive equilibrium M, and M is given by

$$M = \left(\frac{\beta_0 - \gamma}{\gamma}\right)^{1/n} \tag{11.2.23}$$

With eqn (11.2.21) one associates an initial function of the form

$$x(t) = \phi(t) \quad \text{for } -\tau \leqslant t \leqslant 0 \quad \text{where } \phi \in C[[-\tau, 0], \mathbb{R}^+] \text{ and } \phi(0) > 0. \tag{11.2.24}$$

We should point out that eqns (11.2.21) and (11.2.4) are qualitatively different. This is a consequence of the fact that the non-linearity in eqn (11.2.21) is a so-called single-humped function of $x(t - \tau)$ which can lead to chaotic dynamics. On the other hand, the non-linearity in eqn (11.2.4) is a decreasing function of $x(t - \tau)$ which usually gives rise to simpler dynamics.

As in the case of eqn (11.2.1), we can now establish results analogous to Lemma 11.2.1 and Theorem 11.2.1. Their proofs are similar to those given above and for economy we present only their statements.

Lemma 11.2.2.
 (a) *Assume that*

$$\tau > 0, \quad n > 1, \quad \text{and} \quad \beta_0 > \gamma. \tag{11.2.25}$$

Then IVP (11.2.21) and (11.2.23) has a unique solution $x(t)$ which exists and is positive for all $t \geqslant 0$.
 (b) *If $x(t)$ is non-oscillatory about M then it tends to M as $t \to +\infty$.*
 (c) *If*

$$(\beta_0 + \gamma)\tau < 1 \tag{11.2.26}$$

and $x(t)$ is oscillatory about M then there exists $t_1 \geqslant 0$ such that

$$\frac{M(1 - (\beta_0 + \gamma)\tau)}{1 - \beta_0\tau} \leqslant x(t) \leqslant \frac{M}{1 - \beta_0\tau} \quad \text{for } t \geqslant t_1.$$

Remark 11.2.1. *The transformation*

$$x(t) = M + y(t)$$

reduces eqn (11.2.21) to

$$\dot{y}(t) + \gamma y(t) + \beta_0 \left(\frac{M}{1 + M^n} - \frac{M + y(t - \tau)}{1 + (M + y(t - \tau))^n} \right) = 0.$$

Theorem 11.2.2. *Assume that (11.2.24) and (11.2.25) hold. Let*

$$L = \frac{M(1 - (\beta + \gamma)\tau)}{1 - \beta\tau}, \qquad U = \frac{M}{1 - \beta\tau},$$

$$q = \frac{(n - 1)L^n - 1}{(1 + U^n)^2} \qquad and \qquad q_2 = \frac{(n - 1)U^n - 1}{(1 + L^n)^2}.$$

Assume that

$$(n - 1)L^n - 1 > 0 \qquad and \qquad (\gamma + q_2)q_2\tau < \gamma + q_1.$$

Then every positive solution $x(t)$ of eqn (11.2.21) satisfies $\lim_{t \to +\infty} x(t) = M$.

11.3 Global attractivity in a discrete delay logistic model

Our aim in this section is to establish the following global attractivity result about the discrete delay logistic model that we studied in Section 7.9, when the delay k in eqn (7.9.1) is equal to 1.

Theorem 11.3.1. *Assume that*

$$\alpha \in (1, \infty) \qquad and \qquad \beta \in (0, \infty).$$

Then every solution of the difference equation

$$x_{n+1} = \frac{\alpha x_n}{1 + \beta x_{n-1}}, \qquad n = 0, 1, 2, \ldots \qquad (11.3.1)$$

with

$$x_{-1} \geqslant 0 \qquad and \qquad x_0 > 0 \qquad (11.3.2)$$

converges to the positive equilibrium $(\alpha - 1)/\beta$.

Proof. The change of variables

$$x_n = \frac{\alpha - 1}{\beta} + \frac{y_n}{\beta} \qquad \text{for } n \geqslant -1$$

transforms eqn (11.3.1) to

$$y_{n+1} = \frac{\alpha y_n - (\alpha - 1)y_{n-1}}{\alpha + y_{n-1}}, \qquad n = 0, 1, 2, \ldots \tag{11.3.3}$$

where, in view of (11.3.2),

$$\alpha + y_{-1} \geqslant 1 \qquad \text{and} \qquad \alpha + y_n > 1 \qquad \text{for } n \geqslant 0. \tag{11.3.4}$$

Note also that for $n \geqslant 0$,

$$y_{n+1} = \frac{\alpha(y_n - y_{n-1}) + y_{n-1}}{\alpha + y_{n-1}}, \tag{11.3.5}$$

$$y_{n+1} - y_n = -\frac{(\alpha - 1) + y_n}{\alpha + y_{n-1}} y_{n-1}, \tag{11.3.6}$$

$$y_{n+2} = \frac{\alpha y_n}{(\alpha + y_{n-1})(\alpha + y_n)} - \frac{(\alpha - 1)y_{n-1}}{\alpha + y_{n-1}}. \tag{11.3.7}$$

The proof will be complete if we show that

$$\lim_{t \to \infty} y_n = 0. \tag{11.3.8}$$

In view of (11.3.4), by using (11.3.6), we see that (11.3.8) holds for every solution of eqn (11.3.3) that is eventually non-negative or eventually non-positive.

Therefore it remains to establish (11.3.8) for every solution of eqn (11.3.3) that is 'strictly' oscillatory, in the sense that it attains both positive and negative values. Such a solution consists of a 'string' of consecutively negative terms followed by a string of consecutively non-negative terms, etc. We shall call these strings *negative semicycles* and *positive semicycles*. By using (11.3.5) and (11.3.6) we see that every semicycle contains at least three terms and that the third term in a positive semicycle is positive. Also the maximum in a positive semicycle and the minimum in a negative semicycle is equal to the value of the second term in the respective semicycle. Now to complete the proof, consider four consecutive semicycles

$$C_{r-1}, C_r, C_{r+1}, C_{r+2},$$

with C_{r-1} being a negative semicycle as follows:

$$C_{r-1} = \{y_{k+1}, y_{k+2}, \ldots, y_l\} \qquad \text{negative semicycle}$$
$$C_r = \{y_{l+1}, y_{l+2}, \ldots, y_m\} \qquad \text{positive semicycle}$$
$$C_{r+1} = \{y_{m+1}, y_{m+2}, \ldots, y_n\} \qquad \text{negative semicycle}$$
$$C_{r+2} = \{y_{n+1}, y_{n+2}, \ldots, y_o\} \qquad \text{positive semicycle.}$$

We establish the following estimates from which the proof of (11.3.8) will become obvious:

$$|y_{m+2}| \leqslant \frac{(\alpha - 1)^2}{\alpha^2 - \alpha + 1} |y_{k+2}|, \tag{11.3.9}$$

$$y_{n+2} \leqslant \left(\frac{\alpha - 1}{\alpha}\right)^2 y_{l+2}. \tag{11.3.10}$$

To this end, observe that from (11.3.7),

$$y_{l+2} \leqslant -\frac{(\alpha - 1)y_{l-1}}{\alpha + y_{l-1}}$$

and because the function $x/(\alpha + x)$ is strictly increasing in x,

$$y_{l+2} \leqslant \frac{(\alpha - 1)|y_{k+2}|}{\alpha - |y_{k+2}|}. \tag{11.3.11}$$

Also, by using (11.3.7) and (11.3.11), we find

$$|y_{m+2}| < \frac{(\alpha - 1)y_{m-1}}{\alpha + y_{m-1}} \leqslant \frac{(\alpha - 1)y_{l+2}}{\alpha + y_{l+2}}$$

$$\leqslant \left((\alpha - 1)\frac{(\alpha - 1)|y_{k+2}|}{\alpha - |y_{k+2}|}\right) \bigg/ \left(\alpha + \frac{(\alpha - 1)|y_{k+2}|}{\alpha - |y_{k+2}|}\right) = \frac{(\alpha - 1)^2 |y_{k+2}|}{\alpha^2 - |y_{k+2}|}. \tag{11.3.12}$$

As $|y_{k+2}| < \alpha - 1$, it follows that

$$|y_{m+2}| \leqslant \frac{(\alpha - 1)^2}{\alpha^2 - \alpha + 1} |y_{k+2}|$$

and the proof of (11.3.9) is complete. In a similar way we see that

$$y_{n+2} \leqslant -\frac{(\alpha - 1)y_{n-1}}{\alpha + y_{n-1}} \leqslant \frac{(\alpha - 1)|y_{m+2}|}{\alpha - |y_{m+2}|},$$

and by using (11.3.12) and the increasing nature of the function $x/(\alpha - x)$, we find

$$
y_{n+2} \leqslant \left((\alpha - 1)\frac{(\alpha - 1)y_{l+2}}{\alpha + y_{l+2}} \right) \Bigg/ \left(\alpha - \frac{(\alpha - 1)y_{l+2}}{\alpha + y_{l+2}} \right)
$$

$$
= \frac{(\alpha - 1)^2 y_{l+2}}{\alpha^2 + y_{l+2}} < \frac{(\alpha - 1)^2}{\alpha^2} y_{l+2},
$$

which proves (11.3.10) and completes the proof of the theorem.

11.4 Global stability of solutions of linear non-autonomous difference equations

Our aim in this section is to obtain global stability results for linear difference equations of the form

$$
x_{n+1} - x_n + \sum_{i=1}^{m} p_i(n)x_{n-k_i} = 0, \qquad n = 0, 1, 2, \ldots, \qquad (11.4.1)
$$

where

$$
k_i \in \mathbb{N} \quad \text{and} \quad p_i(n) \in \mathbb{R} \quad \text{for } i = 1, 2, \ldots, m \text{ and } n \in \mathbb{N}. \quad (11.4.2)
$$

By investigating the asymptotic behaviour first of the non-oscillatory solutions and then of the oscillatory solutions of eqn (11.4.1), we obtain sufficient conditions for the global asymptotic stability of all solutions of eqn (11.4.1). We may think of eqn (11.4.1) as being a discrete analogue of the delay differential equation

$$
\dot{x}(t) + \sum_{i=1}^{m} p_i(t)x(t - \tau_i) = 0 \qquad (11.4.3)
$$

and in fact the results in this section have similar extensions to solutions of eqn (11.4.3).

Without loss of generality, we assume throughout this section that

$$
0 \leqslant k_1 < k_2 < k_3 < \cdots < k_m. \qquad (11.4.4)
$$

The first result gives sufficient conditions for the non-oscillatory solutions of eqn (11.4.1) to tend to zero as $n \to \infty$.

Lemma 11.4.1. *Assume that there exists a positive constant c such that for n* *sufficiently large,*

$$\sum_{i=1}^{m} p_i(n + k_i - k_m) \geqslant c \tag{11.4.5}$$

and

$$\sum_{i=1}^{m-1} \left(\sum_{j=n-k_m}^{n-k_i-1} p_i(j + k_i)_- \right) \leqslant 1, \tag{11.4.6}$$

where $p_i(j + k_i)_- = \max\{-p_i(j + k_i), 0\}$ *for* $i = 1, 2, \ldots, m - 1$. *Then every non-oscillatory solution of eqn* (11.4.1) *tends to zero as* $n \to \infty$.

Proof. Let $\{x_n\}$ be a non-oscillatory solution of eqn (11.4.1). Without loss of generality we assume that $\{x_n\}$ is an eventually positive solution. Set

$$z_n = x_n + \sum_{i=1}^{m-1} \left(\sum_{j=n-k_m}^{n-k_i-1} p_i(j + k_i)x_j \right). \tag{11.4.7}$$

Then, in view of (11.4.7), for n sufficiently large we have

$$z_{n+1} - z_n = -\left(\sum_{i=1}^{m} p_i(n - k_m + k_i) \right) x_{n-k_m} \leqslant 0. \tag{11.4.8}$$

Hence $\{z_n\}$ is eventually a strictly decreasing sequence. Set $l = \lim_{n \to \infty} z_n$. We claim that l is finite. Otherwise $l = -\infty$. From (11.4.7) it follows that

$$z_n \geqslant -\sum_{i=1}^{m-1} \left(\sum_{j=n-k_m}^{n-k_i-1} p_i(j + k_i)_- x_j \right),$$

which in view of (11.4.6) implies that $\{x_n\}$ is unbounded. Therefore we can choose an $n_0 \geqslant k_m$ such that (11.4.5) and (11.4.6) are satisfied for $n \geqslant n_0$, $z_{n_0} < 0$, and

$$x_{n_0} = \max_{0 \leqslant j \leqslant n_0} x_j.$$

By using (11.4.7) we see that

$$0 > z_{n_0} = x_{n_0} + \sum_{i=1}^{m-1} \left(\sum_{j=n_0-k_m}^{n_0-k_i-1} p_i(j + k_i)x_j \right)$$

$$\geqslant x_{n_0} - x_{n_0} \sum_{i=1}^{m-1} \left(\sum_{j=n_0-k_m}^{n_0-k_i-1} p_i(j + k_i)_- \right)$$

$$= x_{n_0}\left(1 - \sum_{i=1}^{m-1} \left(\sum_{j=n_0-k_m}^{n_0-k_i-1} p_i(j + k_i)_- \right) \right) \geqslant 0.$$

This is a contradiction and so our claim that l is finite is true. Now by taking limits on both sides of (11.4.8) and by using (11.4.5), we see that

$$0 = \lim_{n \to \infty} z_{n+1} - z_n \leqslant -c \lim_{n \to \infty} x_{n-k_m} \leqslant 0.$$

This implies that $\lim_{n \to \infty} x_n = 0$ and the proof is complete.

The next result gives a sufficient condition for the boundedness of all solutions of eqn (11.4.1).

Theorem 11.4.1. *Assume that eventually*

$$\sum_{i=1}^{m} p_i(n + k_i - k_m) > 0, \tag{11.4.9}$$

$$\sum_{i=1}^{m-1} \left(\sum_{j=n-k_m}^{n-k_i-1} |p_i(j + k_i)| \right) \leqslant q_1, \tag{11.4.10}$$

$$\sum_{j=n-k_m}^{n} \left(\sum_{i=1}^{m} p_i(j + k_i - k_m) \right) \leqslant q_2, \tag{11.4.11}$$

where q_1, q_2 are positive constants such that

$$2q_1 + q_2 < 1. \tag{11.4.12}$$

Then every solution of eqn (11.4.1) is bounded.

Proof. Assume, for the sake of contradiction, that eqn (11.4.1) has an unbounded solution $\{x_n\}$. Then we can choose an $n_0 \geqslant k_m$ such that (11.4.9)–(11.4.11) hold for $n \geqslant n_0$ and also

$$\max_{n_0 \leqslant j \leqslant n} |x_j| \geqslant \max_{n-k_m \leqslant j \leqslant n-k_i-1} |x_j|.$$

It follows from (11.4.7) that

$$|z_n| \geqslant |x_n| - \sum_{i=1}^{m-1} \left(\sum_{j=n-k_m}^{n-k_i-1} |p_i(j + k_i)| |x_j| \right) \geqslant |x_n| - (\max|x_j|)q_1.$$

Hence

$$\max_{n_0 \leqslant j \leqslant n} |z_j| \geqslant (1 - q_1) \max_{n_0 \leqslant j \leqslant n} |x_j| \qquad \text{for } n \geqslant n_0, \tag{11.4.13}$$

which shows that $\{z_n\}$ is also unbounded. Therefore there exists a subsequence n_s of positive integers such that

$$\lim_{s \to \infty} n_s = \infty, \qquad \lim_{s \to \infty} |z_{n_s}| = \infty, \qquad n_s \geqslant n_0, \qquad \text{and} \qquad |z_{n_s}| \geqslant \max_{n_0 \leqslant j \leqslant n} |z_j|.$$

$$\tag{11.4.14}$$

From (11.4.7) we find

$$z_{n_s} - z_{n_s-1} = -\left(\sum_{i=1}^{m} p_i(n_s - k_m + k_i - 1)\right)x_{n_s-k_m-1}, \qquad (11.4.15)$$

which implies that

$$(z_{n_s} - z_{n_s-1})x_{n_s-k_m-1} \leqslant 0. \qquad (11.4.16)$$

Furthermore, by using (11.4.14) and (11.4.15) we see that

$$z_{n_s}x_{n_s-k_m-1} \leqslant 0. \qquad (11.4.17)$$

By summing up both sides of (11.4.8) from $n_s - 1 - k_m$ to $n_s - 1$, we obtain

$$z_{n_s} - z_{n_s-1-k_m} = \sum_{j=n_s-1-k_m}^{n_s-1}\left(\sum_{i=1}^{m} p_i(j + k_i - k_m - 1)x_{j-k_m}\right).$$

Now by substituting $z_{n_s-1-k_m}$, from (11.4.7) we find

$$z_{n_s} - x_{n_s-1-k_m} = \sum_{i=1}^{m}\left(\sum_{j=n_s-k_m-1}^{n_s-k_i-2} p_i(j - k_m + k_i)x_{j-k_m}\right)$$

$$- \sum_{j=n_s-k_m-1}^{n_s-1}\left(\sum_{i=1}^{m} p_i(j - k_m + k_i - 1)x_{j-k_m}\right).$$

In view of (11.4.16) we get

$$|z_{n_s}| \leqslant |z_{n_s} - x_{n_s-k_m-1}| \leqslant \sum_{i=1}^{m}\left(\sum_{j=n_s-k_m-1}^{n_s-k_i-2} |p_i(j - k_m + k_i)||x_{j-k_m}|\right)$$

$$+ \sum_{j=n_s-k_m-1}^{n_s-1}\left(\sum_{i=1}^{m} |p_i(j + k_i - k_m - 1)|\right)|x_{j-k_m}|$$

$$\leqslant \max_{n_0 \leqslant j \leqslant n_s} |x_j| \sum_{i=1}^{m}\left(\sum_{n_s-k_m-1}^{n_s-k_i-2} |p_i(j + k_i - k_m)|\right)$$

$$+ \max_{n_0 \leqslant j \leqslant n_s} |x_j| \sum_{j=n_s-k_m-1}^{n_s-1}\left(\sum_{i=1}^{m} p_i(j - k_m + k_i - 1)\right).$$

By using (11.4.13) and (11.4.17) we obtain

$$(1 - q_1)|z_{n_s}| \leqslant (q_1 + q_2)|z_{n_s}|, \qquad \text{which implies that } 1 \leqslant 2q_1 + q_2.$$

This contradicts (11.4.12) and the proof of the theorem is complete.

The next result shows that under the hypotheses of Theorem 11.4.1 every oscillatory solution of eqn (11.4.1) tends to zero as $n \to \infty$.

Lemma 11.4.2. *Assume that the hypotheses of Theorem 11.4.1 are satisfied. Then every oscillatory solution of eqn (11.4.1) tends to zero as $n \to \infty$.*

Proof. Let $\{x_n\}$ be an oscillatory solution of eqn (11.4.1). From Theorem 11.4.1 we know that $\{x_n\}$ is a bounded sequence. Set

$$\mu = \limsup_{n \to \infty} |x_n|.$$

Then $\mu \in \mathbb{R}^+$ and for any $\varepsilon > 0$ there exists a positive integer n_ε such that

$$|x_n| < \mu + \varepsilon \qquad \text{for } n \geqslant n_\varepsilon. \tag{11.4.18}$$

Since $\{x_n\}$ is oscillatory, by using (11.4.18) and (11.4.7) it is easy to see that $\{z_n\}$ is not monotonic. Hence there exists a sub-sequence n_r of positive integers such that

$$n_r \geqslant n_\varepsilon, \qquad (z_{n_r} - z_{n_r - 1})z_{n_r} \geqslant 0, \tag{11.4.19}$$

$$\eta = \lim_{r \to \infty} |z_{n_r}| = \limsup_{n \to \infty} |z_n|. \tag{11.4.20}$$

By using (11.4.7) and (11.4.10), we find

$$|z_n| \geqslant |x_n| - \sum_{i=1}^{m-1} \left(\sum_{j=n-k_m}^{n-k_i-1} |p_i(j + k_i)||x_j| \right) \geqslant |x_n| - (\mu + \varepsilon)q_1. \tag{11.4.21}$$

Hence

$$\limsup_{n \to \infty} |z_n| \geqslant \limsup_{n \to \infty} |x_n| - (\mu + \varepsilon)q_1$$

and so

$$\eta \geqslant \mu - (\mu - \varepsilon)q_1. \tag{11.4.22}$$

As ε is arbitrary,

$$\eta \geqslant \mu - \mu q_1 = (1 - q_1)\mu. \tag{11.4.23}$$

Now by multiplying both sides of (11.4.14) by z_{n_r} we obtain

$$(z_{n_r} - z_{n_r - 1})z_{n_r} = -\left(\sum_{i=1}^{m} p_i(n_r - 1 - k_m + k_i) \right)x_{n_r - 1 - k_m}z_{n_r},$$

which together with (11.4.9) and (11.4.19) implies that

$$z_{n_r}x_{n_r - 1 - k_m} \leqslant 0. \tag{11.4.24}$$

By using an argument similar to that used in the proof of Theorem 11.4.1, we find

$$|z_{n_r}| \leqslant |z_{n_r} - x_{n_r-1-k_m}| \leqslant \max_{n_\varepsilon \leqslant j \leqslant n} |x_j| \sum_{i=1}^{m-1} \left(\sum_{j=n_s-k_m-1}^{n_s-k_i-2} |p_i(j+k_i-k_m)| \right)$$

$$+ \max_{n_\varepsilon \leqslant j \leqslant n_r} |x_j| \sum_{j=n_r-k_m-1}^{n_r-1} \left(\sum_{i=1}^{m} p_i(j-k_m+k_i) \right),$$

In view of (11.4.10), (11.4.11), and (11.4.18) we obtain

$$|z_{n_r}| \leqslant (\mu + \varepsilon)q_1 + (\mu + \varepsilon)q_2. \tag{11.4.25}$$

By taking limits and by using the fact that ε is arbitrary we see that

$$\eta \leqslant \mu(q_1 + q_2). \tag{11.4.26}$$

By combining (11.4.23) and (11.4.26) we obtain $1 \leqslant 2q_1 + q_2$, which is a contradiction and the proof is complete.

By combining Lemmas 11.4.1 and 11.4.2 we obtain the following stability result for eqn (11.4.1).

Theorem 11.4.2. *Assume that conditions (11.4.5), (11.4.6), and (11.4.10)–(11.4.12) are satisfied. Then the trivial solution of eqn (11.4.1) is globally asymptotically stable.*

Remark 11.4.1. *By combining Lemma 11.4.2 with known oscillation results, one may obtain stability theorems for eqn (11.4.1).*

Finally we present conditions for the global asymptotic stability of the delay difference equation

$$x_{n+1} - x_n + p(n)x_{n-k} = 0, \tag{11.4.27}$$

where

$$k \in \mathbb{N} \quad \text{and} \quad p(n) \in \mathbb{R}^+ - \{0\}. \tag{11.4.28}$$

The following result can be established in a way similar to that in the proofs of Theorem 11.4.1 and Lemma 11.4.2. The details of the proof are omitted.

Lemma 11.4.3. *Assume that (11.4.28) holds and that*

$$\limsup_{n \to \infty} \sum_{i=n-k}^{n} p(i) < 1. \tag{11.4.29}$$

Then every oscillatory solution of eqn (11.4.27) tends to zero as $n \to \infty$.

Concerning the non-oscillatory solutions of eqn (11.4.27), we have the following result.

Lemma 11.4.4. *Assume that* (11.4.28) *holds and that*

$$\sum_{n=0}^{\infty} p(n) = \infty. \tag{11.4.30}$$

Then every non-oscillatory solution of eqn (11.4.27) *tends to zero as* $n \to \infty$.

Proof. Let $\{x_n\}$ be a non-oscillatory solution of eqn (11.4.27). Then without loss of generality we may assume that $\{x_n\}$ is eventually positive. From (11.4.27), we find

$$x_{n+1} - x_n = -p(n)x_{n-k} < 0. \tag{11.4.31}$$

Hence $\{x_n\}$ is eventually positive and decreasing, so the limit exists and is non-negative. Set

$$l = \lim_{n \to \infty} x_n.$$

Then by summing up both sides of (11.4.31) from n to ∞ we obtain

$$l = x_n = -\sum_{i=n}^{\infty} p(i)x_{i-k} \leqslant -l \sum_{i=n}^{\infty} p(i),$$

which implies that $l = 0$ and the proof is complete.

Now by combining Lemmas 11.4.3 and 11.4.4 we obtain the following stability results for eqn (11.4.27).

Theorem 11.4.3. *Assume that* (11.4.28)–(11.4.30) *hold. Then the trivial solution of eqn* (11.4.27) *is globally asymptotically stable.*

By combining Lemma 11.4.3 and the oscillation result that was established in Theorem 7.5.1, we obtain the following.

Corollary 11.4.1. *Assume that* (11.4.28) *and* (11.4.29) *hold and that*

$$\liminf_{n \to \infty} \left(\frac{1}{k} \sum_{i=n-k}^{n-1} p(i) \right) > k^k/(k+1)^{k+1}.$$

Then the trivial solution of eqn (11.4.27) *is globally asymptotically stable.*

11.5 Necessary and sufficient conditions for global attractivity in a perturbed linear difference equation

Consider the delay difference equation

$$x_{n+1} - x_n + px_{n-k} = f(x_{n-l}), \qquad n = 0, 1, 2, \ldots \qquad (11.5.1)$$

where

$$p \in (0, \infty) \qquad \text{and} \qquad k, l \in \mathbb{N}. \qquad (11.5.2)$$

The main result in this section is the following necessary and sufficient condition for the global attractivity of the trivial solution of eqn (11.5.1).

Theorem 11.5.1. *Assume that (11.5.2) holds,*

$$0 < p < \frac{k^k}{(k+1)^{k+1}}, \qquad (11.5.3)$$

$$f \in C[\mathbb{R}, \mathbb{R}] \qquad \text{and} \qquad f(u)u \geq 0 \qquad \text{for all } u \in \mathbb{R}. \qquad (11.5.4)$$

Then the following statements are equivalent.

(a) *Every solution of eqn (11.5.1) tends to zero as $n \to \infty$.*

(b) $|f(u)| < p|u| \qquad$ *for all $u \neq 0$.* $\qquad (11.5.5)$

The following corollary is an interesting consequence of Theorem 11.5.1.

Corollary 11.5.1. *Assume $q \in [0, \infty)$ and that (11.5.2) and (11.5.3) hold. Then the trivial solution of the linear difference equation*

$$x_{n+1} - x_n + px_{n-k} - qx_{n-l} = 0, \qquad n = 0, 1, 2, \ldots$$

is globally asymptotically stable if and only if $p > q$.

One should recall from Theorem 7.2.1 that the condition (11.5.3) implies that the homogeneous linear difference equation

$$y_{n+1} - y_n + py_{n-k} = 0, \qquad n = 0, 1, 2, \ldots \qquad (11.5.6)$$

has a non-oscillatory solution. In particular (see Lemma 11.5.2, below), eqn (11.5.6) has a special non-oscillatory solution which will play a fundamental role in the proof of Theorem 11.5.1.

Before we give the proof of Theorem 11.5.1 we establish some lemmas which are interesting in their own rights.

Lemma 11.5.1. *Assume that $k \in \mathbb{N}$ and that (11.5.3) holds. Then exactly k roots of the equation*

$$\lambda^{k+1} - \lambda^k + p = 0 \qquad (11.5.7)$$

lie in the interior of the disk

$$|\lambda| < \frac{k}{k+1} \qquad (11.5.8)$$

and one root lies in the open interval $(k/(k+1), 1)$.

Proof. We will employ Rouche's theorem. Set

$$g(\lambda) = \lambda^{k+1} + p, \qquad h(\lambda) = -\lambda^k, \qquad \text{and} \qquad C = \left\{ \lambda \in \mathbb{C} : |\lambda| = \frac{k}{k+1} \right\}.$$

Then for $\lambda \in C$ we have,

$$|g(\lambda)| \leqslant |\lambda|^{k+1} + p \leqslant \left(\frac{k}{k+1} \right)^{k+1} + \frac{k^k}{(k+1)^{k+1}}$$

$$= \left(\frac{k}{k+1} \right)^k = |\lambda|^k = |h(\lambda)|.$$

Hence by Rouche's theorem, $g(\lambda) + h(\lambda)$ and $h(\lambda)$ have the same number of zeros in the interior of C. This shows that eqn (11.5.7) has exactly k roots, satisfying (11.5.8). Finally observe that eqn (11.5.7) has a root in the interval $(k/(k+1), 1)$ because if we set

$$\phi(\lambda) = \lambda^{k+1} - \lambda^k + p \qquad \text{then } \phi\left(\frac{k}{k+1} \right) \phi(1) < 0.$$

The proof is complete.

An elementary consequence of Lemma 11.5.1 is that if (11.5.3) holds, then every solution of (11.5.6) tends to zero as $n \to \infty$.

The next lemma gives a detailed description of a specific non-oscillatory solution of eqn (11.5.6) which will be useful in the proof of the Theorem 11.5.1.

Lemma 11.5.2. *Assume that $k \in \mathbb{N}$ and that (11.5.3) holds. Then the unique solution $\{z_n\}$ of the initial value problem*

$$\left. \begin{array}{ll} z_{n+1} - z_n + pz_{n-k} = 0, & n = 0, 1, 2, \ldots \\ z_i = 0, & i = -k, \ldots, -1 \\ z_0 = 1 \end{array} \right\} \qquad (11.5.9)$$

satisfies the following properties:

(a) $z_n > 0$ for $n \geq 0$;

(b) $\lim\limits_{n \to \infty} z_n = 0$;

(c) $\sum\limits_{n=0}^{\infty} z_n = \dfrac{1}{p}$;

(d) $\sup\limits_{n \geq 0} \dfrac{z_{n-k}}{z_{n+1}}$
$\begin{cases} < \dfrac{1}{pk} & \text{if } k \neq 0 \\[2ex] = \dfrac{1}{1-p} & \text{if } k = 0. \end{cases}$

Proof. When $k = 0$, (11.5.3) reduces to $0 < p < 1$ and the unique solution of (11.5.9) is

$$z_n = (1 - p)^n, \qquad n = 0, 1, 2, \ldots$$

which, clearly, satisfies (a)–(d).

Next assume that $k \geq 1$. Then in view of (11.5.3),

$$\frac{1}{pk} > \left(\frac{k+1}{k}\right)^{k+1}.$$

Let m be any fixed number such that

$$\left(\frac{k+1}{k}\right)^{k+1} < m < \frac{1}{pk}.$$

We claim that

$$z_{n-k} < m z_{n+1} \qquad \text{for } n \geq 0. \tag{11.5.10}$$

Otherwise, there exists an $n_0 \geq 1$ such that

$$z_{n-k} < m z_{n+1} \qquad \text{for } n < n_0 \text{ and } z_{n_0-k} \geq m z_{n_0+1}. \tag{11.5.11}$$

Then for $n < n_0$,

$$p z_{n-k} < p m z_{n+1} < \frac{1}{k} z_{n+1}$$

and so

$$z_{n+1} - z_n + \frac{1}{k} - z_{n+1} > 0.$$

Hence

$$z_{n+1} > \left(\frac{k}{k+1}\right)z_n, \qquad n < n_0$$

which implies that

$$z_{n_0} > \left(\frac{k}{k+1}\right)^k z_{n_0-k}.$$

Then

$$z_{n_0+1} = z_{n_0} - pz_{n_0-k} > \left[\left(\frac{k}{k+1}\right)^k - p\right]z_{n_0-k} > \left[\left(\frac{k}{k+1}\right)^k - \frac{1}{mk}\right]z_{n_0-k}.$$

By combining this with (11.5.11) we see that

$$m < \frac{1}{\left(\dfrac{k}{k+1}\right)^k - \dfrac{1}{mk}}$$

or equivalently,

$$m \leqslant \left(\frac{k+1}{k}\right)^{k+1}$$

which violates our choice of m. Therefore (11.5.10) is true, from which parts (a) and (d) follow immediately. Part (b) follows from Lemma 11.5.1 and so it remains to establish part (c). To this end, by summing up both sides of the difference equation in (11.5.9) from 0 to N we find

$$z_N - 1 + p \sum_{n=0}^{N} z_{n-k} = 0.$$

By taking limits as $N \to \infty$ and by using (b) we see that (c) holds. The proof of the lemma is complete.

Lemma 11.5.3. *Assume that $k \in \mathbb{N}$ and that (11.5.3) holds. Let $\{w_n\}$ be the unique solution of the initial value problem*

$$\left.\begin{array}{ll} w_{n+1} - w_n + pw_{n-k} = 0, & n = 0, 1, 2, \ldots \\ w_n = 1, & n = -k, \ldots, 0 \end{array}\right\} \tag{11.5.12}$$

and let $\{y_n\}$ be any solution of eqn (11.5.6). Then there exists a positive constant $c = c(y_n)$ such that

$$|y_n| \leqslant cw_n \qquad for\ n \geqslant 0. \tag{11.5.13}$$

Proof. One can easily see that

$$w_n = z_{n+k+1} \qquad \text{for } n \geqslant -k$$

where $\{z_n\}$ is the unique solution of the initial value problem (11.5.9). The proof of the lemma is obvious when $k = 0$. So assume that $k \geqslant 1$. Then by Lemma 11.5.2, $w_n \to 0$ as $n \to \infty$ and

$$q \equiv \sup_{n \geqslant 0} \frac{w_{n-k}}{w_{n+1}} < \frac{1}{pk}. \tag{11.5.14}$$

Set

$$\xi_n = \frac{y_n}{w_n}, \qquad \text{for } n \geqslant 0.$$

It suffices to prove that the sequence $\{\xi_n\}$ is bounded. Observe that

$$\xi_{n+1} - \xi_n = p \frac{w_{n-k}}{w_{n+1}} (\xi_n - \xi_{n-k}) = p \frac{w_{n-k}}{w_{n+1}} \sum_{i=n+k}^{n-1} (\xi_{i+1} - \xi_i)$$

and so

$$|\xi_{n+1} - \xi_n| \leqslant (pqk) \max_{n-k \leqslant i \leqslant n} |\xi_{i+1} - \xi_i|.$$

Set

$$\mu = \max \left\{ 1, \max_{-k \leqslant i \leqslant 0} |\xi_{i+1} - \xi_i| \right\}.$$

Then it follows by induction that

$$|\xi_{n+1} - \xi_n| < \mu(pqk)^{n/k}, \qquad n \geqslant 0$$

and because $pqk \in (0, 1)$,

$$|\xi_{n+1}| \leqslant |\xi_0| + \sum_{i=1}^{n} |\xi_{i+1} - \xi_i| \leqslant |\xi_0| + \mu \sum_{i=0}^{n} (pqk)^{i/k} < \infty.$$

This proves that the sequence $\{\xi_n\}$ is bounded and the proof is complete.

The next lemma is the *variation of constants formula for delay difference equations.*

Lemma 11.5.4. *Let $\{x_n\}$ be the unique solution of eqn (11.5.1) satisfying the initial conditions*

$$x_i = a_i \qquad \text{for } i = -r, \dots, 0$$

where $r = \max\{k, l\}$. Let $\{y_n\}$ be the unique solution of eqn (11.5.6) satisfying *the initial conditions*

$$y_i = a_i \quad for \ i = -k, \ldots, 0$$

and finally, let $\{z_n\}$ be the unique solution of the initial value problem (11.5.9). Then

$$x_n = y_n + \sum_{i=0}^{n-1} z_{n-i-1} f(x_{i-l}) \quad for \ n = 0, 1, 2, \ldots \quad (11.5.15)$$

Proof. The proof follows by a direct substitution of (11.5.15) into (11.5.1).

Proof of Theorem 11.5.1.

(a) $\Rightarrow$ (b). Assume for the sake of contradiction that (11.5.5) is not true. Then in view of (11.5.4) either

$$f(a) = pa \quad \text{for some } a \neq 0 \quad (11.5.16)$$

or

$$f(u) > pu \quad \text{for all } u > 0. \quad (11.5.17)$$

However, (11.5.15) implies that $x_n = a$ is a solution of eqn (11.5.1) which does not tend to zero as $n \to \infty$. Next, assume that (11.5.17) holds and let $\{x_n\}$ be the unique solution of eqn (11.5.1) satisfying the initial conditions

$$x_i = 2 \quad \text{for } -r \leqslant i \leqslant 0$$

where $r = \max\{k, l\}$. The proof that (a) $\Rightarrow$ (b) will be complete if we show that

$$x_n > 1 \quad \text{for } n \geqslant 0.$$

To this end, let $\{y_n\}$ be the solution of eqn (11.5.3) satisfying the initial conditions

$$y_i = 2 \quad \text{for } -k \leqslant i \leqslant 0.$$

Then clearly,

$$y_n = 2w_n \quad \text{for } n \geqslant -k$$

where w_n is the solution of (11.5.12) and by (11.5.15)

$$x_n = 2w_n + \sum_{i=0}^{n-1} z_{n-i-1} f(x_{i-1}), \quad n \geqslant 0. \quad (11.5.19)$$

Assume for the sake of contradiction, that (11.5.18) is false. Then there exists an $n_0 \geqslant 1$ such that

$$x_n > 1 \quad \text{for } n < n_0 \text{ and } x_{n_0} \leqslant 1. \quad (11.5.20)$$

It follows from (11.5.17) and (11.5.19) that

$$x_{n_0} > 2w_{n_0} + p \sum_{i=0}^{n-1} z_{n_0-i-1} x_{i-l} > w_{n_0} + p \sum_{i=0}^{n-1} z_{n_0-i-1}.$$

Set

$$\psi_n = w_n + p \sum_{i=0}^{n-1} z_{n-i-1}, \qquad n = 0, 1, 2, \ldots$$

and observe that

$$\psi_{n+1} = \psi_n \qquad \text{for } n \geqslant 0 \text{ and } \psi_{n_0} = 1 \tag{11.5.21}$$

Hence

$$x_{n_0} > \psi_{n_0} = 1$$

which contradicts (11.5.20).

(b) $\Rightarrow$ (a). Let $\{x_n\}$ be any solution of eqn (11.5.1) and let $\{y_n\}$, $\{z_n\}$, and $\{w_n\}$ be as defined in Lemmas 11.5.4, 11.5.2, and 11.5.3, respectively. Then from (11.5.13) and (11.5.15)

$$|x_n| \leqslant cw_n + p \sum_{i=0}^{n-1} z_{n-i-1} |x_{n-l}|. \tag{11.5.22}$$

Let

$$M > \max\left\{c, \ \max_{-r \leqslant i \leqslant 0} |x_i|\right\}.$$

First we claim that

$$|x_n| < M \qquad \text{for } n \geqslant 0. \tag{11.5.23}$$

Otherwise there exists an $n_0 > 0$ such that

$$|x_n| < M \qquad \text{for } -r \leqslant n < n_0 \qquad \text{and} \qquad |x_{n_0}| \geqslant M.$$

Then from (11.5.21) and (11.5.22) it follows that

$$|x_{n_0}| < M\left(w_{n_0} + p \sum_{i=0}^{n_0-1} z_{n_0-i-1}\right) = M$$

and this contradiction establishes that (11.5.23) is true. Set

$$L = \limsup_{n \to \infty} |x_n|$$

which is finite. It suffices to prove that $L = 0$. Assume, for the sake of contradiction, that $L > 0$ and set

$$\mathscr{L} = \limsup_{n \to \infty} |f(x_{n-l})|.$$

Now we claim that

$$pL \leqslant \mathscr{L}. \qquad (11.5.24)$$

Indeed, let $\varepsilon > 0$ be given and let $N = N(\varepsilon)$ be such that

$$|f(x_{n-l})| \leqslant \mathscr{L} + \varepsilon \qquad \text{for } n \geqslant N.$$

Then for $n \geqslant N$, (11.5.15) yields

$$|x_n| \leqslant cw_n + \sum_{i=0}^{N-1} z_{n-i-1}|f(x_{i-l})| + \sum_{i=N}^{n} z_{n-i-1}|f(x_{i-l})|$$

$$\leqslant cw_n + \sum_{i=0}^{N-1} z_{n-i-1}|f(x_{i-l})| + (\mathscr{L} + \varepsilon) \sum_{i=N}^{\infty} z_{n-i-1}$$

and by taking $\limsup_{n \to \infty}$ on both sides, we find

$$L \leqslant (\mathscr{L} + \varepsilon)/p.$$

Hence (11.5.24) holds. In particular, this shows that $\mathscr{L} > 0$. Let $\{n_s\}$ be a sequence of natural numbers such that

$$\mathscr{L} = \lim_{n_s \to \infty} |f(x_{n_s-l})|.$$

Furthermore, since $\{x_n\}$ is bounded there exists a subsequence $\{n_m\}$ such that $\lim_{n_m \to \infty} |x_{n_m-l}|$ exists. Then

$$\mathscr{L} = \limsup_{n_s \to \infty} |f(x_{n_s-l})| = \left| f\left(\lim_{n_m \to \infty} x_{n_m-l} \right) \right| < p \left| \lim_{n_m \to \infty} x_{n_m-l} \right| \leqslant pL.$$

This contradicts (11.5.24) and completes the proof of the theorem.

11.6 Notes

The results in Section 11.1 are from Kulenovic *et al.* (1989*a, b*). Section 11.2 is extracted from Gopalsamy *et al.* (1990*d*). The results in Section 11.3 are from Kuruklis and Ladas (1990). Section 11.4 is from Ladas *et al.* (1990*c*) and contains discrete analogues of results in Kulenovic *et al.* (1987*c*). Karakostas *et al.* (1990*a, b*) have just obtained some powerful global asymptotic results for general differential and difference equations of the form

$$\dot{x}(t) = -\alpha x(t) + f(x(t - \tau)) \qquad x_{n+1} = \alpha x_n + f(x_{n-k})$$

$$\dot{y}(t) = -\alpha y(t) + y(t - \tau)f(y(t - \tau)) \qquad y_{n+1} = \alpha y_n + y_{n-k}f(y_{n-k}).$$

The results in Section 11.5 are from Györi *et al.* (1990*d*) and are discrete analogues of results in Györi (1990*d, e*).

11.7 Open problems

11.6.1 Extend Theorem 11.1.2 to equations with several delays, of the form

$$\dot{N}(t) = -\delta N(t) + \sum_{i=1}^{m} P_i N(t - \tau_i)\, e^{-a_i N(t - \tau_i)},$$

where, for $i = 1, 2, \ldots, m$,

$$\delta, P_i, \tau_i, a_i \in (0, \infty).$$

11.6.2 Extend Theorem 11.1.2 to equations of the form

$$\dot{N}(t) = -\delta N(t) + P_1 N(t - \tau_1)\, e^{-a_1 N(t - \tau_1)} - P_2 N(t - \tau_2)\, e^{-a N(t - \tau_2)},$$

where

$$\delta, P_1, P_2, \tau_1, \tau_2, a_1, a_2 \in (0, \infty).$$

11.6.3 Extend the results of Section 11.2 to equations with several delays.

11.6.4 Obtain global attractivity results for the equation

$$x_{n+1} = \frac{\alpha x_n}{1 + \beta x_{n-k}},$$

where

$$\alpha \in (1, \infty), \qquad \beta \in (0, \infty), \qquad \text{and} \qquad k \in \{2, 3, \ldots\}.$$

Do the same for the equation in Problem 7.11.7.

11.6.5 Obtain oscillation and global attractivity results for all positive solutions of the difference equation

$$x_{n+1} = \frac{\alpha x_n}{1 + \sum_{i=1}^{m} \beta_i x_{n-k_i}},$$

where

$$\alpha \in (1, \infty), \qquad \beta_i \in (0, \infty), \qquad \text{and} \qquad k_i \in \mathbb{N}.$$

11.6.6 Obtain sharp global attractivity results for the equation

$$\dot{x}(t) + \sum_{i=1}^{n} p_i x(t - \tau_i) + f(x(t - \sigma_1), \ldots, x(t - \sigma_m)) = 0 \qquad (11.7.1)$$

when every solution of the linear equation

$$\dot{y}(t) + \sum_{i=1}^{n} p_i y(t - \tau_i) = 0 \qquad (11.7.2)$$

oscillates and also when eqn (11.6.2) has a non-oscillatory solution (see Györi 1990*d, e*).

12

MISCELLANEOUS TOPICS

Sections 12.1 and 12.2 contain some simple but interesting results on the non-existence of slowly oscillating periodic solutions. Section 12.3 deals with necessary and sufficient conditions for all solutions of a neutral equation with periodic coefficients to oscillate. In Section 12.4 we obtain a result on the growth of rapidly oscillating solutions of some linear equations.

12.1 Non-existence of slowly oscillating periodic solutions

Consider the delay differential equation

$$\dot{x}(t) + f(x(t-1)) = 0, \tag{12.1.1}$$

where $f \in C[\mathbb{R}, \mathbb{R}]$. First we need the following definition.

Definition 12.1.1. *A periodic solution $x(t)$ of eqn (12.1.1) of period p is called slowly oscillating if $(p > 2$ and$)$ there exists a positive constant q such that*

$$q > 1, \qquad p > q + 1,$$

$$x(0) = x(q) = x(p) = 0,$$

$$x(t) > 0 \quad \text{for } 0 < t < q, \qquad x(t) < 0 \quad \text{for } q < t < p,$$

and

$$x(t + p) = x(t) \quad \text{for all } t \in \mathbb{R}.$$

The word 'slowly' refers to the fact that the distance between consecutive zeros of $x(t)$ is greater than the delay in the equation, which in this case is 1.

The main result in this section is the following theorem of Nussbaum (1982).

Theorem 12.1.1. *Assume that $f: \mathbb{R} \to \mathbb{R}$ is a continuous and monotonically increasing function such that $xf(x) > 0$ for $x \neq 0$. Then for any p such that $2 < p \leqslant 3$, eqn (12.1.1) has no slowly oscillating periodic solution of period p which is C^1 and satisfies the equation for all $t \in \mathbb{R}$.*

Proof. Assume, for the sake of contradiction, that eqn (12.1.1) has a slowly oscillating periodic solution $x(t)$ of period $p \in (2, 3]$. Then there exists a $q \in (1, p - 1)$ such that

$$x(0) = x(q) = x(p) = 0,$$

$$x(t) > 0 \quad \text{for } 0 < t < q, \quad x(t) < 0 \quad \text{for } q < t < p,$$

and

$$x(t + p) = x(t) \quad \text{for all } t \in \mathbb{R}.$$

Set

$$q = 1 + b_1 \quad \text{and} \quad p = q + 1 + b_2.$$

Then $b_1, b_2 \in (0, \infty)$ and $b_1 + b_2 \leqslant 1$. Clearly,

$$x(t) \uparrow \text{ on } [0, 1], \quad x(t) \downarrow \text{ on } [1, q + 1] \quad \text{and} \quad x(t) \uparrow [q + 1, p + 1],$$

Set

$$h = x(1), \qquad h_1 = -x(q + 1) \quad \text{and} \quad h_2 = x(p + 1).$$

Then

$$h, h_1, h_2 \in (0, \infty) \quad \text{and} \quad h = h_2.$$

Observe that

$$h = x(1) - x(q) = -\int_1^q \dot{x}(t)\, dt = \int_1^q f(x(t - 1))\, dt$$

$$= \int_0^{q-1} f(x(s))\, ds = -\int_0^{b_1} f(x(s))\, ds < b_1 f(x(b_1)) \quad (12.1.2)$$

and

$$h_1 = x(q) - x(q + 1) = -\int_q^{q+1} \dot{x}(t)\, dt = \int_q^{q+1} f(x(t - 1))\, dt$$

$$= \int_{q-1}^q f(x(s))\, ds = \int_{b_1}^{1+b_1} f(x(s))\, ds > \int_{b_1}^1 f(x(s))\, ds$$

$$\geqslant f(x(b_1))(1 - b_1) \quad \text{and, by using (12.1.2),}$$

$$\geqslant \frac{1 - b_1}{b_1} h. \quad (12.1.3)$$

In a similar way we get

$$h_2 > \frac{1 - b_2}{b_2} h_1. \quad (12.1.4)$$

From (12.1.3) and (12.1.4) we see that

$$h = h_2 > \frac{1 - b_2}{b_2} \frac{1 - b_1}{b_1} h$$

and so

$$1 > \frac{1 - b_2}{b_2} \frac{1 - b_1}{b_1}$$

or equivalently,

$$b_1 + b_2 > 1,$$

which is a contradiction. The proof of the theorem is complete.

12.2 Wright's equation has no solutions of period four

Our aim in this section is to establish that Wright's equation,

$$\dot{y}(t) + \alpha y(t - 1)[1 + y(t)] = 0, \qquad \alpha > 0 \tag{12.2.1}$$

has no non-constant solutions of period four. This result was established by Nussbaum (1990) and is remarkable, for it is known (see Nussbaum 1990) that for each $p > 4$ there exists $\alpha = \alpha(p) > 0$ such that Wright's equation has a slowly oscillating periodic solution of minimal period p.

It should be noted that Wright's equation, named after E. M. Wright (1955), who did some of the earliest significant work on eqn (12.2.1), is obtained from the logistic equation

$$\dot{N}(s) = rN(s)[1 - N(s - \tau)/K]$$

by making the change of variables

$$s = \tau t \qquad \text{and} \qquad N(s) = K[1 + y(t)].$$

Indeed,

$$\dot{N}(s) = \frac{dN(s)}{ds} = \frac{d}{dt}[K(1 + y(t))]\frac{dt}{ds} = K\frac{dy(t)}{dt}\frac{1}{\tau}$$

and

$$N(s - \tau) = K\left[1 + y\left(\frac{s - \tau}{\tau}\right)\right] = K[1 + y(t - 1)],$$

from which the result follows with $\alpha = r\tau$.

It should be noted that Wright's equation has a periodic solution of period four if and only if the logistic equation has a periodic solution of period 4τ.

Theorem 12.2.1. *There does not exist a non-constant periodic function $y(t)$ of period four which is C^1 and satisfied Wright's equation (12.2.1) for all t.*

Proof. Assume, for the sake of contradiction, that Wright's equation (12.2.1) has a C^1 non-constant periodic solution $y(t)$ of period four that satisfies (12.2.1) for all t.

Claim 1. $y(t) + 1 \neq 0$ for all $t \in \mathbb{R}$. Otherwise there exists a $t_0 \in \mathbb{R}$ such that $y(t_0) = -1$. Then in the interval $[t_0, t_0 + 1]$, $y(t)$ satisfies the IVP

$$\left. \begin{array}{c} \dot{y}(t) + \alpha y(t-1)y(t) = -\alpha y(t-1) \\ y(t_0) = -1 \end{array} \right\}.$$

But this IVP has a unique solution. And in fact $y(t) = -1$ is the solution. It follows that $y(t) = -1$ for $t \geqslant t_0$, which contradicts the fact that $y(t)$ is a non-constant periodic function.

Claim 2. $y(t) + 1 > 0$ for all $t \in \mathbb{R}$. Otherwise $y(t) + 1 < 0$ for all $t \in \mathbb{R}$. Then

$$\dot{y}(t) = -\alpha y(t-1)[1 + y(t)] < 0 \qquad \text{for all } t \in \mathbb{R},$$

which contradicts that fact that $y(t)$ is periodic.

Set

$$y(t) + 1 = e^{x(t)}, \qquad t \in \mathbb{R}.$$

Then

$$\dot{x}(t) = -\alpha[e^{x(t-1)} - 1]$$

and $x(t)$ is a non-constant periodic function of period four. Set

$$x_i(t) = x(t-i) \qquad \text{for } i = 1, 2, 3, 4.$$

Then

$$\dot{x}_i(t) = -\alpha[e^{x_{i+1}(t)} - 1] \qquad \text{for } i = 1, 2, 3, 4,$$

where indices are written mod 4 (that is, $x_5 = x_1$, etc.). Set

$$\eta(t) = \sum_{i=1}^{4} [e^{x_i(t)} - 1] \qquad \text{and} \qquad v(t) = \sum_{i=1}^{4} x_i(t).$$

Then η and v are periodic of period four,

$$\dot{\eta}(t) = -\alpha \sum_{i=1}^{4} e^{x_i(t)}[e^{x_{i+1}(t)} - 1]$$

and

$$\dot{v}(t) = -\alpha \eta(t).$$

Next, we need the following result.

Claim 3. Let $u_i \in \mathbb{R}$ for $i = 1, 2, 3, 4$ and set

$$d = \sum_{i=1}^{4} u_i.$$

Then

$$\sum_{i=1}^{4} u_i u_{i+1} \leqslant d^2/4$$

with equality if and only if

$$u_1 + u_3 = u_2 + u_4 = d/2.$$

Indeed, if $u = u_1 + u_3$, then

$$\sum_{i=1}^{4} u_i u_{i+1} = (u_1 + u_3)(u_2 + u_4) = u(d - u) = d^2/4 - (u - d/2)^2,$$

from which the claim follows.

Claim 4.

(a) If $\eta(t_0) < 0$, then $\dot{\eta}(t_0) > 0$.

(b) If $\eta(t_0) = 0$, then $\dot{\eta}(t_0) > 0$ unless

$$|e^{x_1(t_0)} + e^{x_3(t_0)} - 2| + |e^{x_2(t_0)} + e^{x_4(t_0)} - 2| = 0.$$

(c) If $\eta(t_0) \geqslant 0$, then $\eta(t) \geqslant 0$ for $t \geqslant t_0$.

First assume that $\eta(t_0) \leqslant 0$. Set $u_i = e^{x_i(t_0)}$ and $d = \sum_{i=1}^{4} u_i$. Then

$$\dot{\eta}(t_0) = \alpha \left[\sum_{i=i}^{4} u_i - \sum_{i=1}^{4} u_i u_{i+1} \right] \geqslant \alpha(d - d2/4) = \frac{\alpha d(4 - d)}{4}.$$

Also observe that $\eta(t_0) \leqslant 0$ implies that $d \leqslant 4$. The proofs of (a) and (b) are now easy consequences of the above facts and the fact that equality in Claim 3 holds if and only if

$$u_1 + u_3 = u_2 + u_4 = \frac{d}{2}.$$

Finally, if (c) were false, there should exist a $\tau_2 > t_0$ such that $\eta(\tau_2) < 0$. Define τ_1 by

$$\tau_1 = \sup\{t \geqslant t_0 : \eta(t) \geqslant 0\}.$$

Then

$$\tau_1 < \tau_2 \quad \text{and} \quad \eta(t) < 0 \quad \text{for } \tau_1 < t \leqslant \tau_2.$$

By the mean value theorem there exists $t \in (\tau_1, \tau_2)$ such that

$$0 > \eta(\tau_2) - \eta(\tau_1) = \eta'(t)(\tau_2 - \tau_1) \Rightarrow \eta'(t) < 0,$$

which contradicts the Claim 4(a).

Claim 5: $\eta(t) = 0$ for all $t \in \mathbb{R}$. By Claim 4(c), $\eta(t) \geq 0$ for all t. Otherwise $\dot\eta(t) > 0$ for all $t \in \mathbb{R}$, which contradicts the periodicity of $\eta(t)$. Then $\dot{v}(t) = -\alpha\eta(t) \leq 0$ yields a contradiction unless $\eta(t) = 0$ for all t. The proof of the claim is complete.

The proof of the theorem is completed by showing that $x_i(t) = 0$ for all $t \in \mathbb{R}$ and for each $i = 1, 2, 3, 4$. To this end, and by using Claim 4(b) it follows that for all $t \in \mathbb{R}$,

$$e^{x_1(t)} + e^{x_3(t)} = 2 = e^{x_2(t)} + e^{x_4(t)}.$$

Hence

$$\dot{x}_1(t)\, e^{x_1(t)} + \dot{x}_3(t)\, e^{x_3(t)} = 0$$

$$\dot{x}_2(t)\, e^{x_2(t)} + \dot{x}_4(t)\, e^{x_4(t)} = 0$$

and so

$$e^{x_1(t)}[e^{x_2(t)} - 1] + e^{x_3(t)}[e^{x_4(t)} - 1] = 0$$

$$e^{x_2(t)}[e^{x_3(t)} - 1] + e^{x_4(t)}[e^{x_1(t)} - 1] = 0.$$

It follows that

$$e^{x_1(t)}[e^{x_2(t)} - 1] + [2 - e^{x_1(t)}][1 - e^{x_2(t)}] = 0$$

$$[2 - e^{x_4(t)}][e^{x_3(t)} - 1] + e^{x_4(t)}[2 - e^{x_3(t)}] = 0$$

and so

$$[e^{x_1(t)} - 1][e^{x_2(t)} - 1] = 0 = [e^{x_3(t)} - 1][e^{x_4(t)} - 1].$$

Now assume, for the sake of contradiction, that $x_1(t) \neq 0$ in some neighbourhood N of some point x_0. Then from the above identities and the fact that

$$\dot{x}_i(t) = -\alpha[e^{x_{i+1}(t)} - 1]$$

we see that

$$x_2(t) = 0 \qquad \text{for } t \in N.$$

Hence

$$0 = \dot{x}_2(t) = -\alpha[e^{x_3(t)} - 1] \qquad \text{for } t \in N$$

and so

$$x_3(t) = 0 \qquad \text{for } t \in N,$$

which implies

$$0 = \dot{x}_3(t) = -\alpha[e^{x_4(t)} - 1] \qquad \text{for } t \in N.$$

Consequently,

$$x_4(t) = 0 \qquad \text{for } t \in N,$$

and so

$$0 = \dot{x}_4(t) = -\alpha[e^{x_1(t)} - 1] \qquad \text{for } t \in N,$$

which yields

$$x_1(t) = 0 \qquad \text{for } t \in N.$$

This contradiction completes the proof of the theorem.

12.3 Oscillations in neutral equations with periodic coefficients

Consider the neutral delay differential equation

$$\frac{d}{dt}[x(t) + px(t - \tau)] + Q(t)x(t - \sigma) = 0, \tag{12.3.1}$$

where

$$\left.\begin{aligned}
&p \in \mathbb{R}, Q \in C[[0, \infty), \mathbb{R}^+], Q \text{ is } \omega\text{-periodic with } \omega > 0,\\
&Q(t) \not\equiv 0 \text{ for } t \geqslant 0, \text{ and there exist positive integers}\\
&n_1 \text{ and } n_2 \text{ such that } \tau = n_1\omega \text{ and } \sigma = n_2\omega
\end{aligned}\right\} . \tag{12.3.2}$$

Our aim in this section is to obtain a necessary and sufficient condition for the oscillation of all solutions of eqn (12.3.1). More precisely, we establish the following result.

Theorem 12.3.1. *Assume that* (12.3.2) *is satisfied and set*

$$\tau_1 = \int_0^\tau Q(s)\,ds \qquad \text{and} \qquad \sigma_1 = \int_0^\sigma Q(s)\,ds. \tag{12.3.3}$$

Then the following statements are equivalent.

(a) Every solution of eqn (12.3.1) oscillates.

(b) Every solution of the neutral equation with constant coefficients

$$\frac{d}{dt}[y(t) + py(t - \tau_1)] + y(t - \sigma_1) = 0 \tag{12.3.4}$$

oscillates.

This theorem shows that the oscillatory behaviour of eqn (12.3.1) when Q is periodic and (12.3.2) holds is characterized by the oscillatory behaviour of the linear autonomous neutral equation (12.3.4). On the other hand, it is also known (see Theorem 6.3.1) that every solution of eqn (12.3.4) oscillates if and only if its characteristic equation

$$\lambda + p\lambda\,e^{-\lambda\tau_1} + e^{-\lambda\sigma_1} = 0 \qquad (12.3.5)$$

has no real roots.

The way in which one is led to Theorem 12.3.1 is by looking for a solution of eqn (12.3.1) in the form

$$x(t) = \exp\left(\lambda \int_0^t Q(s)\,ds\right). \qquad (12.3.6)$$

By substituting (12.3.6) into eqn (12.3.1), we find

$$\exp\left(\lambda \int_0^t Q(s)\,ds\right)\!\left[\lambda Q(t) + p\lambda Q(t-\tau)\exp\left(-\lambda \int_{t-\tau}^t Q(s)\,ds\right)\right.$$

$$\left. + Q(t)\exp\left(-\lambda \int_{t-\sigma}^t Q(s)\,ds\right)\right] = 0. \quad (12.3.7)$$

Therefore if Q is not identically zero and if Q is ω-periodic with τ and σ integral multiples of ω, then eqn (12.3.7) reduces to eqn (12.3.5) where τ_1 and σ_1 are defined by (12.3.3). The fact that eqn (12.3.5) characterizes the oscillatory behaviour of eqn (12.3.1) is a remarkable results which is not obvious.

Proof of Theorem 12.3.1. When $p = 0$, the proof of Theorem 12.3.1 is a consequence of known results. Indeed, in this case eqn (12.3.1) reduces to

$$\dot{x}(t) + Q(t)x(t - \sigma) = 0, \qquad (12.3.1')$$

where

$$Q \in C[[0, \infty], \mathbb{R}^+] \qquad \text{and} \qquad \sigma > 0.$$

Now it is well known (see Theorem 2.3.1) that every solution of eqn (12.3.1') oscillates provided that

$$\liminf_{t \to \infty} \int_{t-\sigma}^t Q(s)\,ds > 1/e.$$

It is also known (see Theorem 2.3.2) that eqn (12.3.1') has a non-oscillatory solution provided that

$$\sup_{t \geq \sigma} \int_{t-\sigma}^t Q(s)\,ds \leq 1/e.$$

Hence if Q is σ-periodic and if we set

$$\sigma_1 = \int_0^{\sigma} Q(s)\, ds,$$

then every solution of eqn (12.3.1') oscillates if and only if

$$\sigma_1 > 1/e,$$

that is, if and only if every solution of

$$\dot{y}(t) + y(t - \sigma_1) = 0$$

oscillates (see Theorem 2.2.3). In view of the foregoing discussion, we now assume that $p \neq 0$.

For the proof that (a) $\Rightarrow$ (b), assume, for the sake of contradiction, that eqn (12.3.4) has a non-oscillatory solution. Then eqn (12.3.5) must have a real root λ_0 (see Theorem 6.3.1). One can now see by direct substitution into eqn (12.3.1) that

$$x(t) = \exp\left(\lambda_0 \int_0^t Q(s)\, ds \right) \qquad \text{for } t \geqslant \max\{\tau, \sigma\}$$

is a non-oscillatory solution of eqn (12.3.1). This contradicts the hypothesis that every solution of eqn (12.3.1) oscillates and completes the proof that (a) $\Rightarrow$ (b).

The proof that (b) $\Rightarrow$ (a) is quite involved and is accomplished by establishing a series of lemmas. In the sequel we will assume that (b) holds and that, for the sake of contradiction, eqn (12.3.1) has an eventually positive solution which we will denote by $x(t)$. The proof is completed when we reach a contradiction. Set

$$F(\lambda) = \lambda + p\lambda\, e^{-\lambda \tau_1} + e^{-\lambda \sigma_1}.$$

The hypothesis that (b) holds is equivalent to the fact that eqn ((12.3.5) has no real roots (see Theorem 6.3.1). As $F(0) = 1$, it follows that

$$F(\lambda) > 0 \qquad \text{for } \lambda \in \mathbb{R}.$$

Also $F(\infty) = \infty$ and clearly $F(-\infty)$ must be ∞, for otherwise $F(-\infty) = -\infty$ and so eqn (12.3.5) would have a real root.

The following lemma is now an elementary consequence of the above observations.

Lemma 12.3.1. *The following statements are true:*

(a) *There exists a positive number m_0 such that*

$$\lambda + p\lambda\, e^{-\lambda \tau_1} + e^{-\lambda \sigma_1} \geqslant m_0 \qquad \text{for } \lambda \in \mathbb{R},$$

or equivalently,

$$\lambda + p\lambda\, e^{\lambda\tau_1} - e^{\lambda\sigma_1} \leqslant -m_0 \qquad for\ \lambda \in \mathbb{R}.$$

(b) *If $p > 0$ then $\sigma_1 > \tau_1$ and $\sigma > \tau$.*

The following lemma can be easily proved by direct substitution into eqn (12.3.1), and its proof will be omitted.

Lemma 12.3.2. *If $v(t)$ is a solution of eqn (12.3.1) for $t \geqslant t_0 \geqslant 0$, then*

$$w_1(t) = v(t) + pv(t - \tau) \qquad for\ t \geqslant t_0 + \tau$$

and

$$w_2(t) = \int_{t-\sigma}^{t-\tau} Q(s)v(s)\, \mathrm{d}s \qquad for\ t \geqslant t_0 + \max\{\tau, \sigma\}$$

are also solutions of eqn (12.3.1).

Before we state the next lemma, we introduce some notation. Let W^0 be the set of all continuously differentiable solutions $w(t)$ of eqn (12.3.1) with the properties that

$$w(t) > 0 \quad and \quad \dot{w}(t) \leqslant 0 \quad \text{for all large } t \quad and \quad \lim_{t\to\infty} w(t) = 0. \qquad (12.3.8)$$

Also, let W^∞ be that set of all continuously differentiable solutions $w(t)$ of eqn (12.3.1) such that

$$w(t) > 0 \quad and \quad \dot{w}(t) \geqslant 0 \quad \text{for all large } \epsilon\, t \quad and \quad \lim_{t\to\infty} w(t) = \infty. \qquad (12.3.9)$$

For each $w \in W^0$ we define that set

$$\Lambda^0(w) = \{\lambda \in \mathbb{R}^+ : \dot{w}(t) + \lambda Q(t)w(t) \leqslant 0 \qquad \text{for all large } t\}.$$

and for each $w \in W^\infty$ we define the set

$$\Lambda^\infty(w) = \{\lambda \in \mathbb{R}^+ : -\dot{w}(t) + \lambda Q(t)w(t) \leqslant 0 \qquad \text{for all large } t\}.$$

Clearly $0 \in \Lambda^0(w)$ for every $w \in W^0$ and $0 \in \Lambda^\infty(w)$ for every $w \in W^\infty$. It is also easy to see that for any $w \in W^0 \cup W^\infty$, $\Lambda^0(w)$ and $\Lambda^\infty(w)$ are sub-intervals of $\mathbb{R}^+$.

The next three lemmas describe some interesting facts about the sets W^0 and W^∞ and the sets $\Lambda^0(w)$ and $\Lambda^\infty(w)$ with $w \in W^0 \cup W^\infty$.

Lemma 12.3.3. $W^0 \cup W^\infty \neq \varnothing$. *That is, there exists a solution w of eqn (12.3.1) that satisfies either (12.3.8) or (12.3.9).*

Proof. Set

$$y(t) = x(t) + px(t - \tau),$$

where $x(t)$ is an eventually positive solution of eqn (12.3.1). Then

$$\dot{y}(t) = -Q(t)x(t - \sigma).$$

As $Q(t) \geqslant 0$ and $Q(t) \not\equiv 0$, it follows that $\dot{y}(t) \leqslant 0$, so either eventually $y(t) > 0$ or eventually $y(t) < 0$. First, assume that eventually $y(t) > 0$. Then we claim that $y \in W^0$. Otherwise

$$\lim_{t \to \infty} y(t) = l > 0. \tag{12.3.10}$$

Set

$$v(t) = y(t) + py(t - \tau).$$

Then

$$\dot{v}(t) = -Q(t)y(t - \sigma)$$

and so by (12.3.10) we obtain

$$\dot{v}(t) \leqslant -\frac{l}{2}Q(t). \tag{12.3.11}$$

Clearly $\int_0^\infty Q(s)\,\mathrm{d}s = \infty$, so by integrating (12.3.11) from t_0 to ∞, with t_0 sufficiently large, we are led to a contradiction. Next, assume that eventually $y(t) < 0$. Set

$$z(t) = -y(t)$$

and observe that

$$\dot{z}(t) = Q(t)x(t - \sigma).$$

Hence $z(t)$ is eventually positive and increasing, so either

$$\lim_{t \to \infty} z(t) = \infty \tag{12.3.12}$$

or

$$\lim_{t \to \infty} z(t) = L \in (0, \infty). \tag{12.3.13}$$

If (12.3.12) holds, then $z \in W^\infty$. On the other hand, if (12.3.13) holds, set

$$u(t) = -[z(t) + pz(t - \tau)]$$

and observe that eventually

$$\dot{u}(t) = Q(t)z(t - \sigma) \geqslant (L/2)Q(t).$$

From this it is easily seen that $u \in W^\infty$. The proof of Lemma 12.3.3 is complete.

When $W^0 \neq \varnothing$, the next lemma shows how to construct a sequence of functions $\{w_n\}$ in W^0 and a positive number λ^* such that

$$\lambda^* \in \Lambda^0(w_n) \qquad \text{for all } n.$$

A similar construction is also shown when $W^\infty \neq \varnothing$.

Lemma 12.3.4. *The following statements hold*

   (a) *Assume that* $W^0 \neq \varnothing$. *Then*

$$p > -1. \tag{12.3.14}$$

Let $w_o \in W^0$ *and set*

$$w(t) = w_0(t) + pw_0(t - \tau).$$

Then

$$w \in W^0.$$

Furthermore

$$1 \in \Lambda^0(w) \qquad \text{if } -1 < p < 0$$

and

$$\frac{1}{1 + p} \in \Lambda^0(w) \qquad \text{if } p > 0.$$

   (b) *Assume that* $W^\infty \neq \varnothing$. *Then*

$$p < -1. \tag{12.3.15}$$

Let $w_\infty \in W^\infty$ *and set*

$$w(t) = -[w_\infty(t) + pw_\infty(t - \tau)] \qquad \text{if } \sigma \leqslant \tau$$

and

$$v(t) = -[w_\infty(t) + pw_\infty(t - \tau)] + \int_{t-\sigma}^{t-\tau} Q(s)w_\infty(s)\, ds \qquad \text{if } \sigma > \tau. \tag{12.3.16}$$

Then $w, v \in W^\infty$. *Furthermore,*

$$1/(-p) \in \Lambda^\infty(w) \qquad \text{if } \sigma \leqslant \tau$$

and

$$\frac{1}{-p + \sigma_1 - \tau_1} \in \Lambda^\infty(v) \qquad \text{if } \sigma > \tau.$$

Proof.

(a) We have

$$\dot{w}(t) = -Q(t)w_0(t - \sigma) \leqslant 0$$

and $\dot{w}(t)$ is not identically zero for all large t. Also,

$$\lim_{t \to \infty} w(t) = \lim_{t \to \infty} [w_0(t) + pw_0(t - \tau)] = 0.$$

Thus $w(t)$ decreases to zero, which implies that eventually $w(t) > 0$. Hence $w \in W^0$. Observe now that

$$0 < w(t) = w_0(t) + pw_0(t - \tau) \leqslant (1 + p)w_0(t - \tau),$$

which implies that (12.3.14) holds. First, assume that $-1 < p < 0$. Then

$$w(t) = w_0(t) + pw_0(t - \tau) < w_0(t)$$

and so

$$\dot{w}(t) = -Q(t)w_0(t - \sigma) \leqslant -Q(t)w(t - \sigma) \leqslant -Q(t)w(t),$$

which shows that $1 \in \Lambda^0(w)$. Next, assume that $p > 0$. By Lemma 12.3.1(b) we also have $\sigma > \tau$. Therefore

$$w(t) = w_0(t) + pw_0(t - \tau) \leqslant (1 + p)w_0(t - \tau) \leqslant (1 + p)w_0(t - \sigma)$$

and so

$$\dot{w}(t) = -Q(t)w_0(t - \sigma) \leqslant -\frac{1}{1 + p} Q(t)w(t),$$

which show that

$$\frac{1}{1 + p} \in \Lambda^0(w).$$

(b) Set $z(t) = -[w_\infty(t) + pw_\infty(t - \tau)]$. Then

$$\dot{z}(t) = Q(t) w_\infty(t - \sigma)$$

and one can easily see that $z \in W^\infty$. Hence

$$0 < z(t) = -[w_\infty(t) + pw_\infty(t - \tau)] \leqslant -(1 + p)w_\infty(t - \tau),$$

which shows that (12.3.15) holds. First, assume that $\sigma \leqslant \tau$. Then $w = z$, so $w \in W^\infty$. Furthermore,

$$-pw_\infty(t - \sigma) \geqslant -pw_\infty(t - \tau) > w(t)$$

and so

$$0 = -\dot{w}(t) + Q(t)w_\infty(t - \sigma) \geqslant -\dot{w}(t) + \frac{1}{-p} Q(t)w(t),$$

which shows that

$$1/(-p) \in \Lambda^\infty(w).$$

Next, assume that $\sigma > \tau$. Then from (12.3.16) we see that

$$\dot{v}(t) = Q(t)w_\infty(t - \tau), \tag{12.3.17}$$

from which it follows that $v \in W^\infty$. From (12.3.16) and (12.3.2) we see that

$$v(t) < -pw_\infty(t - \tau) + w_\infty(t - \tau) \int_{t-\sigma}^{t-\tau} Q(s)\,\mathrm{d}s = w_\infty(t - \tau)(-p + \sigma_1 - \tau_1).$$

$$\tag{12.3.18}$$

It follows from (12.3.17) and (12.3.18) that

$$-\dot{v}(t) + \frac{1}{-p + \sigma_1 - \tau_1} Q(t)v(t) \leqslant 0,$$

which shows that

$$\frac{1}{-p + \sigma_1 - \tau_1} \in \Lambda^\infty(v).$$

The proof of Lemma 12.3.4 is complete.

Lemma 12.3.5.
(a) *Assume that $W^0 \neq \varnothing$. Then there exists a $\lambda^* \in (0, \infty)$ such that λ^* is an upper bound of $\Lambda^0(w)$ for every $w \in W^0$.*

  (b) *Assume that $W^\infty \neq \varnothing$. Then there exists a $\lambda^* \in (0, \infty)$ such that λ^* is an upper bound of $\Lambda^\infty(w)$ for every $w \in W^\infty$.*

Proof.
   (a) As we proved in Lemma 12.3.4(a)—see (12.3.14)—in this case $p > -1$.
   Case 1: $-1 < p < 0$. Let $w \in W^0$. Then clearly $w(t) + pw(t - \tau)$ decreases to zero, so eventually

$$w(t) > -pw(t - \tau). \tag{12.3.19}$$

Let $\lambda \in \Lambda^0(w)$ and set

$$\phi_\lambda(t) = w(t) \exp\left(\lambda \int_0^t Q(s)\,\mathrm{d}s\right). \tag{12.3.20}$$

Then eventually

$$\dot{\phi}_\lambda(t) \leqslant 0$$

and so $\phi_\lambda(t)$ is eventually decreasing. Hence eventually

$$w(t - \tau) \exp\left(\lambda \int_0^{t-\tau} Q(s)\, ds\right) \geqslant w(t) \exp\left(\lambda \int_0^t Q(s)\, ds\right),$$

and in view of (12.3.3) and (12.3.19),

$$w(t - \tau) \geqslant w(t) \exp\left(\lambda \int_{t-\tau}^t Q(s)\, ds\right) = w(t)\, e^{\lambda\tau_1} > -p\, e^{\lambda\tau_1} w(t - \tau).$$

Hence

$$-p\, e^{\lambda\tau_1} < 1,$$

which shows that

$$\lambda^* = \frac{1}{\tau_1} \ln\left(-\frac{1}{p}\right)$$

is an upper bound of $\Lambda^0(w)$.

   Case 2: $p > 0$. By Lemma 12.3.1(b), in this case $\sigma > \tau$. Set

$$\theta = \sigma - \tau \quad \text{and} \quad k = \int_{t-\theta}^t Q(s)\, ds \quad \text{for } t \geqslant \theta.$$

Clearly $k > 0$. Now for every $t \geqslant \theta$ let $t^* = t^*(t)$ be a point in $(t - \theta, t)$ such that

$$\int_{t-\theta}^{t^*} Q(s)\, ds = \int_{t^*}^t Q(s)\, ds = k/2.$$

Set

$$z(t) = w(t) + pw(t - \tau).$$

Then eventually $z(t) > 0$ and

$$\dot{z}(t) = -Q(t)w(t - \sigma).$$

By integrating both sides of this equation from t^* to t, for t sufficiently large, we obtain

$$z(t) - z(t^*) = -\int_{t^*}^t Q(s)w(s - \sigma)\, ds \leqslant -w(t - \sigma)k/2.$$

Thus

$$(k/2)w(t - \sigma) \leqslant z(t^*) - z(t) \leqslant z(t^*) = w(t^*) + pw(t^* - \tau) \leqslant (1 + p)w(t^* - \tau)$$

and so

$$w(t^* - \tau) \geqslant \frac{k}{2(1 + p)} w(t - \sigma). \qquad (12.3.21)$$

Let $\lambda \in \Lambda^0(w)$. Now by using the fact that the function $\phi_\lambda(t)$ defined in (12.3.20) is decreasing and in view of (12.3.21), we find

$$w(t - \sigma) \geqslant w(t^* - \tau) \exp\left(\lambda \int_{t - \sigma}^{t^* - \tau} Q(s)\, ds\right)$$

$$= w(t^* - \tau) \exp\left(\lambda \int_{t - \theta}^{t^*} Q(s)\, ds\right)$$

$$\geqslant \frac{k}{2(1 + p)} w(t - \sigma)\, e^{\lambda k/2}.$$

Hence

$$\frac{k}{2(1 + p)} e^{\lambda k/2} \leqslant 1,$$

which shows that

$$\lambda^* = \frac{2}{k} \ln \frac{2(1 + p)}{k}$$

is an upper bound of $\Lambda^0(w)$. The proof of part (a) of Lemma 12.3.5 is complete.

(b) As we proved in Lemma 12.3.4(b)—see (12.3.15)—in this case $p < -1$. Let $w \in W^\infty$ and set

$$z(t) = -[w(t) + pw(t - \tau)].$$

Then eventually $z(t) > 0$, so

$$w(t) < -pw(t - \tau).$$

Let $\lambda \in \Lambda^\infty(w)$ and set

$$\psi_\lambda(t) = w(t) \exp\left(-\lambda \int_0^t Q(s)\, ds\right). \qquad (12.3.22)$$

Then eventually

$$\dot{\psi}_\lambda(t) \geqslant 0$$

and so $\psi_\lambda(t)$ is eventually increasing. Hence eventually,

$$w(t) \geqslant \exp\left(\lambda \int_{t - \tau}^{t} Q(s)\, ds\right) w(t - \tau) \geqslant e^{\lambda \tau_1} \frac{1}{(-p)} w(t).$$

Therefore

$$e^{\lambda \tau_1} \frac{1}{(-p)} \leqslant 1,$$

which shows that

$$\lambda^* = \frac{\ln(-p)}{\tau_1}$$

is an upper bound of $\Lambda^\infty(w)$. The proof of Lemma 12.3.5 is complete.

On the basis of the preceding lemmas it suffices to examine each of the following four cases, in each of which it remains to obtain a contradiction:

Case I: $W^0 \neq \varnothing$ and $-1 < p < 0$;

Case II: $W^0 \neq \varnothing$ and $p > 0$;

Case III: $W^\infty \neq \varnothing$ and $\sigma \leqslant \tau$;

Case IV: $W^\infty \neq \varnothing$ and $\sigma > \tau$.

Our strategy in each case is to use Lemma 12.3.4 to construct a sequence of functions $\{w_n\}$ such that for each $n = 1, 2, \ldots$,

$$w_n \in W^0 \quad \text{in Cases I and II}$$

$$w_n \in W^\infty \quad \text{In Cases III and IV}.$$

In each case we shall also find a positive number μ and show that for every $n = 1, 2, \ldots$,

$$\text{if } \lambda \in \Lambda^0(w_n) \qquad \text{then } \lambda + \mu \in \Lambda^0(w_{n+1}) \tag{12.3.23}$$

and

$$\text{if } \lambda \in \Lambda^\infty(w_n) \qquad \text{then } \lambda + \mu \in \Lambda^\infty(w_{m+1}). \tag{12.3.24}$$

In view of Lemma 12.3.5, (12.3.23) and (12.3.24) will eventually lead to the desired contradiction.

Case I: $W^0 \neq \varnothing$ and $-1 < p < 0$. Let $w_0 \in W^0$. Then clearly each of the functions

$$w_n(t) = w_{n-1}(t) + pw_{n-1}(t - \tau) \qquad \text{for } n = 1, 2, \ldots \tag{12.3.25}$$

belongs to W^0. We now show that (12.3.23) holds with $\mu = m_0$, where m_0 is the constant in Lemma 12.3.1(a). To this end, let $\lambda \in \Lambda^0(w_n)$ and set

$$\phi_\lambda(t) = w_n(t) \exp\left(\lambda \int_0^t Q(s)\, ds\right).$$

Then eventually $\dot{\phi}_\lambda(t) \leqslant 0$, so eventually $\phi_\lambda(t)$ is a decreasing function. Finally, observe that

$$\dot{w}_{n+1}(t) + (\lambda + \mu)Q(t)w_{n+1}(t)$$

$$= -Q(t)w_n(t - \sigma) + (\lambda + \mu)Q(t)w_n(t) + (\lambda + \mu)Q(t)pw_n(t - \tau)$$

$$= Q(t)\left[-\phi_\lambda(t - \sigma) \exp\left(-\lambda \int_0^{t-\sigma} Q(s)\,ds \right) \right.$$

$$+ (\lambda + \mu)\phi_\lambda(t) \exp\left(-\lambda \int_0^t Q(s)\,ds \right)$$

$$\left. + (\lambda + \mu)p\phi_\lambda(t - \tau) \exp\left(-\lambda \int_0^{t-\tau} Q(s)\,ds \right) \right]$$

$$\leqslant Q(t)\phi_\lambda(t) \exp\left(-\lambda \int_0^t Q(s)\,ds \right) (-e^{\lambda\sigma_1} + \lambda + \mu + \lambda p\,e^{\lambda\tau_1} + \mu p\,e^{\lambda\tau_1})$$

$$\tag{12.3.26}$$

By using Lemma 12.3.1(a) and the fact that $p < 0$ we see that

$$\dot{w}_{n+1}(t) + (\lambda + m_0)Q(t)w_{n+1}(t) \leqslant Q(t)\phi_\lambda(t) \exp\left(-\lambda \int_0^t Q(s)\,ds \right)(-m_0 + m_0)$$

$$= 0.$$

This shows that $\lambda + m_0 \in \Lambda^0(w_{n+1})$ and the proof in Case I is complete.

Case II: $W^0 \neq \varnothing$ and $p > 0$. By Lemma 12.3.1(b) we know that $\sigma > \tau$. Here we also use the sequence (12.3.25). Finally, we claim that (12.3.23) holds with

$$\mu = \frac{m_0}{1 + p\,e^{\lambda^*\tau_1}}$$

where λ^* is the constant in Lemma 12.3.5(a). Indeed, (12.3.26) implies that

$$\dot{w}_{n+}(t) + (\lambda + \mu)Q(t)w_{n+1}(t)$$

$$\leqslant Q(t)\phi_\lambda(t - \tau) \exp\left(-\lambda \int_0^t Q(s)\,ds \right)(-e^{\lambda\sigma_1} + \lambda + \mu + \lambda p\,e^{\lambda\sigma_1} + \mu p\,e^{\lambda\tau_1})$$

$$\leqslant Q(t)\phi_\lambda(t - \tau) \exp\left(-\lambda \int_0^t Q(s)\,ds \right)[-m_0 + \mu(1 + p\,e^{\lambda\tau_1})]$$

$$\leqslant Q(t)\phi_\lambda(t - \tau) \exp\left(-\lambda \int_0^t Q(s)\,ds \right)[-m_0 + \mu(1 + p\,e^{\lambda^*\tau_1})]$$

$$= 0.$$

This shows that $\lambda + \mu \in \Lambda^0(w_{n+1})$ and the proof in Case II is complete.

Case III: $W^\infty \neq \emptyset$ and $\sigma \leq \tau$. In this case we arrive at a contradiction by using the sequence

$$w_n(t) = -[w_{n-1}(t) + pw_{n-1}(t - \tau)] \qquad \text{for } n = 1, 2, \ldots,$$

where w_0 is some fixed element of W^∞, and by taking

$$\mu = \frac{m_0}{-p - 1}.$$

Of course, $p < -1$ in this case. The proof is as in Cases I and II, but here we utilize the substitution

$$\psi_\lambda(t) = \exp\left(- \lambda \int_0^t Q(s)\, ds\right) w_n(t). \tag{12.3.27}$$

Case IV: $W^\infty \neq \emptyset$ and $\sigma > \tau$. In this case we use the sequence

$$w_n(t) = -[w_{n-1}(t) + pw_{n-}(t - \tau)] + \int_{t-\sigma}^{t-\tau} Q(s)\, ds \qquad \text{for } n = 1, 2, \ldots,$$

where w_0 is some fixed element of W^∞. We also take

$$\mu = \frac{m_0}{-p-1+(1/\lambda_0)} \qquad \text{with } \lambda_0 = \frac{1}{-p + \sigma_1 - \tau_1},$$

which by Lemma 12.3.4(b) lies in $\Lambda^\infty(w_n)$ for every $n = 1, 2, \ldots,$ In this case we also use (12.3.27) with $\lambda \geq \lambda_0$.

The proof of Theorem 12.3.1 is complete.

12.4 Asymptotic behaviour of rapidly oscillating solutions

Consider the delay differential equation

$$\dot{x}(t) + p(t)x(t - \tau) = 0, \qquad t \geq 0, \tag{12.4.1}$$

where

$$p \in C[[0, \infty), \mathbb{R}] \qquad \text{and} \qquad \tau > 0. \tag{12.4.2}$$

In this section we obtain bounds for the growth of rapidly oscillating solutions of eqn (12.4.1).

Definition 12.4.1. *We say that a solution x of eqn (12.4.1) is* rapidly oscillating *if there exists $T_0 \geq \tau$ such that for every $t_1 \geq T_0$, $x(t)$ has a zero in the open interval $(t_1 - \tau, t_1)$.*

The main result in this section is the following:

Theorem 12.4.1. *Assume that* (12.4.2) *holds and that there exists* $\alpha \in (0, 1]$ *and* $T_1 \geqslant \tau$ *such that*

$$\int_{t-\tau}^{t} |p(s)| \, ds \leqslant 2\alpha \qquad \text{for } t \geqslant T_1. \tag{12.4.3}$$

Let x be a rapidly oscillating solution of eqn (12.4.1). *Then there exists a constant* $K = K(x)$ *such that*

$$|x(t)| \leqslant K \, e^{-\beta t} \qquad \text{for } t \geqslant -\tau, \tag{12.4.4}$$

where

$$\beta = \frac{1}{2\tau} \ln \frac{1}{\alpha}. \tag{12.4.5}$$

Proof. Let $T_2 = \max\{2\tau, T_0, T_1\}$, where the T_0 is as given by Definition 12.4.1 for the rapidly oscillating solution x. Define a constant K such that

$$K > \max\{|x(t)| \, e^{\beta t} \colon -\tau \leqslant t \leqslant T_2\}.$$

We now claim that (12.4.4) is true. Otherwise there exists a $\bar{t} > T_2$ such that

$$|x(t)| \leqslant K \, e^{-\beta t} \qquad \text{for } -\tau \leqslant t < \bar{t} \qquad \text{and} \qquad |x(\bar{t})| = K \, e^{-\beta \bar{t}}. \tag{12.4.6}$$

Set

$$Q = \sup\{t \in [-\tau, \bar{t}] \colon x(t) = 0\}$$

and

$$R = \inf\{t \geqslant \bar{t} \colon x(t) = 0\}.$$

Then clearly,

$$R - \tau < Q < \bar{t} < R, \quad x(Q) = x(R) = 0 \quad \text{and} \quad x(t) \neq 0 \quad \text{for } Q < t < R. \tag{12.4.7}$$

Also, by integrating eqn (12.4.1) first from Q to $\bar{t}$ and then from $\bar{t}$ to R we obtain

$$|x(\bar{t})| = \left| \int_{Q}^{\bar{t}} p(s) x(s - \tau) \, ds \right| \leqslant \int_{Q}^{\bar{t}} |p(s)| \, |x(s - \tau)| \, ds$$

and

$$|x(\bar{t})| = \left| \int_{\bar{t}}^{R} p(s) x(s - \tau) \, ds \right| \leqslant \int_{\bar{t}}^{R} |p(s)| \, |x(s - \tau)| \, ds.$$

By adding these inequalities and then by using (12.4.6), (12.4.3) and (12.4.5), we find

$$K e^{-\beta \bar{t}} = |x(\bar{t})| \leqslant \frac{1}{2} \int_Q^R |p(s)| |x(s - \tau)| \, ds \leqslant \frac{1}{2} \int_{R-\tau}^R |p(s)| |x(s - \tau)| \, ds$$

$$\leqslant \frac{1}{2} \int_{R-\tau}^R |p(s)| K e^{-\beta(s-\tau)} \, ds \leqslant \frac{1}{2} K e^{-\beta(R-2\tau)} \int_{R-\tau}^R |p(s)| \, ds$$

$$\leqslant K\alpha e^{-\beta R} e^{2\tau\beta} = K e^{-\beta R} < K e^{-\beta \bar{t}}.$$

This is a contradiction and the proof is complete.

12.5 Notes

The results of Sections 12.1 and 12.2 are due to Nussbaum (1982, 1990). See also Chapin and Nussbaum (1984) and Walther (1981). Section 12.3 is from Ladas *et al.* (1990*b*). Theorem 12.4.1 is from Myskis (1972) and Györi (1989*a*).

12.6 Open problems

12.6.1   Obtain discrete analogues of Theorems 12.1.1, 12.2.1 and 12.4.1.

12.6.2   Extend Theorem 12.3.1 by relaxing the hypothesis that $Q(t)$ is non-negative.

12.6.3   Obtain conditions for the existence (or non-existence) of rapidly oscillating solutions of eqn (12.4.1)

12.6.4   Extend Theorem 12.4.1 to systems of neutral equations.

BIBLIOGRAPHY

Aftabizadeh, A. R. and Wiener, J.
(1985). Oscillatory properties of first order linear functional differential equations. *Applicable Analysis* **20**, 165–87.

(1988). Oscillatory and periodic solutions for systems of two first order linear differential equations with piecewise constant arguments. *Applicable Analysis* **26**, 327–33.

Aftabizadeh, A. R., Wiener, J., and Xu, J.-M.
(1987). Oscillatory and periodic properties of delay differential equations with piecewise constant argument. *Proceedings of the American Mathematical Society* **99**, 673–9.

Arino, O. and Ferreira, J. M.
(1989). Total oscillatory behavior globally in the delays. *Portugaliae Mathematica* **46**, 71–85.

Arino, O. and Györi, I.
(1990). Necessary and sufficient condition for oscillation of neutral differential system with several delays. *Journal of Differential Equations* **81**, 98–105.

Arino, O. and Kimmel, M.
(1986). Stability analysis of models of cell production systems. *Mathematical Modelling* **17**, 1269–300.

Arino, O., Györi, I., and Jawhari, A.
(1984). Oscillation criteria in delay equations. *Journal of Differential Equations* **53**, 115–23.

Arino, O., Ladas, G., and Sficas, Y. G.
(1987). On oscillations of some retarded differential equations. *SIAM Journal on Mathematical Analysis* **18**, 64–73.

Atkinson, F. V. and Haddock, J. R.
(1983). Criteria for asymptotic constancy of solutions of functional differential equations. *Journal of Mathematical Analysis and Applications* **91**, 410–23.

Bainov, D. D., Myshkis, A. D., and Zahariev, A. I.
(1988). Oscillatory properties of the solutions of linear equations of neutral type. *Bulletin of the Australian Mathematical Society* **38**, 255–61.

Barbalat, I.
(1959). Systèmes d'équations différentielles d'oscillations non-linéaires. *Revue Roumaine de Mathematiques Pures et Appliquées* **4**, 267–70.

Bellman, R. and Cooke, K. L.
(1963). *Differential-difference equations*. Academic Press, New York.

Braddock, R. D. and van den Driessche, P.

(1980).   A population model with time delay. *Mathematical Scientist* **5**, 55–6.

(1983).   On a two lag differential delay equation. *Journal of the Australian Mathematical Society, Series B* **23**, 292–317.

Brayton, R. K. and Willoughby, R. A.

(1967).   On the numerical integration of a symmetric system of difference-differential equations of neutral type. *Journal of Mathematical Analysis and Applications* **18**, 182–9.

Buchanan, J.

(1974).   Bounds on the growth of a class of oscillatory solutions of $y'(x) = my(x - d(x))$ with bounded delays. *SIAM Journal on Applied Mathematics* **27**, 539–43.

Burton, T. A.

(1983).   *Volterra integral and differential equations.* Academic Press, New York.

Busenberg, S., Fisher, D., and Martelli, M.

(1990).   Minimal periods of discrete and smooth orbits. *American Mathematical Monthly.* (To appear.)

Carvalho, L. A. V. and Cooke, K. L.

(1988).   A nonlinear equation with piecewise continuous argument. *Differential and Integral Equations* **1**, 359–67.

Chapin, S. A. and Nussbaum, R. D.

(1984).   Asymptotic estimates for the periods of periodic solutions of differential-delay equation. *Michigan Mathematical Journal* **31**, 215–29.

Chuanxi, Q. and Ladas, G.

(1989*a*).   Existence of positive solutions for neutral differential equations, *Journal of Applied Mathematics and Simulation* **2**, 267–76.

(1989*b*).   Oscillations of neutral differential equations with variable coefficients. *Applicable Analysis* **32**, 215–28.

(1990*a*).   Oscillation in differential equations with positive and negative coefficients. *Canadian Mathematical Bulletin.* (To appear.)

(1990*b*).   A comparison theorem for odd-order neutral differential equations. (To appear.)

(1990*c*).   Linearized oscillations for odd-order neutral delay differential equations. *Journal of Differential Equations.* (To appear.)

(1990*d*).   Linearized oscillations for even order neutral differential equations. *Journal of Mathematical Analysis and Applications.* (To appear.)

(1990*e*).   Existence of positive solutions. (To appear.)

Chuanxi, Q., Kulenovic, M. R. S., and Ladas, G.

(1989).   Oscillations of neutral equations with variable coefficients. *Radovi Matematicki* **5**, 321–31.

Chuanxi, Q., Kuruklis, S. A., and Ladas, G.
(1990*a*). Oscillations of systems of difference equations with variable coefficients, *Journal of Mathematical and Physical Sciences*. (To appear.)

(1990*b*). Oscillations of linear autonomous systems of difference equations. *Applicable Analysis* **36**, 51–63.

Chuanxi, Q., Ladas, G., Zhang, B. G., and Zhao, T.
(1990*c*). Sufficient conditions for oscillation and existence of positive solutions. *Applicable Analysis* **35**, 187–94.

Chuanxi, Q., Ladas, G., and Yan, J.
(1990*d*). Oscillations of higher order neutral differential equations. *Portugaliae Mathematica*. (To appear.)

Clark, C. W.
(1976). A delayed-recruitment model of population dynamics, with an application to Baleen whale populations. *Journal of Mathematical Biology* **3**, 381–91.

Cooke, K. L. and van den Driessche, P.
(1986). On zeroes of some transcendental equations. *Funkcialaj Ekvacioj* **29**, 77–90.

Cooke, K. L. and Wiener, J.
(1984). Retarded differential equations with piecewise constant delays. *Journal of Mathematical Analysis and Applications* **99**, 265–94.

(1987*a*). An equation alternately of retarded and advanced type. *Proceedings of the American Mathematical Society* **99**, 726–32.

(1987*b*). Neutral differential equations with piecewise constant argument. *Bolletino, Unione Matematica Italiana* **1-B**, 321–45.

Cooke, K. L. and Witten, M.
(1986). One-dimensional linear and logistic harvesting models. *Mathematical Modelling* **7**, 301–40.

Coppel, W. A.
(1965). *Stability and asymptotic behavior of differential equations.* D. C. Heath, Boston, Massachusetts.

Corduneanu, C.
(1971). *Principles of differential and integral equations.* Allyn and Bacon, Boston.

Cushing, J. M.
(1977). Integrodifferential equations and delay models in population dynamics. *Lecture Notes in Mathematics*, **20**, Springer-Verlag, Berlin and New York.

Dahiya, R. S.
(1982). Oscillation theorems for nonlinear delay differential equations. *Bulletin of the Faculty of Science, Ibaraki University, Series A: Mathematics* **14**, 13–18.

Dahiya, R. S., Kusano, T., and Naito, M.
(1984). On almost oscillation of functional differential equations with deviating arguments. *Journal of Mathematical Analysis and Applications* **98**, 332–40.

Domshlak, Y. I.

(1982).   Comparison theorems of Sturm type for first and second order differential equations with sign variable deviations of the argument. *Ukrainskii Matematicheskii Zhurnal* **34**, 158–63, 267.

Domshlak, Y. I. and Aliev, A. I.

(1988).   On oscillatory properties of the first order differential equations with one or two retarded arguments. *Hiroshima Mathematical Journal* **18**, 31–46.

Driver, R. D.

(1977).   *Ordinary and delay differential equations.* Springer-Verlag, Berlin and New York.

(1978).   *Introduction to ordinary differential equations.* Harper and Row, New York.

(1984).   A mixed neutral system. *Nonlinear Analysis: Theory, Methods and Applications* **8**, 155–8.

Elbert, A.

(1976).   Comparison theorem for first order nonlinear differential equations with delay. *Studia Scientiarum Mathematicarum Hungarica* **11**, 259–67.

Erbe, L. H. and Zhang, B. G.

(1989).   Oscillation of discrete analogues of delay equations. *Differential and Integral Equations* **2**, 300–9.

Farrell, K.

(1990a).   Necessary and sufficient conditions for oscillation of neutral equations with real coefficients. *Journal of Mathematical Analysis and Applications.* (To appear.)

(1990b).   Necessary and sufficient conditions for oscillation of a mixed neutral equation. (To appear.)

(1990c).   Bounded oscillations of neutral differential equations. *Radovi Matematicki* **6**, 21–40.

Farrell, K., Grove, E. A., and Ladas, G.

(1988).   Neutral delay differential equations with positive and negative coefficients. *Applicable Analysis* **27**, 181–97.

Farrell, K., Ladas, G., Wang, Z., and Zou, X.

(1990).   Necessary and sufficient conditions for the oscillation of all unbounded solutions. *Journal of Mathematical and Physical Sciences* **24**, 351–61.

Ferreira, J. M.

(1990).   Total oscillatory behavior globally in the delays. (To appear.)

Ferreira, J. M. and Györi, I.

(1987).   Oscillation behaviour in linear retarded functional differential equations. *Journal of Mathematical Analysis and Applications* **128**, 332–46.

Finizio, N. and Ladas, G.
(1982). *An introduction to differential equations.* Wadsworth Publishing, Belmont, California.

Fink, A. M. and Kusano, T.
(1983). Nonoscillation theorems for differential equations with general deviating arguments. *Lecture Notes in Mathematics* no. 1032, 224–39, Springer, Berlin.

Freedman, H. I. and Rao, S. H.
(1983). The trade-off between mutual interference and time lags in predator-prey systems. *Bulletin of Mathematical Biology* **45**, 991–1004.

Georgiou, D. A., Grove, E. A., and Ladas, G.
(1990). Oscillation of neutral difference equations. *Applicable Analysis.* (To appear.)

Glass, L. and Mackey, M. C.
(1977). Oscillation and chaos in physiological control systems. *Science* **197**, 287–9.

(1979). Pathological conditions resulting from instabilities in physiological control systems. *Annals of the New York Academy of Sciences* **316**, 214–35.

Gopalsamy, K.
(1983). Stability, instability, oscillation and nonoscillation in scalar integro-differential systems. *Journal of the Australian Mathematical Society* **28**, 233–46.

(1984). Oscillation in linear systems of differential-difference equations, *Bulletin of the Australian Mathematical Society* **29**, 377–88.

(1986a). Oscillations in a delay-logistic equation, *Quarterly of Applied Mathematics* **44**, 447–61.

(1986b). Oscillations in systems of integro-differential equations. *Journal of Mathematical Analysis and Applications* **113**, 78–87.

(1987). Oscillatory properties of first order linear delay differential inequalities, *Pacific Journal of Mathematics* **128**, 299–305.

(1990). Oscillations in a delay logistic equation. *Quarterly of Applied Mathematics.* (To appear.)

Gopalsamy, K. and Zhang, B. G.
(1990). On a neutral delay-logistic equation. (To appear.)

Gopalsamy, K., Grove, E. A., and Ladas, G.
(1988a). Neutral delay differential equations with variable delays. *Differential Equations and Applications, Proceedings of the International Conference on Theory and Applications of Differentiial Equations, March 21–25, 1988.* Ohio University Press, Athens, pp. 343–7.

Gopalsamy, K., Kulenovic, M. R. S., and Ladas, G.

(1988*b*). Oscillations in a periodic delay differential equation. *Differential Equations and Applications, Proceedings of the International Conference on Theory and Applications of Differential Equations, March 21–25, 1988.* Ohio University Press, Athens, pp. 348–52.

(1988*c*). Time lags in a 'food limited' population model. *Applicable Analysis* **31**, 225–37.

(1989*b*). Oscillations and global attractivity in respiratory dynamics. *Dynamics and Stability of Systems* **4**, 131–9.

(1990*a*). On a logistic equation with piecewise constant argument. *Differential and Integral Equations.* (To appear.)

(1990*b*). Environmental periodicity and time delays in a 'food limited' population model. *Journal of Mathematical Analysis and Applications* **147**, 545–55.

(1990*c*). Oscillations of a system of delay logistic equations. *Journal of Mathematical Analysis and Applications* **146**, 192–202.

(1990*d*). Oscillations and global attractivity in models of haematopoiesis. *Journal of Dynamics and Differential Equations.* (To appear.)

Gopalsamy, K., Györi, I., and Ladas, G.

(1989*a*). Oscillations of a class of delay equations with continuous and piecewise constant arguments. *Funkcialaj Ekvacioj* **32**, 395–406.

Grace, S. R. and Lalli, B. S.

(1982). A note on Ladas' paper: Oscillatory effects of retarded actions. *Journal of Mathematical Analysis and Applications* **88**, 257–64.

(1990). Oscillation theorems for second order neutral differential equations. (To appear.)

Grace, S. R., Lalli, B. S., and Yeh, C. C.

(1986). Comparison theorems for difference inequalities. *Journal of Mathematical Analysis and Applications* **113**, 468–72.

Graef, J. R., Grammatikopoulos, M. K., and Spikes, P. W.

(1989). Behavior of the nonoscillatory solutions of first order neutral delay differential equations. *Differential Equations: Proceedings of the 1987 Equadiff Conference.* Dekker, New York, 265–72.

Grammatikopoulos, M. K., Ladas, G., and Meimaridou, A.

(1985). Oscillations of second order neutral neutral delay differential equations. *Radovi Matematicki* **1**, 267–74.

Grammatikopoulos, M. K., Grove, E. A., and Ladas, G.

(1986*a*). Oscillations of first order neutral delay differential equations. *Journal of Mathematical Analysis and Applications* **120**, 510–20.

(1986*b*). Oscillation and asymptotic behavior of neutral differential equations with deviating arguments. *Applicable Analysis* **22**, 1–19.

Grammatikopoulos, M. K., Ladas, G., and Sficas, Y. G.
(1986*c*). Oscillation and asymptotic behavior of neutral equations with variable coefficients. *Radovi Matematicki* **2**, 279–303.

(1987*a*). Oscillation and asymptotic behavior of second order neutral differential equations with deviating arguments. *Canadian Mathematical Society Conference Proceedings* **8**, 153–61.

(1987*b*). Oscillation and asymptotic behavior of second order neutral differential equations. *Annali di Matematica Pura ed Applicata* **148**, 29–40.

(1987*c*). Necessary and sufficient conditions for oscillation of delay equations with constant coefficients. *Czechoslovak Mathematical Journal* **37**, 262–70.

Grammatikopoulos, M. K., Ladas, G., and Meimaridou, A.
(1988*a*). Oscillation and asymptotic behavior of higher order neutral equations with variable coefficients. *Chinese Annals of Mathematics* **9B**, 322–38.

Grammatikopoulos, M. K., Sficas, Y. G., and Stavroulakis, I. P.
(1988*b*). Necessary and sufficient conditions for oscillations of neutral equations with several coefficients. *Journal of Differential Equations* **76**, 294–311.

Grove, E. A., Kulenovic, M. R. S., and Ladas, G.
(1987*a*). Sturm comparison theorem for neutral differential equations. *Canadian Mathematical Society Conference Proceedings* **8**, 163–69.

Grove, E. A., Ladas, G., and Meimaridou, A.
(1987*b*). A necessary and sufficient condition for the oscillation of neutral equations. *Journal of Mathematical Analysis and Applications* **126**, 341–54.

Grove, E. A., Ladas, G., and Schultz, S. W.
(1988*a*). Oscillations and asymptotic behavior of first-order neutral delay differential equations. *Applicable Analysis* **27**, 67–78.

Grove, E. A., Ladas, G., and Schinas, J.
(1988*b*). Sufficient conditions for the oscillation of delay and neutral delay equations. *Canadian Mathematical Bulletin* **31**, 459–66.

Grove, E. A., Györi, I., and Ladas, G.
(1990*a*). On the characteristic equations for equations with continuous and piecewise constant arguments. *Radovi Matematicki* **5**, 271–81.

Grove, E. A., Kulenovic, M. R. S., and Ladas, G.
(1990*b*). A Myskis-type comparison result for neutral equations. *Mathematische Nachrichten* **146**, 195–206.

Gurney, W. S. and Nisbet, R. M.
(1976). Population dynamics in a periodically varying environment. *Journal of Theoretical Biology* **56**, 459–75.

Gurney, W. S., Blythe, S. P., and Nisbet, R. M.
(1980). Nicholson's blowflies revisited. *Nature* **287**, 17–21.

Györi, I.

(1984). On the oscillatory behaviour of solutions of certain nonlinear and linear delay differential equations. *Nonlinear Analysis: Theory, Methods and Applications* **8**, 429–39.

(1986). Oscillation conditions in scalar linear delay differential equations, *Bulletin of the Australian Mathematical Society* **34**, 1–9.

(1987). Asymptotic behaviour and oscillation of solutions of a logistic delay equation. *Proceedings of the Eleventh International Conference on Nonlinear Oscillations, Budapest, August 17–23, 1987*. Janos Bolyai Mathematical Society, Budapest, 406–8.

(1988a). Asymptotic stability and oscillations of linear non-autonomous delay. *Differential Equations and Applications, Proceedings of the International Conference on Theory and Applications of Differential Equations, March 21–25, 1988*. Ohio University Press, Athens, pp. 389–97.

(1988b). Oscillations of nonlinear delay differential equations—general linearized oscillations. Technical Report No. 91, University of Rhode Island, Kingston, 26 pp.

(1989a). Existence and growth of oscillatory solutions of first order unstable type differential equations. *Nonlinear Analysis: Theory, Methods and Applications* **13**, 739–51.

(1989b). Oscillations of neutral delay differential equations arising in population dynamics. *Differential Equations, Proceedings of the 1987 Equadiff Conference*. Dekker, New York, pp. 291–301.

(1989c). Oscillation and comparison results in neutral differential equations and their applications to the delay logistic equation. *Computers and Mathematics with Applications* **18**, 883–906.

(1989d). Oscillations of retarded differential equations of the neutral and the mixed types. *Journal of Mathematical Analysis and Applications* **141**, 1–20.

(1990a). Comparison results for integro-differential equations of Volterra-type. *Nonlinear Analysis: Theory, Methods and Applications* **14**, 245–52.

(1990b). Global attractivity in delay differential equations using a mixed monotone technique. *Journal of Mathematical Analysis and Applications* **152**, 131–55.

(1990c). Interaction between oscillations and global asymptotic stability in delay differential equations. *Differential and Integral Equations* **3**, 181–200.

(1990d). Global attractivity in a perturbed linear delay differential equation. *Applicable Analysis* **34**, 167–82.

(1990e). Sharp conditions for existence of nontrivial invariant cones of nonnegative initial values of difference equations. *Applied Mathematics and Computation* **36**, 89–112.

(1990f). Invariant cones of positive initial functions for delay differential equations. *Applicable Analysis* **35**, 21–42.

Györi, I. and Witten, M.
(1990). Cell population models with a growth rate law. (To appear.)

Györi, I. and Ladas, G.
(1988). Oscillations of systems of neutral differential equations. *Differential and Integral Equations* **1**, 281–6.

(1989). Linearized oscillations for equations with piecewise constant arguments. *Differential and Integral Equations* **2**, 123–31.

(1990). Positive solutions of integro-differential equations with unbounded delay. (To appear.)

Györi, I. and Stavroulakis, I. P.
(1988). Positive solutions of functional differential equations. *Bolletino Unione Italiana Matematica, Series A* **7**, 2–18.

Györi, I. and Wu, J.
(1990). A neutral equation arising from compartmental systems with pipes. *Journal of Dynamics and Differential Equations.* (To appear.)

Györi, I., Ladas, G., and Pakula, L.
(1989). On oscillations of unbounded solutions. *Proceedings of the American Mathematical Society* **106**, 785–92.

(1990*a*). Conditions for oscillation of difference equations with applications to equations with piecewise constant arguments. *SIAM Journal on Mathematical Analysis.* (To appear.)

(1990*b*). Oscillation theorems for delay differential equations via Laplace transforms. *Canadian Mathematical Bulletin* **33**, 323–6.

Györi, I., Ladas, G., and Qian, C.
(1990*c*). Existence of positive solutions of difference equations. (To appear.)

Györi, I., Ladas, G., and Vlahos, P. N.
(1990*d*). Necessary and sufficient conditions for global attractivity in a difference equation. *Journal of Nonlinear Analysis: Theory, Methods and Applications.* (To appear.)

Hadeler, K. P.
(1979). Delay equations in biology. *Functional differential equations and approximation of fixed points, Lecture Notes in Mathematics* no. 730. Springer-Verlag, New York, pp. 136–56.

Hale, J. K.
(1977). *Theory of functional differential equations.* Springer-Verlag, New York.

Hall, L. M. and Trimble, S. Y.
(1990). Asymptotic behavior of solutions of Poincaré difference equations. *Differential Equations and Applications.* (To appear.)

Hallam, T. G. and De Luna, J. T.
(1984). Effects of toxicants on populations: a qualitative approach, III. Environmental and food chain pathways. *Journal of Theoretical Biology* **109**, 411–29.

Hethcote, H. W., Stech, H. W., and Van den Driessche, P.
(1987). Nonlinear oscillations in epidemic models. *SIAM Journal on Applied Mathematics* **40**, 1–9.

Hille, E.
(1963). *Analytic function theory*. Blaisdell Publishing, New York.

Hooker, J. W., Man Kam Kwong, and Patula, W. T.
(1987). Oscillatory second order linear difference equations and Riccati equations. *SIAM Journal on Applied Mathematics* **18**, 54–63.

Huang, Y. K.
(1988). On a system of differential equations alternately of advanced and delay type. *Proceedings of the International Conference on Theory and Applications of Differential Equations, Ohio University, 1988*. Dekker, New York, pp. 455–65.

(1989). A nonlinear equation with piecewise constant argument. *Applicable Analysis* **33**, 183–90.

Hunt, B. R. and Yorke, J. A.
(1984). When all solutions of $x' = \sum_{j=1}^{n} q_j(t) x(t - T_j(t))$ oscillate. *Journal of Differential Equations* **53**, 139–45.

Hutchinson, G. E.
(1948). Circular causal systems in ecology. *Annals of the New York Academy of Science* **50**, 221–40.

Israelsson, D. and Johnsson, A.
(1967). A theory of circumnutations of Helianthus annus. *Physiology of Plants* **20**, 957–76.

Ivanov, A. F. and Shevelo, V. N.
(1981). On the oscillation and asymptotic behavior of solutions of first order functional differential equations. *Ukrainskii Matematicheskii Zhurnal* **33**, 745–51.

Ivanov, A. F., Kitamura, Y., Kusano, T., and Shevelo, V. N.
(1982). Oscillatory solutions of functional differential equations generated by deviations of arguments of mixed type. *Hiroshima Mathematical Journal* **12**, 645–55.

Jaros, T. and Kusano, T.
(1988). Oscillation theory of higher order linear functional differential equations of neutral type. *Hiroshima Mathematical Journal* **18**, 509–31.

Jones, G. S.

(1962a). The existence of periodic solutions of $f'(x) = -af(x-1)[1 + f(x)]$. *Journal of Mathematical Analysis and Applications* **5**, 435–50.

(1962b). On the nonlinear differential-difference equation $f'(x) = \alpha f(x-1)[1 + f(x)]$. *Journal of Mathematical Analysis and Applications* **3**, 440–69.

Kakutani, S. and Markus, L.

(1958). On the nonlinear difference-differential equation $\dot{y}(t) = [A - By(t-\tau)]y(t)$. *Contributions to the theory of nonlinear oscillations*, Vol. 4. Princeton University Press, Princeton, NJ, pp. 1–18.

Karakostas, G. and Staikos, V. A.

(1984). μ-Like-continuous operators and some oscillation results. University of Ioannina, Greece, TR 104.

Karakostas, G., Philos, Ch. G., and Sficas, Y. G.

(1990a). Stable steady state of some population models. (To appear.)

(1990b). *The dynamics of some discrete population models.* (To appear.)

Kartsatos, A. G.

(1969). On oscillation of solutions of even-order nonlinear differential equations. *Journal of Differential Equations* **6**, 232–7.

Kelly, W. and Peterson, A.

(1991). *Difference equations. An introduction with applications.* Academic Press, New York.

Kitamura, Y. and Kusano, T.

(1980). Oscillation of first order nonlinear differential equations with deviating arguments. *Proceedings of the American Mathematical Society* **78**, 64–9.

Kolmogorov, A. N. and Fomin, S. V.

(1970). *Introduction to real analysis.* Prentice-Hall, Englewood Cliffs, NJ.

Koplatadze, R. G. and Chanturia, T. A.

(1982). On the oscillatory and monotone solutions of the first order differential equations with deviating arguments. *Journal of Differential Equations* **18**, 1463–5.

Kosmalla, W. A.

(1986). Oscillation theorems for higher order delay equation. *Applicable Analysis* **21**, 45–53.

Kreith, K. and Ladas, G.

(1985). Allowable delays for positive diffusion processes. *Hiroshima Mathematical Journal* **15**, 437–43.

Kuang, Y.

(1990a). On neutral delay logistic Gauss-type predator–prey systems. (To appear.)

(1990b). Global stability and oscillation in delay-differential equations for single-species population growths. (To appear.)

Kuang, Y. and Feldstein, A.
(1990).   Monotonic and oscillatory solutions of a linear neutral delay equation with infinite lag. (To appear.)

Kulenovic, M. R. S. and Ladas, G.
(1987*a*).   Linearized oscillation theory for second order differential equations. *Canadian Mathematical Society Conference Proceedings* **8**, 261–7.

(1987*b*).   Linearized oscillations in population dynamics. *Bulletin of Mathematical Biology* **49**, 615–27.

(1988).   Oscillations of the sunflower equation. *Quarterly of Applied Mathematics* **46**, 23–8.

Kulenovic, M. R. S., Ladas, G., and Meimaridou, A.
(1987*a*).   Necessary and sufficient condition for oscillations of neutral differential equations. *Journal of the Australian Mathematical Society, Series B* **28**, 362–75.

(1987*b*).   On oscillation of nonlinear delay equations. *Quarterly of Applied Mathematics* **45**, 155–64.

(1987*c*).   Stability of solutions of linear delay differential equations. *Proceedings of the American Mathematical Society* **100**, 433–41.

Kulenovic, M. R. S., Ladas, G., and Sficas, Y. G.
(1988).   Oscillations of second order linear delay differential equations, *Applicable Analysis* **27**, 109–23.

(1989*a*).   Global attractivity in population dynamics. *Computers & Mathematics with Applications* **18**, 925–8.

(1989*b*).   Global stability in population dynamics. *Computers and Mathematics with Applications* **18**, 925–8.

(1990).   Comparison results for oscillations of delay equations. *Annali di Matematica Pura ed Applicata* **CLVI**, 1–14.

Kuruklis, S. A. and Ladas, G.
(1990).   Oscillations and global attractivity of a delay logistic model. (To appear.)

Kusano, T.
(1984).   Oscillation of the even order linear differential equations with deviating arguments of mixed type. *Journal of Mathematical Analysis and Applications* **98**, 341–7.

Kusano, T. and Onose, H.
(1974).   Oscillations of functional differential equations with retarded argument. *Journal of Differential Equations* **15**, 269–77.

Kusano, T. and Singh, B.
(1984).   Positive solutions of functional differential equations with singular nonlinear terms. *Nonlinear Analysis: Theory, Methods and Applications* **8**, 1081–90.

356   BIBLIOGRAPHY

Kwong, M. K.
(1990).   Oscillation of first-order delay equations. (To appear.)

Kwong, M. K. and Patula, W. T.
(1990).   Comparison theorems for first order linear delay equations. (To appear.)

Ladas, G.
(1971).   Oscillation and asymptotic behavior of solutions of differential equations
          with retarded argument. *Journal of Differential Equations* **10**, 281–90.

(1977).   Oscillatory effects of retarded actions. *Journal of Mathematical Analysis and
          Applications* **60**, 410–16.

(1979).   Sharp conditions for oscillations caused by delays. *Applicable Analysis* **9**,
          93–8.

(1988*a*). Oscillations of equations with piecewise constant mixed arguments. *Differen-
          tial Equations and Applications, Proceedings of the International Conference
          on Theory and Applications of Differential Equations, March 21–25, 1988.*
          Ohio University Press, Athens, pp. 64–9.

(1988*b*). Oscillations of difference equations with positive and negative coefficients.
          *Proceedings of the Geoffrey J. Butler Memorial Conference on Differential
          Equations and Population Biology, June 20–25, 1988.* University of Alberta,
          Edmonton.

(1989).   Linearized oscillations for neutral equations. *Differential Equations, Pro-
          ceedings of the 1987 Equadiff Conference.* Dekker, New York, pp. 379–87.

(1990*a*). Explicit conditions for the oscillation of difference equations. *Journal of
          Mathematical Analysis and Applications* **153**, 276–87.

(1990*b*). Recent developments in the oscillation of delay difference equations.
          Difference equations: stability and control. Dekker, New York, pp. 321–32.

(1990*c*). Necessary and sufficient conditions for oscillation. Open problems and
          conjectures. *Proceedings of the International Symposium on Functional
          Differential Equations and Related Topics, August 30–September 2, 1990,
          Kyoto, Japan.*

Ladas, G. and Lakshmikantham, V.
(1974).   Oscillations caused by retarded actions. *Applicable Analysis* **4**, 9–15.

Ladas, G. and Schultz, S. W.
(1989).   On oscillations of neutral equations with mixed arguments. *Hiroshima
          Mathematical Journal* **19**, 409–29.

Ladas, G. and Sficas, Y. G.
(1984).   Oscillations of delay differential equations with positive and negative
          coefficients. *Proceedings of the International Conference on Qualitative
          Theory of Differential Equations, University of Alberta, June 18–20, 1984,*
          pp. 232–40.

Ladas, G. and Sficas, Y. G. (*continued*)

(1986*a*). Oscillations of neutral delay differential equations. *Canadian Mathematical Bulletin* **29**, 438–45.

(1986*b*). Oscillations of higher-order neutral equations. *Journal of the Australian Mathematical Society, Series B* **27**, 502–11.

(1988). Asymptotic behavior of oscillatory solutions. *Hiroshima Mathematical Journal* **18**, 351–9.

Ladas, G. and Stavroulakis, I. P.

(1982*a*). On delay differential inequalities of first order. *Funkcialaj Ekvacioj* **25**, 105–13.

(1982*b*). On delay differential inequalities of higher order. *Canadian Mathematical Bulletin* **25**, 348–54.

(1982*c*). Oscillations caused by several retarded and advanced arguments. *Journal of Differential Equations* **44**, 143–52.

(1984*a*). Oscillations of differential equations of mixed type. *Journal of Mathematical and Physical Sciences* **18**, 254–62.

(1984*b*). On differential inequalities with several deviating arguments. *Proceedings of the IXth International Conference on Nonlinear Oscillations.* Naukova Dumka, pp. 215–19.

Ladas, G., Ladde, G., and Papadakis, J. S.

(1972*a*). Oscillations of functional-differential equations generated by delays. *Journal of Differential Equations* **12**, 385–95.

Ladas, G., Lakhsmikantham, V., and Papadakis, J. S.

(1972*b*). Oscillations of higher-order retarded differential equations generated by the retarded argument. *Delay and functional differential equations and their applications.* Academic Press, New York, pp. 219–31.

Ladas, G., Sficas, Y. G., and Stavroulakis, I. P.

(1983*a*). Asymptotic behavior of solutions of retarded differential equations. *Proceedings of the American Mathematical Society* **88**, 247–53.

(1983*b*). Necessary and sufficient conditions for oscillations. *American Mathematical Monthly* **90**, 247–53.

(1984*a*). Necessary and sufficient conditions for oscillations of higher order delay differential equations. *Transactions of the American Mathematical Society* **285**, 81–90.

(1984*b*). Nonoscillatory functional differential equations. *Pacific Journal of Mathematics* **115**, 391–8.

(1984*c*). Functional differential inequalities and equations with oscillating coefficients. *Proceedings of the Vth International Conference on Trends in Theory and Practice of Nonlinear Differential Equations.* Dekker, New York, pp. 277–84.

Ladas, G., Partheniadis, E. C., and Schinas, J.

(1989*a*). Existence theorems for second order differential equations with piecewise constant argument. *Differential Equations, Proceedings of the 1987 Equadiff Conference.* Dekker, New York, pp. 389–95.

(1989*b*). Oscillation and stability of second order differential equations with piecewise constant argument. *Radovi Matematicki* **5**, 171–8.

Ladas, G., Partheniadis, E. C., and Sficas, Y. G.

(1989*c*). Oscillations of second order neutral equations. *Differential Equations, Proceedings of the 1987 Equadiff Conference.* Dekker, New York, pp. 397–401.

(1989*d*). Necessary and sufficient conditions for oscillations of second order neutral equations. *Journal of Mathematical Analysis and Applications* **138**, 214–31.

(1989*e*). Oscillations of second order neutral equations. *Canadian Journal of Mathematics* **41**.

Ladas, G., Philos, Ch. G., and Sficas, Y. G.

(1989*f*). Sharp conditions for the oscillation of delay difference equations. *Journal of Applied Mathematics and Simulations* **2**, 101–12.

(1989*g*). Necessary and sufficient conditions for the oscillation of difference equations. *Libertas Mathematica* **9**, 121–5.

(1990*a*). Oscillations of integro-differential equations. *Differential and Integral Equations.* (To appear.)

(1990*b*). Oscillation in neutral equations with periodic coefficients. *Proceedings of the American Mathematical Society.* (To appear.)

Ladas, G., Qian, C., Vlahos, P. N., and Yan, J.

(1990*c*). Stability of solutions of linear nonautonomous difference equations. *Applicable Analysis.* (To appear.)

Ladde, G. S.

(1977). Oscillation caused by retarded perturbations of first order linear ordinary differential equations. *Atti Accademia Nazionale dei Lincei, Rend. C.I. Sci. Fis. Mat. Natur.* (8) **63**, 351–9.

Ladde, G. S., Lakshmikantham, V., and Zhang, B. G.

(1987). *Oscillation theory of differential equations with deviating arguments.* Dekker, New York.

Levin, S. A. and May, R. M.

(1976). A note on difference-delay equations. *Theoretical Population Biology* **9**, 178–87.

Li, T.-Y. and Yorke, J. A.

(1975). Period three implies chaos. *American Mathematical Monthly* **82**, 985–92.

Mackey, M. C.

(1978*a*). Dynamic haematological disorders in stem cell origin. In *Cellular mechanisms of reproduction and aging* (ed. J. Vassileva-Popova). Plenum Press, New York, pp. 373–409.

(1978*b*). A unified hypothesis for the origin of aplastic anaemia and periodic haematopoiesis. *Blood* **51**, 941–56.

(1979). Periodic auto-immune hemolytic anemia; an induced dynamical disease. *Bulletin of Mathematical Biology* **41**, 829–34.

(1981). Some models in hemopoiesis: predictions and problems. In *Biomathematics and cell kinetics* (ed. M. Rotenberg). Elsevier/North Holland, Amsterdam, pp. 23–38.

Mackey, M. C. and an der Heiden, U.

(1982). Dynamical diseases and bifurcations: understanding functional disorders in physiological systems. *Func. Biol. Med.* **156**, 156–64.

Mackey, M. C. and Glass, L.

(1977). Oscillations and chaos in physiological control systems. *Science* **197**, 287–9.

Mackey, M. C. and Milton, J. G.

(1979). Dynamical diseases. *Annals of the New York Academy of Science* **316**, 214–35.

Mahfoud, W. E.

(1979). Comparison theorems for delay differential equations. *Pacific Journal of Mathematics* **83**, 187–97.

Mallet-Paret, J. and Nussbaum, R. D.

(1986). Global continuation and asymptotic behavior for periodic solutions of a differential-delay equation. *Annali di Matematica Pura ed Applicata* **145**, 33–128.

Marusiak, P. and Norkin, S. B.

(1981). The oscillatory nature of systems with deviating argument. *Prace a Studie Vysokej Skoly Dopravy a Spjov Ziline, Ser. Mat. Fyz.* **3**, 13–25.

May, R. M.

(1975). Biological populations obeying difference equations: stable points, stable cycles and chaos. *Journal of Theoretical Biology* **51**, 511–24.

May, R. M. and Oster, G. F.

(1976). Bifurcations and dynamic complexity in simple ecological models. *American Naturalist* **110**, 573–99.

Mickens, R. E.

(1987). *Difference equations.* Van Nostrand Reinhold, New York.

Mingarelli, A. B.

(1983). *Volterra–Stieltjes integral equations and generalized ordinary differential expressions*, Lecture Notes in Mathematics No. 989. Springer-Verlag, New York.

Minorsky, N.
(1962).   *Nonlinear oscillations.* Van Nostrand Reinhold, New Jersey.

Mishev, D. P. and Bainov, D. D.
(1983).   Asymptotic behavior of the nonoscillating solutions of functional differential equations of nth order. *Mathematics Reports Toyama University* **6**, 83–94.

Myskis, A. D.
(1972).   *Linear differential equations with retarded argument* (in Russian). Izdatel'stvo 'Nauka', 2nd Edn, Moscow.

Naito, M.
(1981).   On strong oscillation of retarded differential equations. *Hiroshima Mathematical Journal* **11**, 553–60.

(1984).   Nonoscillatory solutions of linear differential equations with deviating arguments. *Annali di Matematica Pura ed Applicata* (4) **136**, 1–13.

Nicholson, A. J.
(1954).   An outline of the dynamics of animal populations. *Australian Journal of Zoology* **2**, 9–65.

Nussbaum, R. D.
(1982).   Cyclic differential equations and period three solutions of differential-delay equations. *Journal of Differential Equations* **46**, 379–408.

(1989).   Wright's equation has no solutions of period four. *Proceedings of the Royal Society of Edinburgh, Section A* **113** (3/4), 281–8.

Olah, R.
(1982).   Note on the oscillation of differential equations with deviating argument. *Casopis pro Pestovani Matematiky* **107**, 380–7, 428.

Onose, H.
(1981).   Nonlinear oscillations of functional differential equations with complicated deviating argument. *Bulletin of the Faculty of Science, Ibaraki University* **13**, 29–43.

(1983).   Oscillatory properties of the first order differential inequalities with deviating argument. *Funkcialaj Ekvacioj* **26**, 189–95.

(1984).   Oscillatory properties of the first order nonlinear advanced and delayed differential inequalities. *Nonlinear Analysis: Theory, Methods and Applications* **8**, 171–80.

(1986).   Nonoscillation of nonlinear first order differential equations with forcing term. *Hiroshima Mathematical Journal* **16**, 617–24.

Partheniadis, E. C.
(1988).   Stability and oscillation of neutral delay differential equations with piecewise constant argument. *Differential and Integral Equations* **1**, 459–72.

Peterson, A. C. and Ridenhour, J.
(1989). Atkinson's superlinear oscillation theorem for matrix difference equations. *Proceedings of the International Conference on Differential Equations: Theory and Applications in Stability and Control, Colorado Springs, Colorado, June 7–10, 1989.*

Philos, C. G.
(1980). On the existence of nonoscillatory solutions tending to zero at ∞ for differential equations with positive delays. *Archiv der Mathematik* **36**, 168–78.

(1990). Oscillations in difference equations with periodic coefficients. (To appear.)

Philos, C. G. and Sficas, Y. G.
(1990). Positive solutions of difference equations. *Proceedings of the American Mathematical Society* **108**, 107–15.

Philos, C. G., Sficas, Y. G., and Staikos, V. A.
(1982). Some results on the asymptotic behavior of nonoscillatory solutions of differential equations with deviating arguments. *Journal of the Australian Mathematical Society, Series A* **32**, 295–317.

Pianka, E. R.
(1974). *Evolutionary ecology.* Harper and Row, New York.

Pielou, E. C.
(1969). *An introduction to mathematical ecology.* Wiley, New York.

(1974). *Population and community ecology.* Gordon and Breach, New York.

Rosen, G.
(1987). Time delays produced by essential nonlinearity in population growth models. *Bulletin of Mathematical Biology* **28**, 253–6.

Rudin, W.
(1966). *Real and complex analysis.* McGraw-Hill, New York.

Schauder, J.
(1930). Der Fixpunktsatz in Funktionalräumen. *Studia Mathematica* **2**, 171–80.

Schultz, S. W.
(1990). Necessary and sufficient conditions for the oscillation of bounded solutions. *Applicable Analysis* **30**, 47–63.

Seifert, G.
(1986). Oscillation of solutions of a population equation with delay. *Nonlinear Analysis and Applications: Proceedings of the 7th International Conference at UTA, 28 July–1 August, 1986* (ed. V. Lakshmikantham). Dekker, New York, pp. 549–54.

(1990). Oscillatory solutions for certain delay-differential equations. *Proceedings of the American Mathematical Society.* (To appear.)

Sficas, Y. G. and Stavroulakis, I. P.
(1987). Necessary and sufficient conditions for oscillations of neutral differential equations. *Journal of Mathematical Analysis and Applications* **123**, 494–507.

Shah, S. M. and Wiener, J.
(1983). Advanced differential equations with piecewise constant argument deviations. *International Journal of Mathematics and Mathematical Sciences* **6**, 671–703.

Shevelo, V. N. and Varekh, N. V.
(1979). *Asymptotic methods in the theory of nonlinear oscillations* (in Russian). Naukova Dumka, Kiev.

Shevelo, V. N., Varekh, N. V., and Gritsai, A. G.
(1982). Oscillatory properties of solutions of systems of differential equations with retarded argument. *Akademiya Nauk Ukrainskoi SSR, Institut Matematiki, Preprints* No. 2; 48 pp.

Simmons, G. F.
(1963). *Introduction to topology and modern analysis*. McGraw-Hill, New York.

Singh, B.
(1979). Necessary and sufficient condition for maintaining oscillations and non-oscillations in general functional equations and their asymptotic properties. *SIAM Journal on Mathematical Analysis* **10**, 18–31.

Slemrod, M. and Infante, E. F.
(1972). Asymptotic stability criteria for linear systems of difference-differential equations of neutral type and their discrete analogues. *Journal of Mathematical Analysis and Applications* **38**, 399–415.

Smith, F. E.
(1963). Population dynamics in *Daphnia magna. Ecology* **44**, 651–63.

Smith, H.
(1976). Bounded oscillation in a class of functional differential equations. *Journal of Mathematical Analysis and Applications* **56**, 223–32.

(1990). Monotone semiflows generated by functional differential equations. *Journal of Differential Equations*. (To appear.)

Snow, W.
(1965). Existence, uniqueness and stability for nonlinear differential-difference equations in the neutral case. *New York University, Courant Institute of Mathematical Sciences, Report No. IMM NYU 328.*

Somolinos, A. S.
(1978). Periodic solutions of the sunflower equation: $\ddot{x} + (a/r)\dot{x} + (b/r)\sin x(t - r) = 0$. *Quarterly of Applied Mathematics* **35**, 465–77.

Staikos, V. A.
(1980). Basic results on oscillation for differential equations with deviating arguments. *Hiroshima Mathematical Journal* **10**, 495–515.

Stavroulakis, I. P.
(1982). Nonlinear delay differential inequalities. *Nonlinear Analysis: Theory, Methods and Applications* **6**, 389–96.

Sugie, J.
(1988). Oscillating solutions of scalar delay-differential equations with state dependence. *Applicable Analysis* **27**, 217–27.

Swanson, C. A.
(1968). *Comparison and oscillation theory of linear differential equations*. Academic Press, New York and London.

Takano, K.
(1983). On global properties of solutions of a certain delay differential equation. *Journal of the London Mathematical Society* **28**, 519–30.

(1985). On oscillatory solutions of nonlinear delay differential equations. *Journal of Mathematical Analysis and Applications* **105**, 491–501.

Tarski, A.
(1955). A lattice theoretical fixed point theorem and its applications. *Pacific Journal of Mathematics* **5**, 285–309.

Tramov, M. I.
(1975). Conditions for oscillatory solutions of first order differential equations with a delayed argument. *Izvestiya Vysshikh Uchebnykh Zavedenii, Seriya Matematika* **19**, 92–6.

(1982). The oscillatory nature of solutions of differential equations with deviating argument. *Differentsial'nye Uravneniya* **18**, 245–53, 364.

Walther, H. O.
(1981). Density of slowly oscillating solutions of $\dot{x}(t) = -f(x(t-1))$. *Journal of Mathematical Analysis and Applications* **79**, 127–40.

Waltman, P.
(1968). A note on an oscillation criterion for an equation with a functional argument. *Canadian Mathematical Bulletin* **11**, 593–5.

Wang, Z.
(1988*a*). Necessary and sufficient conditions for oscillations of the neutral equation

$$\frac{d}{dt}[x(t) + px(t-r)] + qx(t-s) - hx(t-v) = 0.$$

Kexue Tongbao **33**, 1452–4.

(1988*b*). Necessary and sufficient conditions for oscillations of neutral delay equations with several positive and negative coefficients. *Advances in Mathematics* **17**, 330–1.

Wazewska-Czyzewska, M. and Lasota, A.
(1988).   Mathematical problems of the dynamics of the red blood cells system. *Annals of the Polish Mathematical Society, Series III, Applied Mathematics* **17**, 23–40.

Widder, D. V.
(1971).   *An introduction to transform theory.* Academic Press, New York.

Wiener, J.
(1990).   Partial difference equations with piecewise constant delays. *Abstracts of Papers Presented, American Mathematical Society* **11**, 61.

Wright, E. M.
(1955).   A nonlinear difference-differential equation. *J. Reine und Angewandte Mathematik* **194**, 66–87.

Yan, J.
(1990).   Necessary and sufficient conditions for oscillations of neutral differential equations. *Journal of Mathematical Analysis and Applications.* (To appear.)

Zahariev, A. I. and Bainov, D. D.
(1980).   Oscillating properties of the solutions of a class of neutral type functional differential equations. *Bulletin of the Australian Mathematical Society* **22**, 365–72.

Zhang, B. and Gopalsamy, K.
(1990).   Oscillation and nonoscillation in a nonautonomous delay-logistic equation. (To appear.)

AUTHOR INDEX

SUBJECT INDEX